Albion Thomas Snell

Electric Motive Power

The Transmission and Distribution of Electric power by Continuous and Alternate

Currents

Albion Thomas Snell

Electric Motive Power
The Transmission and Distribution of Electric power by Continuous and Alternate Currents

ISBN/EAN: 9783337024512

Printed in Europe, USA, Canada, Australia, Japan

Cover: Foto ©berggeist007 / pixelio.de

More available books at **www.hansebooks.com**

ELECTRIC MOTIVE POWER:

THE TRANSMISSION AND DISTRIBUTION OF ELECTRIC POWER BY CONTINUOUS AND ALTERNATE CURRENTS.

WITH A SECTION ON THE APPLICATIONS OF ELECTRICITY TO MINING WORK.

BY

ALBION T. SNELL,

Associate Member of the Institution of Civil Engineers;
Member of the Institution of Electrical Engineers;
Member of the Federated Institutes of Mining Engineers.

LONDON:
"THE ELECTRICIAN" PRINTING AND PUBLISHING COMPANY,
LIMITED,
SALISBURY COURT, FLEET STREET, E.C.

1894.

Printed and Published by
" THE ELECTRICIAN " PRINTING AND PUBLISHING CO., LIMITED.
1, 2, and 3, Salisbury Court, Fleet Street,
London, E.C.

PREFACE.

"ELECTRIC MOTIVE POWER" is designed to be a practical treatise for Mechanical and Mining Engineers and other Students of Applied Electricity. Taking for granted an elementary knowledge of the theory of Electricity and Magnetism, I have endeavoured to give a clear exposition of the principles governing Electric Transmission of Power, and to develop them in accordance with the best practice of the present time. In doing this I have deemed it advisable to discuss the design of Dynamos, Alternate and Continuous Current Motors, Alternators and Transformers, as well as the various systems in vogue for transmitting and distributing power.

A large part of the book is devoted to a careful consideration of the alternate current systems, both single and polyphase. This is necessitated by their growing importance, as it is now well recognised by the leading electrical engineers of Europe and America that for power transmission they are superior to continuous current systems. In a short Appendix are given some figures with reference to the prime cost of plants for long-distance transmission.

The last two Chapters deal exhaustively with the applications of electricity to mining work, particularly with reference to collieries and coal-getting. It is hoped they will prove to be of interest to the large number of Mining Engineers who are now using or are advising the use of electric plants for lighting and power purposes.

B

The contents are arranged in ten chapters, each of which is, as far as possible, a monograph on one portion of the subject. Cross references enable the engineer or student to find required information with the greatest possible facility. The mathematics employed are of the simplest kind, entirely practical in character, and inch-pound-minute units are used throughout. Most of the tables of data and tests are new, and all have been compiled from recent and trustworthy information. The illustrations, about 250 in number, are chiefly diagrams rather than perspective drawings, as the former are more useful in practice, and the majority have been prepared specially for this work.

In endeavouring to treat the subject as comprehensively as possible, I have necessarily drawn on the experience of my *confrères* on both sides of the Atlantic, and have to some extent availed myself of the store of information contained in the columns of the Technical Press and in the Proceedings of the Technical Societies.

I am much indebted to Dr. W. E. SUMPNER for many instructive hints with reference to the alternate current sections; and my thanks are also due to many firms and companies who have kindly supplied me with information referring to special features in their manufactures.

Although written primarily for Engineers and Students, I hope the book will be found of value to users of electric plant generally.

ALBION T. SNELL.

November, 1894.

CONTENTS.

(A Complete Synopsis of the Contents of each Chapter appears under each Chapter Heading.)

INTRODUCTION.

CIVIL ENGINEERING has been defined as "the art of directing the great sources of power in Nature for the use and convenience of man." This definition was apt in the days which saw the incorporation of the Institution of Civil Engineers; it is doubly so now. Then gravitation, the expansion of vapours and gases, and the pressure of wind, formed the sum total of forces at the disposal of man for engineering purposes. The study of Electricity and Magnetism was in its infancy, and neither physicist nor engineer dreamed of the possibilities of the new force. It was not until nearly half a century later that the principles of electro-dynamics were developed, giving a fresh meaning to the old definition of Civil Engineering by the inclusion of another force, all-pervading as gravitation, and powerful and manageable in a measure suggesting boundless possibilities of application.

As yet we do not know what Electricity is, but its utilisation is by no means limited by our want of insight into its nature. We do not know what Gravity is, but we have a clear conception of weight, and we measure horse-power in gravitation units, estimating the effects of gravity, though its real nature is still a mystery. In like manner the electrician defines electricity in terms of the magnetic effects produced by it, having no exact knowledge of the causes, although empirical suggestions, which fit the phenomena more or less completely, may give "educated guesses" as to their nature. Thus the utilisation of electricity demands no closer acquaintance with its nature than does that of gravity, which is made use of without direct reflection in almost every contrivance devised by man.

At any rate, the laws governing the production and regulation of electricity have been very completely investigated, and in most practical cases are capable of treatment by the simplest mathematics. Electric Power is measured in definite units bearing an easily defined ratio to the mechanical expressions for horse-power and work. In fact, electric quantities are gauged with a degree of readiness and accuracy impossible with the corresponding quantities in steam, compressed air, or any system involving the transference of matter as well as of power. It is to this transmission of power without the conveyance of matter that electricity owes its superiority to all other means of transmitting power where distance is a factor in the problem. Since neither gravity nor material friction enters into the question, power is transmitted equally well horizontally or vertically, round the sharpest curves and by the most devious routes. The advantages of electricity from this point of view are at once apparent, and have already been widely recognised, both at home and abroad ; more especially where natural water power is available, and the distance between the source of the prime power and the point of application is considerable. On the Continent, water power is largely used for generating electricity; and at home, although less abundant, is steadily gaining favour.

As might be expected, Electricity has found ready favour with mining engineers, and in more than one case the application of E'ectricity to Mining has resulted in profit where previously such a result was unattainable. But the importance of electric power in this connection has not yet been fully appreciated. So far back as the year 1889, the author pointed out in the *Engineer* that in districts where coal pits are in close proximity to each other considerable economy of power might be obtained by the establishing of central power stations on a basis similar to that used for electric lighting. The idea has not commended itself to mine owners as yet, but it is certain to do so as the conditions for economic working become better understood. At present it is not customary to keep so close an account of the cost of power at mines as is done at electric supply stations, and therefore it is not generally recognised how much waste may be involved in this item. Experience in lighting stations shows that the cost of power per unit is largely reduced when the load curve is kept fairly constant during the night and day. Now the conditions of many collieries are such as to permit of a nearly straight line load curve; for pumping, &c., can often be done at times when coal is not being drawn, and hence the one load can, in great measure, be made to balance the other.

At the central power station alternators could be kept running day and night. The power could be transmitted by duplicate mains at high pressure to transformers at each colliery, where the current could be transformed to a working pressure of, say, 500 volts. The power consumed by each colliery could be measured by recording wattmeters, or by ordinary current meters if the pressure were kept approximately constant. The mains between the central station and the collieries could generally be carried on poles, but in a few cases it would pay to lay them underground.

Such an arrangement would have the great advantage of localising all the generating plant under one roof, and in the charge of a special staff, whose only duty would be to supply power at all times at the lowest possible cost. It is evident that the cost of a unit of power would be much less under such conditions than when it is produced by a number of small plants, and that the cost of upkeep would be less because of more efficient supervision, while the reserve plant could be reduced very much below that absolutely necessary with independent plants.

When a number of adjacent pits belong to the same proprietors, the introduction of a central power station would be simple. If the pits belong to different owners it would be easy to arrange an equitable system of charging on the basis of actual power consumed. The station might be financed by a single proprietor, or by an independent company, whose function would be to supply power exclusively.

In the future developments of coal working some modification of this scheme will, no doubt, find a place.

An electric power plant is more easily erected and controlled, and is generally more flexible, than any other. It is no more experimental than the steam engine and boiler, and, if built and erected by competent men, is cheap and trustworthy.

CHAPTER I.

THE GENERAL PLAN OF ELECTRICAL TRANSMISSION OF POWER.

§ 1. Electricity not a Prime Power. Steam and Water Power. The Dynamo. The Line. The Motor. The Machines Driven.

BEFORE entering on the general subject of the Electrical Transmission of Power it will be well to define what is understood by the phrase, and also to note at this early stage that electricity is not a prime power (*see* foot-note on page 2). The problem will be at once understood by considering a hypothetical case. On a hillside there is a water-fall, with its power running to waste. A few miles from this is a mine, where power is urgently required to drive the winding engines, pump the dips, and work the fans. A turbine or waterwheel is so arranged as to rotate by means of the falling water, and a dynamo is coupled to it. The electrical power generated by the dynamo is led by two copper conductors to the mine, and there runs the motors which reconvert the electrical into mechanical power and drive the mining machinery.

Now, it is clear that electricity is simply the medium by which the power of the water is carried to the mine, and, therefore, is not a prime, but a secondary power. Natural water power is not often available, and, therefore, in the majority of cases, a steam engine is employed to drive the dynamo ; but the function of electricity remains the same.

C

The essential factors of an electrical power plant may be summarised as follows :—

(*a*) Prime Motors ....
$$\begin{cases} \text{Steam.} \begin{cases} \text{Boilers.} \\ \text{Engines.} \end{cases} \\ \text{Water... Turbine or waterwheel.} \\ \text{Gas} \ ... \begin{cases} \text{Gas Producers.} \\ \text{Engines.} \end{cases} \end{cases}$$

(*b*) Dynamos to convert the mechanical power of the prime motors into electrical power.

(*c*) Copper conductors to transmit the power from the dynamos to the motors.

(*d*) Motors* to reconvert the electrical into mechanical power.

(*e*) Machines to absorb the work given out by the motors.

We will now briefly examine these before developing the main scheme of the book.

(*a*) **Prime Source of Power.**—The consideration of this part of the working plant belongs more properly to a treatise on prime motors ; but the requirements of electrical work have introduced quite a revolution into the building of both steam engines and turbines, and probably a few hints as to the class of plant most suitable will not be unacceptable or out of place.

The dynamo is essentially a high-speed machine, one of the fundamental factors of both output and efficiency being a high peripheral speed of the armature. Practical considerations confine this to about 3,000ft. per minute ; and hence, for a given output, economy of material and labour will limit the diameter and length of the armature so as to make the speed a maximum. Now the stationary engine of 15 or 20 years ago was, on the contrary, a slow-speed machine, *i.e.*, the flywheel made relatively few revolutions per minute, although the piston velocity may have been in some cases as high as 300 or even 400ft.

---

* Probably most installations will be used for lighting as well as power purposes. By suitable arrangements, either the line or the distributing mains can be tapped at any part, and lamps of the proper voltage can be coupled across them. The power absorbed by the lamps, however, will not be available for use at the motors, and so a margin must be made in the output of the dynamos to allow for the lighting work.

per minute. The cylinders were of large diameter and long stroke, being designed to work with low boiler pressures. Therefore, when it became necessary to drive dynamos by this type of engine, difficulties arose in belting the dynamo pulley to the flywheel; and it soon became evident that the engine was in many ways unsuitable. Firstly, it was inefficient for the class of work; secondly, it governed badly; thirdly, it was costly in relation to its output; and, fourthly, it required foundations of a strength out of all proportion to the load. These weak points in steam-engine design were gradually forced on the notice of mechanical engineers by the introduction of the efficient and high-speed dynamo. But improvements were made before long. Boilers were built for higher pressures, enabling engines to be designed with smaller cylinders and to rotate at higher speeds. Next, governors were improved, and soon the tell-tale pilot lamp scarcely showed the variations of speed, and now it is by no means uncommon to find engines governing to within 2 per cent. or even less. Compound and triple expansion engines have now become general. The vertical type of engine, with inverted cylinders, finds most favour for driving dynamos. It is used either single, compound, or triple, with or without condensers, according to output and circumstances. It is coupled direct or by belting, requires small foundations, gives a minimum of trouble, and is cheap and efficient. Compared with the beam or horizontal engines of from 20 to 30 years ago, it is as David to Goliath.

The change in boiler design has been equally remarkable. Fifty pounds to the square inch used to be considered a fair working pressure with boilers of the Lancashire type; now from 80lbs. to 120lbs. is not uncommon; and with multitubular and watertube designs 140lbs. or even a greater pressure frequently occurs; and water-heaters, economisers, and mechanical stokers are now thoroughly appreciated. All these improvements, it is needless to say, tend to raise efficiency and decrease the running charges.

In choosing steam plant for an electrical transmission installation, therefore, the rules and precautions adopted in laying down a central station for lighting should be followed. First choose a convenient unit of power, *i.e.*, select the best size of dynamo to give the minimum work required during the twenty-four hours, and then if this be of convenient size keep to it as the standard, and simply multiply the unit as the work increases. If the minimum output be very small, it will probably be best to have a separate plant to deal with this part of the work, and to select a larger unit for the main installation. Having fixed the output of the dynamo, the size of the engine which has to run it, either direct or by belting, is easily determined, and the boiler power is also known. The boiler unit may be sufficient to run one or two engines; this will depend on their size, and cannot be definitely fixed without a knowledge of each case. The boilers should each be coupled with a stop-valve to the steam main, so that any boiler can have the fires banked or be shut down separately. One boiler and an engine and dynamo should be held in reserve in all large installations.

If water-power be used, the turbine or water-wheel can advantageously drive the dynamo direct, or belting may be used. The general considerations for determining the size of dynamo unit will hold just the same as with the steam plant; and, indeed, the whole station arrangements will be similar, except that the turbine will replace the boiler and engine. It is unnecessary to point out the simplicity of such a power station, or to refer to the advantages of the utilisation of waste water power by electricity—they are self-evident. It is not surprising that magnificent transmission plants have sprung into existence during the past few years; the wonder is that the number of such plants is not tenfold greater.

Surprise may be felt that the author includes gas engines among the prime motors for electric transmission plants, but experience most decidedly points to them as a feature of future installations for small outputs, especially if the load be intermittent, and the load factor consequently low. It has been thoroughly demonstrated that it is more economic as regards the quantity of light obtainable to consume coal gas in an engine, and to utilise the energy for driving a lighting dynamo, than to burn the gas direct in the ordinary way;

and the cost of power may be still further reduced if producer gas be employed. The gas-making apparatus is to the gas engine what the boiler is to the steam engine, and the cost of wages and repairs in each case is about the same. As regards the cost of producing this gas, the Dowson Economic Gas and Power Company state that, on a moderate scale, with coke costing 12s. per ton, the average cost of the volume of Dowson gas required to give the same heat as 1,000 cubic feet of ordinary town gas is only 10d., while in large plants the cost falls as low as $6\frac{1}{2}$d. Allowing for the variations of price of coal gas and fuel, it may be fairly assumed that producer gas will be about one-quarter as costly. Considerable improvements have of late years been introduced in the design of gas engines; and double-cylinder engines of 100 B.H.P. are now running successfully, and prove more economic for intermittent and varying loads than steam plant of similar capacity. It may be assumed, therefore, that gas engines will find a field for transmission of power work in certain cases.

At Schwabing, a suburb of Munich, a lighting plant with gas engines and Dowson gas runs ten arc lamps and 270 glow lamps of 16 c.p. each, and 30 of 32 c.p. each, at a cost for fuel of 1d. per brake horse-power per hour, the fuel consumption being $2\frac{1}{4}$lb. of mixed coke and anthracite per brake horse-power hour. This result compares favourably with the performance of good steam plant. And in England the Morecambe Electric Light Company use gas engines, with Dowson gas, indicating 110 H.P. each.

(*b*) **The Dynamo.**—The function of the dynamo is to convert mechanical into electrical power, and the machine, if well made, accomplishes this end in a most satisfactory manner. The ratio of conversion of a good dynamo at full load varies between 90 and 94 per cent., and this high efficiency is maintained through a considerable range of output; at half load it is about 85 to 90 per cent., and at one quarter load will not fall below 75 per cent.; while running on open circuit the power absorbed is only sufficient to overcome friction. If the plant be divided into suitable units, it will always be possible to run at nearly the full output of the dynamo, and so secure a high working efficiency. In addition to the high-speed dynamo already referred to, there is a slow-speed type, which is used when it is desired to couple dynamos directly to the engine shaft.

The output of a dynamo is roughly proportional to the length of the armature into the square of its diameter = say l D^2. Let the surface velocity be fixed. If D be large then l will be small, and *vice versâ.* In order to design a slow-speed dynamo it is necessary, then, to take a large diameter and a short length for the armature proportions. This type of machine is heavy and costly, in consequence of the large field-magnets, yokes, and extended beds, and the increased cost of machining and labour; but these disadvantages are, in the opinion of some designers, more than counterbalanced by the decreased speed.

American practice favours dynamos running at from 800 to 1,000 revolutions per minute, coupled with belts to high-pressure engines making from 200 to 350 revolutions. Continental engineers are leaning to the slow-speed steam dynamos, and recently huge plants, with units of 500 H.P. and upwards, and triple expansion condensing engines, have been laid down for town lighting. English engineers are very much divided in opinion upon the subject; but there certainly seems to be a tendency to use steam dynamos of about 100 unit capacity running at from 175 to 250 revolutions per minute.

Viewing the question of the electrical power station from all sides, and laying particular stress on the high pressure that will be required, it appears that the dynamo will generally be belt-driven, and therefore it will be permissible to use high-speed dynamos. A convenient size for a high-pressure transmission dynamo for use in mines is 50 or 100 H.P.; the speed may be from 450 to 600 revolutions per minute.

(*c*) **The Line.**—The success of an installation will very largely depend on the care and forethought shown in the selection of the gauge of copper and the erection of the line. As the general conditions to be observed will be discussed hereafter

seriatim, it need only be noted here that the power wasted
in the conductor may be made as small as desired, for the
loss varies directly as the length and inversely as the area
of cross-section. There are, of course, limits both to gauge
and length of conductor ; but these are determined as much
by financial as by practical considerations. The essential conditions of a successful line are :—Firstly, a continuous metallic
circuit between the dynamo and the motor ; and, secondly,
high insulation resistance, so that the current does not leak to
earth ; for defective insulation means not only a direct loss of
power but may also entail a risk of fire.

(*d*) **The Motor.**—The motor is similar in appearance to the
dynamo, though its function is the converse, viz., to convert
electrical into mechanical energy. It is, in fact, a reversed
dynamo, is governed by precisely the same laws, and its action
is accompanied by similar phenomena ; and, moreover, it is
quite as efficient. The reversibility of the dynamo is said to
have been discovered by an accident, though more probably it
was the result of the steady and laborious study of early inventors. Be this as it may, the fact has been thoroughly recognised, and full advantage has been taken of the opportunities
afforded by it.

The motor is as varied in the details of construction as the
dynamo, the design being modified to suit the work expected
from it. It is made to give either high or low speeds, and for
high or low pressures.

(*e*) **The Machines Driven.**—The machines to be driven need
no special reference, but care must be taken to couple them to
the motors in a suitable manner. High-speed machines may
often be connected directly to the motor-shaft ; or they may
be coupled by gearing or belts ; and countershafts, friction
clutches, and other devices may be used. But whatever
arrangement be adopted, care must be taken that the motor
speed, pressure, and current be approximately those designed
by the makers. If any of these points be neglected, the motor
will not be working under the best conditions, and there will
be a loss in efficiency.

The importance of this will be made clear in the sequel. Attention will be drawn to the different types of electric motors and the kind of work for which they are specially adapted. And an electric plant will be shown to be more flexible and suitable for a greater variety of conditions of working than any other means of transmitting and distributing power now in vogue.

The general idea involved in the expression "an electrical transmission plant" has now been explained, and the way is cleared for a more minute study of the electrical part of the subject, which is the main theme under consideration.

CHAPTER II.

THE DYNAMO AND MOTOR.

§2. INTRODUCTORY.

THE author has not thought it advisable to enter into an elementary discussion of the principles underlying the construction of dynamo-electric machines. This is the province of text-books on electricity and magnetism. Here the object aimed at is rather to develop the lines on which these principles are applied in practice, and to give useful information to those using electric plant. Although machines suitable for the transmission of power are chiefly considered, yet the method advanced is equally applicable to dynamos intended for lighting and other purposes.

First, it is necessary to distinguish between the dynamo and the motor. In the introductory remarks it was mentioned that the dynamo was a reversible machine, the motor being the converse of the dynamo. The functions of the two machines must now be compared. The chief conditions of working in the two cases are widely different, and these distinctions must be duly considered in the design. The dynamo will usually be driven at a speed as nearly constant as possible, and will be required to give a variable output at a constant pressure. The motor will usually be run at a fairly constant pressure of supply, and will be required to give a variable torque at a constant speed. The dynamo in most cases will be placed in a dry engine-room, and will meet with at least as much attention and care as ordinary steam plant. The motor, on the other hand, must be in close proximity to the machine driven, and will frequently be in a dusty, dirty, and more or less exposed position; and the attendance will be most likely a few visits daily at the most. The dynamo will probably run with fairly steady loads, and not be subjected to large sudden variations of output. The motor must, in the nature of things, meet with rougher usage. If the work be intermittent, the stoppages frequent, or overloading common, the motor must meet the demand, no matter how great the strain. It follows, therefore, that the motor should be built with a larger factor of safety than is usually necessary with a dynamo; and that extra care should be given to the insulation and general details. To ensure due prominence being given to these conditions, it is the author's practice to base the design of the dynamo on the total armature E.M.F., and to give special attention in the building of the motor both to the maximum and the average torque.

§3. UNITS AND SYMBOLS.

To make the equations suitable for the drawing office, inch-pound-minute units will be used. The author has for some time endeavoured to use the C.G.S. system in dynamo design; but he has invariably found it necessary to convert the measurements into English quantities before the drawings could be used in the shops, because the workman uses a foot-rule and thinks in eighths and quarters of an inch. The decimal system has not met with much favour in this country; and until it is in more general use there is nothing to be gained by making the duplicate measurements. To the pure physicist the C.G.S. system of units offers a real advantage, for it simplifies his work; but to the construction engineer, at present at any rate, it only means an increased number of figures with a corresponding chance of error from clerical slips. The symbols will be as nearly as possible those used in Prof. Silvanus Thompson's "Dynamo-Electric Machinery," because these are, no doubt, already widely used, and so will be readily followed by the majority of men interested in the subject.

SYMBOLS USED IN THE FOLLOWING CALCULATIONS AND EQUATIONS.

A = area in square inches.

AT_s = ampere turns with armature on open circuit. ⎫ Both refer to field-
CT_s = compensating ampere turns for full load. ⎰ magnets.

B''_a = magnetic induction per square inch of armature cross section.

B''_f = magnetic induction per square inch of field-magnet cores.

B''_g = magnetic induction per square inch of area of air gap.

N_a, N_f, N_g = total magnetic fluxes severally in armature, field-magnet, and air gap.

C_a = total number of conductors in armature, counted all round the external periphery, and therefore applicable to either drum or Gramme winding.

C_m, C_s = total number of turns of wire on series and shunt coils respectively of field-magnets.

D = outer diameter of armature,
d = inner diameter of armature,
r = radial depth of iron of armature, ⎱ in inches.
l = length of armature reckoned over iron,

E = total electromotive force generated in armature, either ⎫ in volts.
   dynamo or motor, ⎰

e = difference of potential from terminal to terminal,
i = current in external circuit,
i_a = current in armature, ⎱ in amperes.
i_s = current in shunt coils,
i_m = current in series or main coils,

n = number of revolutions per minute.
p = number of pairs of poles.

R = resistance of external circuit,
r_a = resistance of armature coils, ⎱ in ohms.
r_s = resistance of shunt coils,
r_m = resistance of series or main coils,

T = torque or turning moment.
v = coefficient of magnetic leakage.

δ = length of air gap *on one side only*,
w = width of polar cavity,
L_g = length of polar cavity, ⎱ in inches.
m_a, m_f, m_g = mean length of magnetic paths, respectively, in armature, cores and pole pieces, and yoke,

FOR ALTERNATE CURRENTS.

V_{mp} = total impressed electromotive force of armature,
v_{imp} = impressed difference of potential from terminal to terminal, ⎱ in effective volts.
v_r = resultant electromotive force,
v_s = self induction electromotive force,

i_{imp} = impressed current,
i_d = dynamic or working current, ⎱ in effective amperes.
i_c = wattless or condenser current,

L = coefficient of self-induction in henries.

§4. FIRST APPROXIMATION TO THE SIZE OF ARMA-
TURE FOR A GIVEN OUTPUT AND SPEED.

The first point to determine in designing a dynamo is the
size of armature that is best suited to meet all the requirements of the problem. Usually the volts, amperes and speed
are fixed, and frequently the selling price too. The latter item
cannot be specially discussed, but a few hints gathered from
practice will be given when possible. We know, then, e, i and
n. The size of machine can generally be selected by reference
to a price list, and by making a few rough calculations. But,
if the type be new to the shop, no data are available, and it is
necessary to resort to a few figures and to solve a few simple
equations.

The output of any dynamo is proportional to the volume of
the armature, the number of revolutions per minute, and the
magnetic flux. Hence output $\propto l\,D^2 n\,\mathbf{N}_a k$, where k is a dimensional constant to reduce the product to watts or horse-power.
The author first published empirical equations of this form
in 1888, and coupling $\mathbf{N}_a$ and k, he based the resultant value
on an average magnetic density and radial depth of core. Such
assumptions are found to be fairly correct for the greater
number of bipolar machines with outputs of from 10 to
100 H.P. ; but below and above these limits the equations are
not so nearly in accord with actual results. Small machines
realise smaller outputs than this theory indicates, owing to the
disproportionate resistance of the air gap ; while in large
machines the output is so largely affected by the design of the
field-magnets and the arrangement of the poles that separate
calculations are needed for every type. Nevertheless, similar
equations can easily be determined for any size and type, but
the constant must be adjusted to suit the conditions. For
Gramme armature dynamos the author finds the coefficient
0·01 to give fair results between the limits stated ; and for
drum machines 0·015, the output in each case being in watts.
Thus for dynamos :—

$$\text{Watts} = l\,D^2 n\,0\cdot01 \quad \text{(Gramme)} \quad . \quad . \quad . \quad (1)$$

$$\text{Watts} = l\,D^2 n\,0\cdot015 \quad \text{(drum)} \quad . \quad . \quad . \quad (2)$$

and for motors, estimating the output in brake horse-power :—

$$B.H.P. = l D^2 n\, 0\, 00001 \text{ (Gramme)} \qquad (3)$$
$$B.H.P. = l D^2 n\, 0.000015 \text{ (drum)}. \qquad (4)$$

It should be noted that a drum armature is capable of giving an output which is 50 per cent. in excess of a Gramme ring of the same over-all dimensions. This enormous advantage in the weight efficiency of drum armature machines is well recognised, and where the conditions of running admit of the design it should be invariably used. Indeed, if it were not for the difficulties in the winding and insulation of drums, Gramme armatures would soon be obsolete. The superiority of the drum lies chiefly in the following points : (a) it has no inside wires to cause an internal field, as well as to increase the dead resistance of the winding ; (b) there is less inequality of E.M.F. in the coils, since each turn of wire embraces the whole armature field ; and (c) the iron can be carried right to the shaft, and therefore the magnetic induction per unit of area may be made low, though the total flux be very large.

The Gramme winding, although causing an internal field and a more irregular distribution of the lines of force, is better adapted for rough work, since coils of very different potentials are not in close contact, as occurs with ordinary drum windings. There are arrangements of drum bar windings with spiral or crank-shaped end connections which obviate this objection to some extent ; but experience has all along pointed to the Gramme type of winding as the safer for hard work. The author has tried both methods, and has no hesitation in advising the Gramme winding for high-tension transmission dynamos and motors. For lighting dynamos of, say, from 100 to 200 volts, the drum type, with bar windings, and spiral strip connections, is the better, and is generally adopted now when other considerations do not prevent its use.

The rough dimensions of the armature for dynamo or motor may now be assumed to be known, and the calculation of the winding may be proceeded with.

§5. FUNDAMENTAL EQUATION CONNECTING THE TOTAL E.M.F. WITH THE ARMATURE CONSTANTS.

The factor of prime importance in armature design is E, the total E.M.F. The value of E is determined by the relationship

$$E = e \pm i_a r_a. \qquad \ldots \quad \ldots \quad (5)$$

The positive sign must be used for the armature of a generator, and the negative for that of a motor. The armature current i_a is known from the conditions of the problem; but r_a has to be determined, as the winding is not yet chosen. An assumed value must be substituted in (5), which can be amended when the gauge and number of turns of wire are definitely fixed. The numerical value of $i_a r_a$ cannot yet be fixed, but it can be conveniently expressed as a percentage of E. The following table of approximate values will be of use at this stage :—

Table A.

Output of dynamo or motor in H.P.	$i_a r_a$ expressed as a percentage of E.	
2·5	6	per cent.
5	5·5	,,
10	5	,,
15	4	,,
20	3	,,
25	2·75	,,
30	2·5	,,
50	2·25	,,
100	2	,,

The total armature E.M.F. is given by

$$E = C_a \, \mathbf{N}_a \, n \, \frac{10^{-8}}{60}. \qquad \ldots \quad \ldots \quad (6)$$

C_a, the number of turns of wire counted all round the periphery, is the most important variable; n is the number of revolutions per minute, and is given by the conditions of running. $\mathbf{N}_a$ is the total magnetic flux, and is numerically equal to the product of the number of square inches in the cross section of the armature iron and the magnetic density per square inch, or

$$\mathbf{N}_a = \mathbf{B}_a'' \, 2 \, r \, l \, 0 \cdot 85. \qquad \ldots \quad \ldots \quad (7)$$

The coefficient 0·85 is introduced to allow for the insulation between the plates of the armature.

The value of l was fixed when the over-all dimensions of the armature were chosen; and so in practice would be that of r; for the radial depth of core has presumably a fixed ratio to the diameter of armature for a given type of field-magnet. Various considerations determined by experience fix the magnetic flux per unit area at from 90,000 to 110,000 lines of force. Therefore N_a can readily be determined as soon as the size of armature is selected.

§6. THE RATIO BETWEEN RADIAL DEPTH OF CORE AND DIAMETER OF ARMATURE.

The following ratios between radial depth of core and diameter in Gramme armatures have been taken from practice, and illustrate the difference of opinion on the subject. They all refer to bipolar machines. It will be noted that they vary from $\frac{1}{5}$ to $\frac{1}{7}$.

Table B.

Diameter of core = D.	Radial depth of core = r.
7 inch.	1·0 inch.
8 ,,	1·25 ,,
8·875 ,,	1·125 ,,
10 ,,	1·5 ,,
10 ,,	2·0 ,,
12 ,,	1·875 ,,
12 ,,	2·25 ,,
13 ,,	2·0 ,,
14 ,,	2·5 ,,
15 ,,	2·25 ,,
15 ,,	3·0 ,,
16·875 ,,	3·0 ,,
18 ,,	4·0 ,,
24 ,,	4·5 ,,

In multipolar design the radial depth will be given approximately by $\frac{r}{p}$, where r is the depth for a bipolar machine and p is the number of pairs of poles.

§7. DETERMINATION OF THE NUMBER OF CONDUCTORS ON THE ARMATURE.

The factor neither known nor approximated to as yet is the number of conductors. This is found by putting

$$C_a = \frac{E.60}{N_a n 10^{-8}}, \quad \cdot \quad \cdot \quad \cdot \quad \cdot \quad \cdot \quad (8)$$

which determines the trial value of C_a. Dividing the necessary number of turns into the space for winding on the periphery gives the width over the insulation of one turn. Deducting the insulation thickness, say, from 7 to 8 mils on each side of the wire, the remainder gives the net width of copper. C_a may be a number that will conveniently form one or two complete layers on the armature; if such be the case, a round wire may be generally chosen; but if the number will not fit into one or two layers, it will be advisable to use a wire of rectangular section, the width being such as to fill up compactly the winding space, and the depth such as to give a suitable air gap.

The current density has been thus far purposely neglected, as the determination of a suitable number of turns is of the first importance; and therefore special stress has been laid upon it. The relation of cross-section of copper to the current must now be carefully considered, since the heating of the armature ($i_a^2 r_a$), as well as the fall of potential ($i_a r_a$), and the efficiency of the machine depend on this. No current density can be said to be the best in all cases; it is a question of size of machine, cooling surface of armature, character of output, peripheral velocity, ventilation, and time of duration of maximum load. The practical limits are between 1,500 amperes per square inch in large machines, and 3,000 with outputs as small as 5 H.P. The permissible total heat waste for a given rise of temperature in degrees Fahrenheit may be roughly estimated for any ordinary armature by the following empirical equation, which is based on a surface velocity of about 3,000 feet per minute :—

$$\text{Rise of temp. Fahr. deg.} = \frac{\text{Watts } 64}{\text{Surface in sq. in.}} \quad . \quad . \quad (8\text{A})$$

The constant is determined for the exterior surfaces only; hence, for Grammes, the surface is given by the perimeter × the length + the area of the two end winding spaces; and for drums the length may be measured from the commutator along the extreme length of the winding in order to allow for the two ends which are usually covered up—if they are open their areas should be included in the cooling surface.

The usual rise of temperature arranged for by designers is about 70°F., and then the equation allows an area of rather more than one square inch per watt. This is an ample allowance for most small and medium size armatures, but it is not nearly sufficient with very large machines having massive iron cores. Internal ventilation and, perhaps, a "forced draught" have then to be resorted to—as is seen in the case of the modern bar wound drum armatures. The watts must be understood to comprise the heat from all sources, hysteresis, eddies, and $i_a^2 r_a$ loss in the copper. If the number of volts to be lost over the armature resistance be determined from Table A, then r_a is known, and the cross-section of copper to give this value with the calculated number of turns can be found. It is advisable also with Gramme armatures to consider the depth of copper in relation to the radial depth of armature core, since the output of the machine will vary as the depth of the copper up to a certain practical limit. This point is reached where the air gap becomes so deep that the resistance is too high for the field excitation to give the necessary magnetic density; or else when the circumflux,* $\frac{i_a C_a}{2}$, on the armature causes cross and back fields of such magnitude that sparking ensues. It should be remarked, however, that the output of the armature is not greatly affected by even considerable differences in the length of the gap. For, with the standard sizes of fields the excitation is practically fixed, and so a large gap with deep copper implies a weak magnetic density; and a small gap with less copper means a corresponding gain in the number of lines of force.

Two empirical equations, due to Mr. W. B. Esson, connecting the circumflux and the armature diameter, are important, for they give at once the permissible value of C_a.

---

* This expression for the circumflux of an armature is applicable to machines with any number of poles, and must not be confounded with the ampere turns. The numeral occurs simply because the symbol i_a is chosen to represent the total armature current and not that in each conductor.

For Gramme armatures—

$$\text{Circumflux} = \frac{i_s C_s}{2} = D\ 1000 ; \quad . \quad . \quad . \quad (9)^*$$

for drum armatures—

$$\text{Circumflux} = \frac{i_s C_s}{2} = D\ 1500, \quad . \quad . \quad . \quad (10)$$

D in each case being the diameter measured over the winding in inches.

Practically, with Gramme ring armatures, it is found that the best all-round results are obtained with a ratio of radial depth of copper to iron of from $\frac{1}{8}$ to $\frac{3}{16}$. With drums some 50 per cent. more copper may be allowed, as is evident from the empirical equations (1), (2), (9), and (10).

This ratio is important in Gramme armatures, because the narrow section of iron must be magnetised to at least $B_s'' = 100,000$. In drums this ratio has no meaning, since the iron is made as deep as possible, reaching even to the shaft itself, and the depth of copper is determined simply from the heating limits and the possible length of gap, having regard to an economical field-magnet and excitation.

The thickness of the copper being known, the length of the air gap can be determined approximately. It may have to be adjusted when the field-magnet excitation is calculated from the trial equations.

§8. DESIGN OF FIELD-MAGNETS.

The function of the field-magnets is simply to supply the necessary magnetic flux, with due regard to an economical distribution of iron and copper. The shape does not directly enter into the question, and is really a matter of convenience, and perhaps fancy. Broadly speaking, field-magnets may be divided into the bipolar and multipolar types, each of which is subdivided into a number of forms, sometimes pos-

---

* In (9) and (10) the diameter is expressed in inches instead of centimetres, and so the coefficients have different values from those given by Mr. Esson.

sessing real advantages, but often only expressing individual taste. Some of the more general designs are shown in Figs. 1 to 9.

(*a*) **The Single Magnetic Circuit Bipolar Field-Magnet.**— This popular shape of magnet is simply a development of

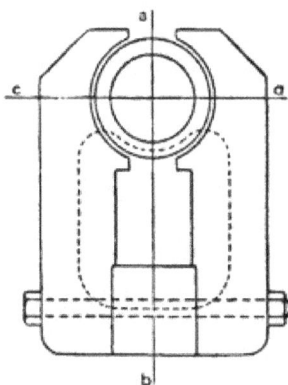

FIG. 1.—Wrought Cores, Cast Yoke, Upright Type.

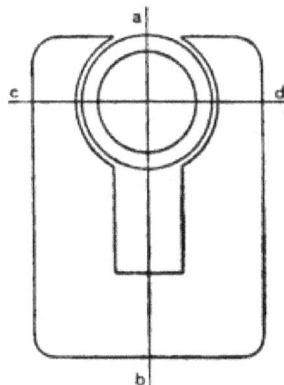

FIG. 2.—Cast Iron.

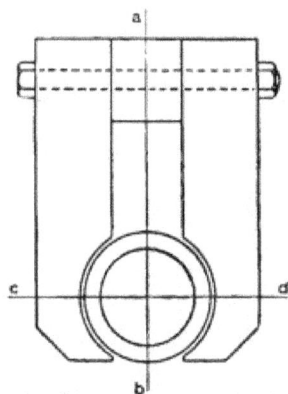

FIG. 3.—Wrought Iron Inverted Type.

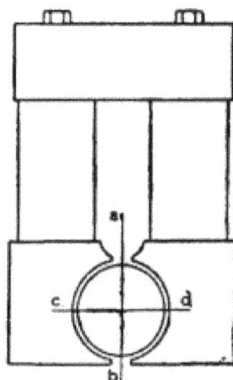

FIG. 3A.—Wrought Iron Cores and Yoke, Cast Iron Pole Pieces.

the primitive horseshoe magnet, and is largely used in both the upright and inverted forms (Figs. 1 to 3A). The former arrangement has a lower leakage coefficient than the latter, and

is generally adopted for small armatures ; but since the centre
of gravity of the rotating part is high there is a tendency to
instability, and, consequently, with large diameter armatures
the latter is more generally adopted.

The single magnetic circuit field-magnet is cheap to build,
and is excited with a relatively small quantity of wire.   Its
principal disadvantage lies in the unequal distribution of the
magnetic flux in the polar space, and hence inequality of induction in the armature wire.   For this reason drum armatures are most frequently used in these fields.   Large Gramme
rings are practically impossible with the single magnetic circuit. The largest size, in the author's opinion, that is advisable
is a diameter of about 12in.; if this be exceeded there is a
tendency to spark with large outputs, so that the armature
cannot be loaded to its proper limit.   Larger sizes should be
run in double magnetic circuit bipolar fields, or, if very large,
in multipolar fields.

(*b*) **The Double Magnetic Circuit Bipolar Field-Magnet.**—
Figs. 4, 5, and 6 illustrate this well-known form of field-
magnet, and show the variety of ways in which it is built.

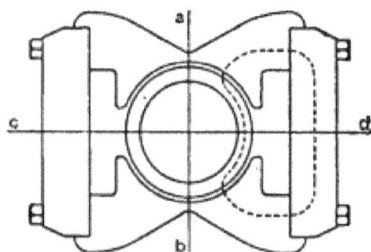

Fig. 4.—" Snell " Type of Double Magnetic Circuit Magnet. Wrought
Iron Cores and Cast Iron Pole Pieces.

The resulting distribution of magnetism is symmetrical, though
the "field" as a whole may be rotated considerably (*see* Fig. 14A,
page 51).   Therefore a Gramme armature as large as, say, 15in.
in diameter may be profitably excited in this type.   The chief
objection is that the number of ampere-turns in the exciting

coils has to be about twice as great as in the single-circuit type for a given induction in the armature. This type has also a larger coefficient of leakage than the single magnetic circuit.

If the armature be 18in. or more in diameter, it will be advisable to increase the number of pairs of poles, although by doing so the exciting turns will be somewhat more numerous.

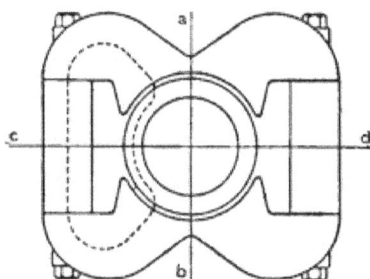

FIG. 5.—Wrought Iron Cores and Cast Iron Pole Pieces.

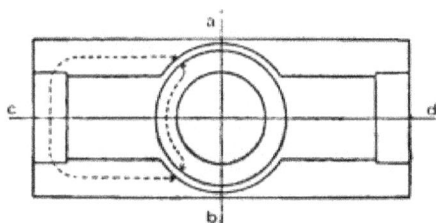

FIG. 6.—Wrought Iron.

(*c*) **Multipolar Field-Magnets.**—The two types most in vogue are shown in Figs. 7 and 8. It is not easy to state exactly at what diameter of armature it is commercially profitable to depart from the bipolar type. But there seems to be some connection between the angular width of the polar cavity and the diameter of the armature. The ratio of interpolar space to the width of the poles, with Gramme rings, is usually about 1:3; and the width of the pole, *w*, may be expressed in terms of the diameter, as

$$w = \frac{\pi D}{3}.$$

If the maximum economic diameter with a single magnetic circuit field be 12in. the width of pole will be $\frac{\pi\,12}{3} = 12\cdot5$ inches approximately. With a double magnetic circuit the diameter may be as large as, say, 15in.; and the polar width will be $\frac{\pi\,15}{3} = 15\cdot75$. The limiting width of the polar

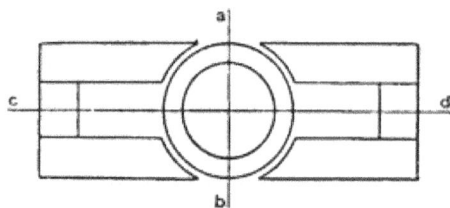

FIG. 7.—Wrought Iron. Four Poles.

cavities therefore varies between 12·5in. and 15·75in. These limits accord well with practice, poles seldom being found shorter than the former or longer than the latter in multipolar designs. Continental engineers have lately developed large Gramme rings with multipolar field-magnets, and have attained considerable success. Recently this design has been

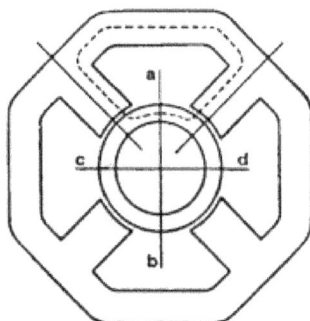

FIG. 8.—Multipolar Design in Cast Iron.

appreciated in England and America. Some four years ago the author selected it for large transmission-of-power machines, and also for central-station dynamos; Mr. Kapp has developed the same idea, using, however, drum armatures; and the Thomson-Houston Company have also adopted the device for their traction dynamos.

(*d*) Armour-clad fields, as in Fig. 9, are coming into use for certain purposes, particularly for motors which have to work in very exposed positions. For tramcars, they are often built with only one exciting coil, one pole piece being made very short.

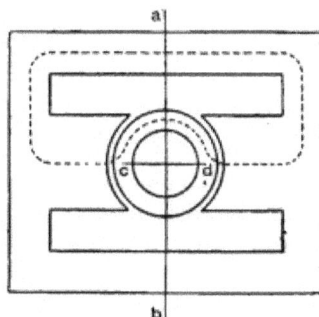

Fig. 9.—Made in Wrought or Cast Iron or with Wrought Poles and Cast Yokes. Armour-clad Type.

This type (*see* Fig. 10) offers no special advantages other than the excellent protection it affords to the armature and field winding.

(*e*) **General Conclusions with reference to Field-Magnet Design.**—A point to be specially observed before discussing the proportions of iron and copper is that no one design of field is equally well adapted for every size of armature. If cost of production be the ruling factor the single-magnetic circuit type is preferable for armatures up to, say, 12in. in diameter; from this to 15in. the double circuit may be used; but from 18in. and upwards one of the multipolar forms should always be employed.

This feature of dynamo design has received little attention outside of the workshop. The author has been led to investigate it when designing large machines; and he some time ago discussed at length the relation of weight to cost, with varying

types of machines.* He finds that the curves connecting
weight and output, and cost per kilowatt with output, follow
very different laws with distinct types of machines.   The

Fig. 10.—Thomson-Houston Tramcar Motor.

subject is interesting, but rather outside the scheme of this
book, and therefore only two diagrams will be given, which
show the general shape and order of the curves.   It will be
seen that the weight - output curve follows practically a

* *Electrical Review*, July, 1692.

straight line law, and that the output of a dynamo or motor is proportional to the cube of the linear dimensions of the machine, practically of the armature diameter (*see* Figs. 11 and 11ᴀ).

The question of material has not yet been mentioned, because this is rather a matter of cost and convenience. The necessary sections and corresponding inductions with wrought and cast iron are given in the following Table C, and the student can estimate for himself the relative advantages. Practically, with bipolar machines at the inductions usually employed, viz., $B''_t =$ from 77,000 to 100,000 in wrought iron and from 45,000 to 50,000 in cast iron, the relative magnet cross-sections should be in the ratio of $1\cdot6:3$, the effective armature section being taken as unity. Of late, cast steel magnets have been used with some success. The induction curve for cast steel lies nearer to that of wrought than to that of cast iron, and so there seems a probability of its use becoming more frequent.

Table C.

Table *giving Area of Wrought and Cast Iron to correspond to given inductions in air, assuming all the lines to pass through the iron and none to leak.*

Wrought Iron.		Cast Iron.		Air.
Area.	$B'_a.$	Area.	$B''_a.$	B.
1	70,750	2·2	36,500	100
1	90,750	2·17	42,000	200
1	100,250	2·17	46,000	300
1	100,575	2·07	48,500	400
1	100,650	1·9	52,500	500
1	100,770	1·9	54,500	600
1	110,050	1·93	57,250	700
1	110,150	1·83	59,750	800
1	110,250	1·8	61,750	900
1	110,265	1·74	63,500	1,000
1	110,500	1·7	65,250	1,100
1	110,650	1·7	66,500	1,200
1	110,750	1·6	67,750	1,300
1	110,800	1·6	69,250	1,400
1	110,900	1·57	70,250	1,500
1	120,000	1·4	71,250	1,600

From George Halliday's " Notes on Designs of Small Dynamos," p. 33.

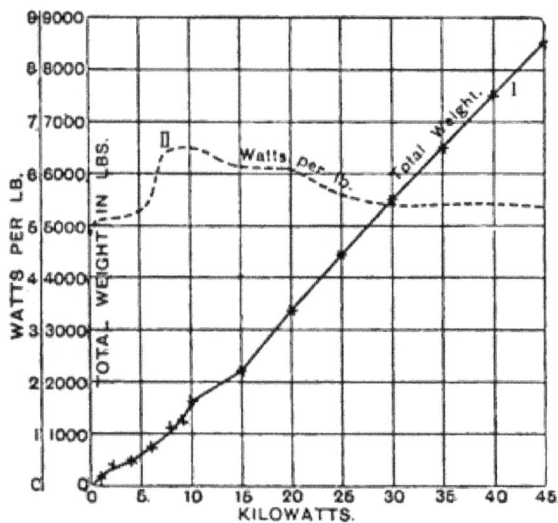

Fɪɢ. 11.

Curve I. connects total weight with output.
Curve II. connects watts per pound with output.

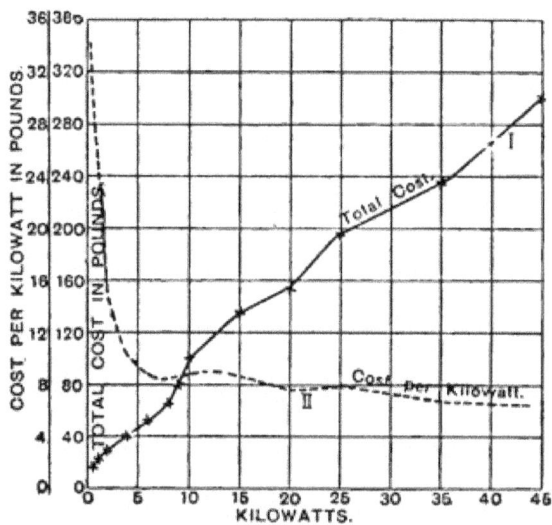

Fɪɢ. 11ᴀ.

Curve I. connects total cost with output.
Curve II. connects cost per kilowatt with output.

Annealed mild steel has met with some favour, but is not likely to supersede wrought iron at present. Good samples show a permeability equal to that of Swedish iron, and therefore better than that of ordinary magnet forgings, which are simply heavy machine scrap, and often of doubtful quality.

The shape of the magnet cores is of importance, because the length of the mean turn of the winding is affected thereby. The circle has the minimum perimeter for a maximum area enclosed, and therefore possesses the ideal section ; and the square gives, from this point of view, better results than any other rectangular form. Structural considerations, however, carry great weight, and it is found in many designs that a rectangular core, with length equal to twice or three times the depth, most nearly meets all the conditions, in spite of the extra expense in copper.

§ 9. CALCULATION OF THE MAGNET DETAILS.

(*a*) **Winding Space.**—In the preceding section the usual areas of cross-section of armature and field-core were shown to be in the ratio of $1:1.6$ for wrought-iron magnets, and $1:3$ for cast-iron. These ratios are not absolutely the best in all circumstances, but in practice they give economical results. It may be assumed, then, that a type of field is chosen, and that the material has been decided on. The area of the field-magnet cores will also be known from its relation to the area of the armature iron. The open factor at present is the length of core to carry the exciting coils. This cannot be settled without calculation ; but a few trial-and-error determinations will show the minimum length required to carry the necessary turns of wire, allowing at the same time the proper area of surface for the watts spent. For bipolar single magnetic circuit machines 0.75 D, for double circuit 0.9 D, and for multipolar fields from 0.4 D to 0.5 D will be very nearly correct for all ordinary designs. And when the other quantities are properly balanced, the number of square inches in the surface of the winding, not including the ends, will be equal to about twice the number of watts wasted in heat ; or each watt of energy wasted in the magnet windings will be dissipated through a superficial area of two square inches. This is found to be sufficient in machines with windings not exceeding 2·5in. in depth; but for large machines a greater margin is preferable if it can be obtained. A useful equation for determining the probable rise in temperature in degrees Fahrenheit is—

$$\text{Rise of temp. Fahr. deg.} = \frac{\text{Watts } 100}{\text{Surface in sq. in.}} \quad . \quad (10\text{A})$$

Here, the cooling surface does not include the ends of the shapes, but is expressed by the perimeter of the coils × their height. The permissible rise of temperature is generally about 60°F.; under these conditions the empirical equation allows a radiating surface of nearly 1·7 sq. in. per watt. This is sufficient for small coils with shallow winding; but if the depth be more than 2·5in., the margin must be increased.

(b) **Poles and Polar Extensions.**—The next point to determine is the length of the polar cavity. This has already been referred to in §8, division (c), and the practical limits have been given. It may be further noted here that since the magnetic resistance is lowered by increasing the width of the polar cavity, it appears advisable to make the arc embraced as large as possible. But this is limited by the strength of the cross field, which is directly proportional to the extent of the pole pieces. With drum windings the lead of the brushes is comparatively little, as the field disturbance is small, and so the polar angle may be somewhat larger than is permissible if a gramme ring be used. A safe rule is to make the angle 135deg. for drums and 125deg. for Grammes; but both of these figures may require to be slightly modified by the peculiar conditions of the design.* In a well-proportioned dynamo, when the armature is carrying the maximum circumflux for its diameter,† the brushes will require a lead just sufficient to bring them under the tips of the pole piece, so that the lead and the width of the polar cavity roughly depend on each other. The ratio of area of polar cavity to the cross-section of the magnet core is determined mainly by the induction B''_g in the gap. It is not found economical to push the flux beyond $B''_g = 33,000$; and, therefore, neglecting leakage, the ratio of area of core section to polar

---

* *See* § 8, division (c).   † *See* equations (9) and (10), p. 22.

cavity will be, approximately, $1:2.75$ if wrought iron be used, and $1:1.6$ for cast iron. But there is always a certain loss of magnetism from leakage, and it is necessary to allow for this in estimating the total field core flux. Calling the leakage coefficient v, the relation between the field lines and the armature lines is expressed by

$$N_f = v\, N_a \text{ or } v\, N_g.$$

The assumption that the armature induction is equal to the induction in the gap will not introduce a large error for ordinary designs, and it simplifies matters. The stray field is not constant for every type of magnet, but varies with the armature induction, increasing in value as B''_a approaches its maximum. The practical values of v in the types of machines most in vogue are given in Table **D**.

Table D.

TABLE *giving Approximate Values of v, the leakage coefficient, for various types of Field-Magnets.*

Type of Magnet.	Open Circuit.	Full Load.
Single magnetic circuit, upright type, Figs. 1 and 2	$v = 1.3$	$v_1 = 1.32$
Single magnetic circuit, inverted form, Figs. 3 and 3a	$v = 1.31$	$v_1 = 1.33$
Double magnetic circuit, Figs. 4, 5 and 6	$v = 1.4$	$v_1 = 1.42$
Four-pole machines, Fig. 7	$v = $ from 1.4 to 1.45	$v_1 = $ from 1.42 to 1.47
„ „ Fig. 8	$v = $ from 1.45 to 1.5	$v_1 = $ from 1.47 to 1.6
Armour-clad field, Fig. 9	$v = 1.2$	$v_1 = 1.22$

(c) **The Yoke.**—The yoke is that part of the magnet which connects the cores. It is not usually covered with windings, and is made either in wrought or cast iron so as to suit the design. Generally cast iron is preferred, since the bed and yoke can, in many cases, be combined in the same pattern. If this device be adopted, as in Fig. 1, p. 23, care must be taken that the path through which the magnetic flux will principally pass is at least as wide as the area of cross-section of the cores, and usually it is found advisable to make it about 1·2 times larger, if the cores and yoke are of the same material. If the cores be made in wrought and the yoke in cast iron, the area of the cross-section at right angles to the lines of force should be about 1·6 to 2·0 times as great for the latter as for the former. Attention must be given to the joints between the cores and the yoke, that they be truly machined and be in close contact over the entire face. In machines of the double-magnet circuit type there is no yoke, since the pole pieces are connected directly to the magnet cores ; these joints must be carefully made.

In some designs the polar cavity can be bored from the solid iron (*see* Figs. 2, 3, and 6) ; but in others it is necessary to attach extension pieces, so as to increase the polar cavity (*see* Figs. 1, 4, and 5). But the dimensions of the pole-pieces will in all cases be governed by the rules already given.

§ 10. CALCULATION OF THE FIELD EXCITATION.

(*a*) The following points have now been determined :—The size of armature, the gauge of wire, and the number of turns for its winding ; the shape of field-magnets, and length of the winding space ; the bore of polar cavity and the width of the poles ; the magnetic flux, and the densities in armature, gap, field cores and yokes.

The excitation necessary to give these fluxes and the corresponding winding have now to be calculated. The method adopted by the author does not take into direct consideration the permeability of the iron, this being understood in the proposed limiting values of B. It is based on curves (*see* Fig. 12) connecting induction per square inch of core with ampere-turns per linear inch of distance through which the induction has to be maintained. (The principle of these curves was first suggested by Drs. J. and E. Hopkinson in 1886.) The co-ordinates are the mean values of the ascending and descending curves of the magnetisation. The variation of induction due to hysteresis is most marked with cast iron, and is of little importance with well annealed Swedish plates, such as are used for armature cores, especially if the induction be about $B'' = 100,000$. Therefore in practice no trouble is caused by hysteresis other than the loss of energy due to the cyclic change of magnetism, which is rendered evident by a development of heat. The calculation for this is given in Table L and footnote on page 173.

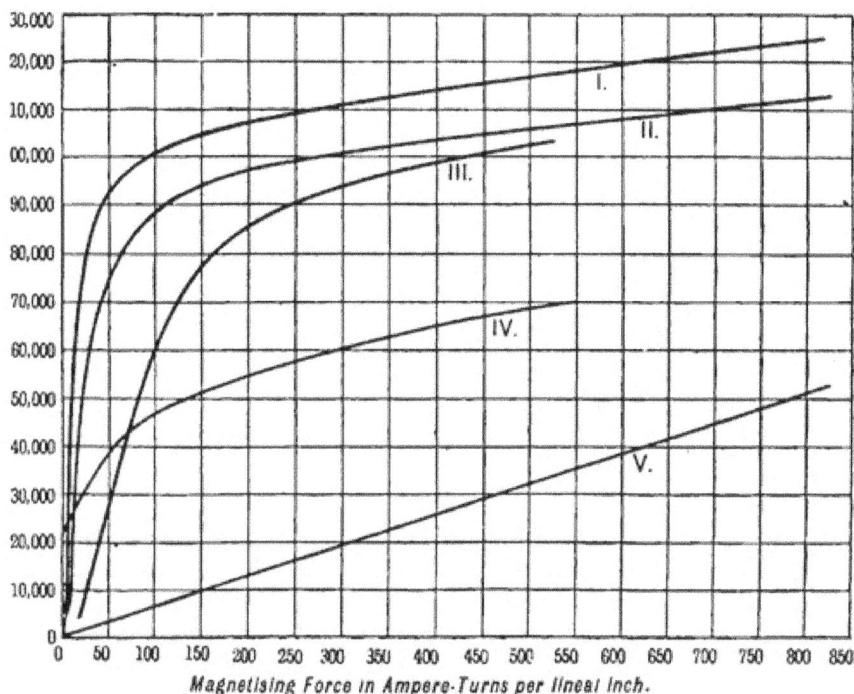

FIG. 12.—Curves of Magnetic Induction in Iron and Air.

Scale of Ampere-Turns for Iron :—One division = 50 Ampere-Turns per inch.
    I. Annealed Swedish Charcoal Plates, as used for Armatures.
    II. Wrought Iron Forgings, as used for Field-Magnets.
    III. Cast Steel, as used for Field-Magnets.
    IV. Cast Iron, as used for Field-Magnets.
    V. Air—To read Ampere-Turns multiply abscissæ by 20.
All the Readings have been determined from the Mean of the Ascending and Descending Curves of Magnetisation.

In Fig. 12, Curve I. refers to annealed charcoal iron plates, from 22 to 24 B.W.G. The data are taken from experiments made by Mr. George Halliday.

Curve II. relates to magnet iron, large scrap, hammered and allowed to cool slowly after forging. The values of the co-ordinates were originally taken from Drs. J. and E. Hopkinsons' Paper on the characteristics of the dynamo ; but they have been altered to suit the samples of iron used by the author, most of which have been supplied from Yorkshire forges.

Curve III. is taken from the recent experiments of Mr. Steinmetz with annealed cast steel, which is likely to be largely used for magnets.

Curve IV. is for the class of grey cast-iron generally supplied in London for magnets. The samples which have recently come under the author's notice have perhaps a slightly higher permeability than that implied in this curve. However, it is best to be on the safe side with cast iron, as the cooling so largely affects the results, and with these values there will be no doubt about the output of the machine.

Curve V. shows the number of ampere-turns required to send the induction through the gap. Its abscissæ are calculated from the equation

$$\mathbf{B}''_{g} = \frac{\text{ATs}}{0\cdot31328} \cdot \frac{1}{\delta}, \quad \cdots \quad (11)$$

where $\delta = 1$ in., and $0\cdot31328$ is a dimensional coefficient to convert the absolute measures into practical units. To use this curve find the ampere-turns corresponding to the density, and multiply by the length of the gap expressed in decimals of an inch.

(b) **Excitation for Open Circuit.**—The data must now be collected in order to make the calculations for the number of ampere-turns. It is necessary to know for the armature :—

B''_{a} = the induction per unit area.

$A_{a}0\cdot85$ = the area of cross section, less 15 per cent., which will be occupied by the insulating material between the plates.

m_{a} = mean length of magnetic path in armature core.

For the gap :—

B''_g = the induction per unit area.

δ = length of the air gap on one side.

A_g = area of one of the polar cavities.

For the field-magnets :—

B''_f = the induction per unit area in the cores.

B''_n = the induction per unit area in the yoke.

A_f = area of core in square inches.

A_n = area of yoke in square inches.

m_f = mean length of magnetic path in cores and pole-pieces.

m_n = mean length of magnetic path in yoke.

Since the design is symmetrical about the perpendicular line $a\,b$ (*see* Figs. 1 and 4), it is only necessary to measure the path of the lines of force around one-half of the circuit. The excitation thus determined refers to one limb only, and must be doubled to give the full number of ampere-turns. This method is usually adopted because it both saves time and gives smaller figures.

The excitation for the several parts of the circuit is :—

Armature :— $m_a\, f\, (B''_a)$.

Field cores :— $m_f\, f\, (B''_f)$.

Field yoke :— $m_n\, f\, (B''_n)$.

Gap :— $\delta\, f\, (B''_g)$.

And, summing up—

$$\text{ATs} = m_a\, f\, (B''_a) + m_f\, f\, (B''_f) + m_n\, f\, (B''_n) + \delta\, f\, (B''_g). \quad . \quad (12)$$

(*c*) **Excitation to be added at Full Load.**—In the preceding equation the demagnetising effect of the armature is neglected, and no allowance is made for the fall of pressure over the internal resistance of the armature and series coils.

The demagnetising and cross-magnetising fields due to the armature have been already referred to, and it has been shown that these effects are most pronounced with the Gramme type of armature. The subject has been exhaustively treated during the past two years, and need not be discussed in detail here. Suffice it to say, that the band of conductors

E

lying between the pole tips will, as a whole, tend to produce a field in the opposite direction to that due to the main excitation ; while the coils lying under the pole-pieces, *i.e.*, in the polar cavity, will tend to cause a cross magnetism. The latter is not of serious importance if the pole-pieces be very thin near the centre, as at a and b in Figs. 4 and 6. If, however, the pole-pieces be very massive in the region of the centre line there will be a cross field, which will have a tendency to twist the main field by crowding out some of the forward induction. But this can be almost entirely obviated with the relative proportions and shapes of magnets given in § 8.

The demagnetising or back ampere-turns are of more importance, and must be allowed for by an equal forward induction.   The disturbance is, with Gramme rings, proportional to the armature current multiplied into the number of turns included in the angle of lead ; and this product must be multiplied by the coefficient of leakage between the field and armature at full load, since a certain quantity of the field induction is lost in space.   Therefore the compensating turns are

$$\text{BTs} = C_a \cdot \frac{2\,\theta}{360} i_a n_1 = C_a \frac{\theta}{180} i_a n_1 \quad . \quad . \quad (13)$$

where θ = the angle of lead in degrees.

With drum armatures the disturbance is about one-half as great, and so the compensating turns are

$$\text{BTs} = C_a \frac{\theta}{360} i_a n_1. \quad . \quad . \quad . \quad (14)$$

To find the turns to compensate for the fall of potential, it is necessary to make two calculations for the ATs, see (12), one with no current flowing in the main circuit, and hence no armature reactions, and one at full load.   The former will determine the strength of the shunt coils if the machine be compound wound ; and the latter will give the shunt excitation plus extra turns to be added for the fall of potential due to the internal resistance.   The total excitation is :—

ATs for open circuit + BTs for demagnetising effects + CTs for fall of potential.

Practically, with large Gramme rings the two latter quantities will be about one-third of the first.

If the machine be shunt wound the winding must be determined for full load, and the BTs must be added before calculating the gauge of wire. The same remark also applies to a series machine. A few trial windings will soon illustrate the point, and facility will come with experience.

Mr. W. B. Sayers' device* for the "prevention and control of sparking" is likely to introduce important modifications in dynamo and motor design. It is well recognised that the limits of output of a given machine are determined simply by the heating and the sparking effects. The former is controlled by the current density, the eddies in the conductors, and hysteresis; the latter depends mainly on the circumflux and the relative proportions of the machine, or, briefly, on the armature reactions. It is well known that a dynamo requires a forward lead and a motor a backward lead to prevent sparking, and it is also recognised that the amount of the lead is proportionate to the load. The effect of the angular displacement of the diameter of commutation is to cause the convolutions, included in the angle of lead to be traversed by currents in an opposite direction to the E.M.F. induced in them, and in a direction such as to tend to demagnetise the field-magnets. The demagnetising effect of this belt of back-ampere turns is merely an accidental condition, and is compensated for by extra excitation as already shown; but the reversed E.M.F. is an absolute necessity in order to check the self-induction of the current in the coils short-circuited beneath the brushes at the moment of current reversal. The explanation is simple. If the current in these coils is acted upon by an E.M.F. in the same direction as itself, or even if the coils are in a neutral field, it is clear that the current cannot be stopped and reversed without a self-induction spark; but if the coils, at the moment of reversal, are in a magnetic field of such direction and intensity as to induce an E.M.F. just sufficient to counter-balance the self-induction of the current, then the reversal may be made without a spark, and the collection of the current will be

_____

* *Proceedings* of Institution of Electrical Engineers, Vol. XXII., Part 107.

sparkless. This apparently complicated action is easily accomplished in practice by simply giving the brushes sufficient lead ; and it is obvious that since the armature reactions are proportionate to the lead that the best designed machines will have the least lead. The converse of the problem is also suggested for consideration, for if a forward lead tends to demagnetise the field-magnets of a dynamo it may be reasonably inferred that a backward lead would tend to strengthen them. This is found to be so in practice, and many attempts

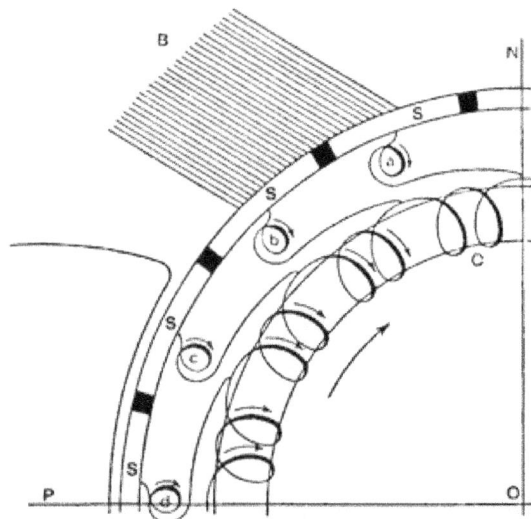

Fig. 12a.—P, Trailing Horn of Pole-piece. B, Positive Brush. *a, b, c, d,* Commutator Coils to furnish the E.M.F. necessary for Sparkless Reversal. S, S, S, S, Segments of Commutator. The current in main coil C has to be reversed in passing under the brush ; this is effected by the E.M.F. in the coil *b,* which is actively cutting Magnetic Lines in the Dense Field under the Pole Horn, and hence can overcome the Self-Inductive E.M.F. of the current in C. O P, the Axis of Poles, and O N the Neutral Axis at right angles to O P. The large arrow shows the direction of armature rotation, and the small ones the direction of the E.M.F. acting in the several coils. The current is flowing round both sides of the armature to the positive brush, B.

have been made to assist the field excitation by the armature reactions, but without success, until Mr. Sayers invented his *Commutator Coils,* as he calls them. Their application will be seen at once by reference to Fig. 12A, and a little study will make clear the theory of their action. In the diagram, which represents a Gramme ring dynamo, the commutator bars, S, are drawn, for convenience, outside the armature, the main windings are seen to be continuous, and each coil to be coupled to a segment of the commutator through a commutator coil, *a, b, c, d,* the order of connecting up being one main coil backwards. It is of special importance to mark that these coils only carry current when touching the brushes, and then the whole armature current passes through them momentarily; also the brush, B, is made sufficiently wide to always bridge over two segments. The direction of motion of the armature is clockwise, as indicated by the large arrow, the brush has, therefore, a *backward lead,* and the *active* commutator coil, *b,* is, practically, just beneath the edge of the trailing pole piece, while the main current in C is commutated at a diameter not far removed from a line at right angles to the axis of the pole pieces. In this position the field due to the armature magnetism is partly added to that of the field-magnets, and sparking is prevented by the E.M.F. induced in the commutator coils, which are cutting lines of force in the dense magnetic field at the trailing horn of the pole piece. These effects will obviously increase in intensity as the brush is brought nearer to the pole piece. The importance of this device lies not so much in the application, just illustrated, as in the new line of investigation suggested by it. In the first place, it permits a much smaller air gap, and, therefore, less excitation in the field coils, which means smaller magnets; secondly, since sparking can be obviated, or its magnitude restricted, a greater output may be expected from a given armature, heating being the only limiting condition. A host of applications at once present themselves for immediate consideration, amongst which may be instanced:—*Series-wound motors* for traction purposes, to which light compact machines are essential; and *regulators,* for compensating the fall of potential on feeders, which, carrying widely different loads, are always subject to sparking troubles (*see* § 25, page 137).

The winding details present some difficulties in practice, especially as regards insulation; and generally the device

lends itself most readily to slotted cores in which the main
winding can be wound underneath and the commutator coils
on the top, so as to be all but flush with the surface.   In one
form Mr. Sayers has used the commutator coils as keys to
hold the main winding in position, as shown in Fig. 12B, which
refers to a slotted drum.   In drum windings the ends of the
main windings are brought to the back end of the armature
away from the commutator, and the commutator coils are
simply lengths of wire sufficiently long to couple up to the
bars.   In Gramme rings the commutator coils are wound in a

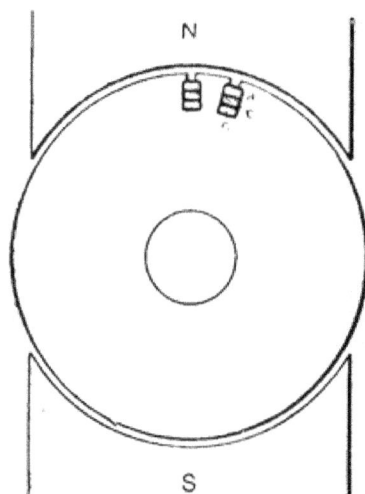

Fig. 12B.—Sayers' Winding as applied to Drums.
a, Commutator Coils.   c, Main Coils.

plane at right angles to the main coils, and may therefore lie
on the outside of the armature.   Various methods of carrying
out the device will suggest themselves to different designers,
but it is too soon to predict the probable results.

§ 11. SPECIAL POINTS TO BE OBSERVED IN DESIGNING MOTORS.

So far the author has considered the design of either dynamo or motor simply from the point of view of the total E.M.F. of the armature, and has given no special attention to the torque. But it was mentioned in §2 that in designing motors it was essential to give due weight to both the maximum and average couples likely to obtain. The importance of this will be best seen from an inspection of the equation for the torque in terms of the armature constants :—

$$\text{Torque} = T = \frac{C_a \, i_a \, N_a}{k}, \quad \ldots \quad (15)$$

or the torque is directly proportional to the product of the number of turns of wire, the magnetic flux, and the armature current; k being simply a dimensional coefficient. The numerical value of k is $852 \cdot 3 \times 10^6$, if T be in pound feet, so that, adopting the practical unit,

$$T = \frac{C_a \, i_a \, N_a}{852 \cdot 3 \times 10^6} \text{ pound feet.} \quad \ldots \quad (16)$$

Now, in all ordinary motors the number of turns of wire in the armature, C_a, is constant, and only the current i_a, and the magnetic flux N_a vary. There are two simple cases for consideration in practice. First, when both i_a and N_a vary. This occurs with a series-wound motor, and it should be noticed that a relatively small increment of current may cause a considerable increase in the value of the torque, for the magnetism will be greatly varied at the straight part of the characteristic with low values of **B**, and therefore N_a will at this point increase much more rapidly than i_a; whilst after the knee of the curve is passed the rate of increase of N_a for increments of i_a will slowly decrease, until a point is reached where the torque varies simply as the current.

The second case is where N_a is constant, and i_a alone varies. This condition is given with a shunt-wound motor, if it be assumed that the armature reactions are negligible. The torque, therefore, will simply vary as the armature current; it will be a minimum with small loads, but will finally attain the same maximum value as it would if the motor were series wound. Both of these windings give special advantages to a motor, and hence they are severally selected for particular work. The compound-wound motor is a combination of the two methods, and has not met with much favour as yet. The uses of the three types of winding are more fully discussed in §14.

It remains now to investigate the best lines upon which to build armatures for special turning moments.

If the desired brake horse-power and the speed of the motor be known, the work per revolution and the torque can be easily determined.

The work per revolution in foot pounds

$$= \frac{\text{B.H.P. } 33000}{n} \text{ foot pounds . . . (17)}$$

and the torque $T = \dfrac{\text{B.H.P. } 33000}{2\,\pi\,n}$ pound feet, . . . (18)

where $n =$ revolutions per minute.

It should be noticed that T is independent of the diameter of the armature.

The peripheral pull exerted by the armature

$$= \frac{\text{B.H P. } 33000}{2\,\pi\,n} \cdot \frac{1}{g} \text{ pounds . . . (19)}$$

where $g =$ the radius of the armature in feet.

This force will act tangentially on the winding, *practically* at two parts diametrically opposite each other, and just underneath the leading horns of the pole-pieces. If the value represented by B.H.P. be the average load, we can thus determine the average strain on the wire ; and if it equal the maximum output the maximum *running* torque is known. But suppose the current to fluctuate largely, and at times to exceed that corresponding to the maximum output—at starting, for example, the momentary rush of current may easily be twice or even three times as great. Then, from what has been said, it is clear that the torque will be increased also, and it may be so sufficiently to endanger the stability of the winding, or even to bend the shaft. Therefore, in motors it is usual to allow a very large factor of safety by supporting the winding with specially-devised driving horns, and by the use of much stiffer shafts than are necessary for dynamos of the same average output. In designing motors, then, it is advisable first to examine the limits of the turning couple, and then to take precautions to ensure sufficient mechanical strength. For tramcar and coal-cutting motors it is usual to provide shafts from $2\frac{1}{2}$in. to 3in. in diameter for outputs of from 15 to 20 H.P.

§15. PRACTICAL EXAMPLE OF PRECEDING METHOD OF DESIGN.

The application of the principles of dynamo and motor design, described in the preceding sections, is fairly illustrative of the ordinary methods followed in most drawing offices, but the details are varied to suit the fancy or necessities of each engineer.

The design has been treated in a general way, so as to include as many cases as possible, and hence appears to be complicated. In practice, however, owing to fixed types of field magnets, definite sizes of armature plates, and data from machines already built, it is very simple. A machine can be designed for a given output at the required speed in a very short time ; indeed, nearly all of the calculations will be in simple proportions that can be seen at a glance on a slide rule.

However, to illustrate the use of the equations, the following calculations for a 25-unit dynamo and a 25·5-B.H.P. motor are appended.

It is assumed that the problem is to build a pair of machines to transmit 25·5 B.H.P. a distance of one mile from the dynamo. Since there is only one motor, the series winding will be most suitable for both machines,* this arrangement giving fairly good

---
* *See* Chap. IV., § 23.

speed regulation of the motor for variable load if the generator
speed be kept constant.

The efficiency of the motor may be safely taken at 90 per
cent., and, therefore, the watts absorbed by it at full load
will be

$$\frac{25\cdot5}{0\cdot9} = 28\cdot32 \text{ E.H.P.}$$

If the line loss be fixed at 5 E.H.P., the output of the dynamo
will be 33·32 E.H.P. = 24,857 watts. On referring to Fig. 13,
which gives the volts and horse power lost per mile of cable,
with different currents and areas, it is seen that a wire of
sectional area corresponding to No. 6 S.W.G. will carry 35·75
amperes with a drop of 52 volts and a loss of 2·5 E.H.P. per
mile. This is a convenient size of wire,* as small as it is practically safe to erect for a permanent line and is not too costly;
it weighs 589lb. per mile, and has a resistance of 1·4694 ohms
for the same length. The two miles of line wire will therefore
have a resistance of nearly 3 ohms and the fall of potential will
be 104 volts, and the watts lost in it 3,720. Dividing the
dynamo output by the current (35·75 amperes), the potential
difference is seen to be approximately 695 volts. Since the
pressure is high the armature winding will be of the Gramme
type, and the double magnetic circuit two-pole field (Fig. 4)
will be selected for both machines. The armatures will be
made interchangeable in order that one "spare" may cover
the two machines; and the field carcases will also be identical
save in respect of the number of ampere-turns, and perhaps the
area of the magnet cores.

Determination of Armature Dimensions :—

From $l \, D^2 \, n \, 0\cdot01$ . . . . . . (1)

$$15 \times 15^2 \times 800 \times 0\cdot01 = 27,000,$$

where $l = 15,$ $D = 15,$ $n = 800.$

This will give a safe margin in output.

Now, $E = e + i_a \, (r_a + r_m)$ . . . . (5)

and, also $E = 695 + (0\cdot06 \times 695) = 736.$. *From Table A.*

(Take one and a-half times the percentage, since the field resistance must
be included.)

---

* The economic area is purposely neglected in this case, as the subject has
not yet been referred to in the present work.

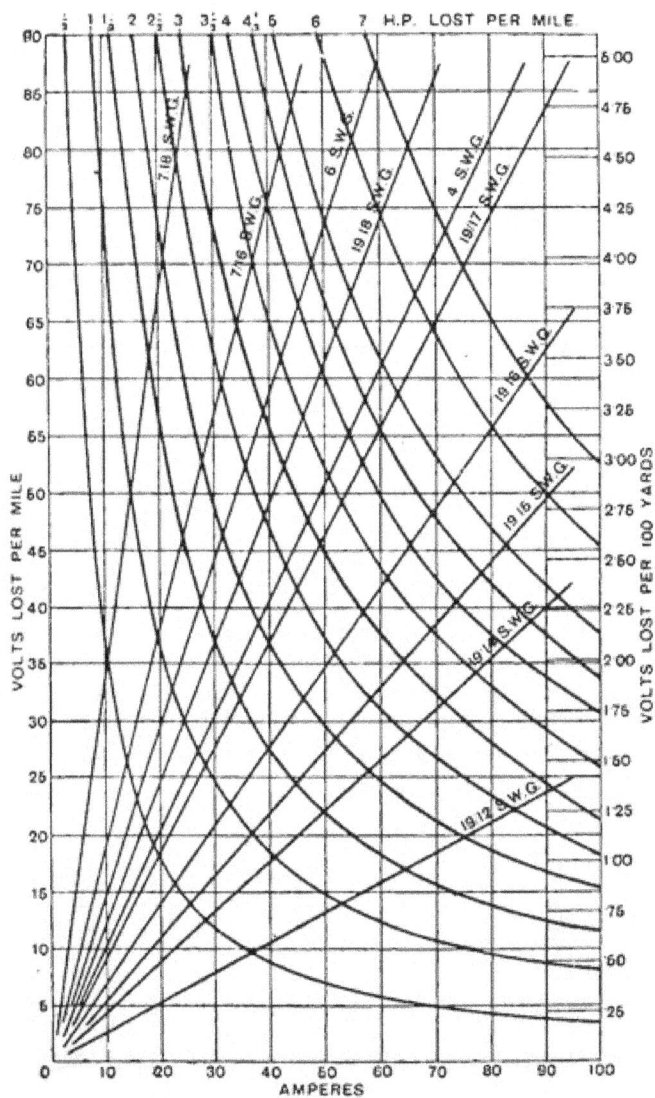

FIG. 13.—Curves showing Volts lost per Mile and per 100 Yards with different Currents and usual Sizes of Cable. Calculated from the formula E = (0·0427/area) C. Also H.P. lost per Mile.

$$N_a = B''_a \, 2 \, r \, l \, 0\cdot85 \quad . \quad . \quad . \quad . \quad . \quad . \quad . \quad . \quad (7)$$
(See Table B.)

$$l = 15, \quad \text{let } r = 2\cdot5, \quad \text{and } B''_a = 110,000.$$
$$\therefore \; N_a = 110,000 \times 5 \times 15 \times 0\cdot85 = 7\cdot0 \times 10^6.$$

Put this trial value of N_a in (8)

$$C_a = \frac{E \, 60}{N_a n 10^{-8}} = \frac{736 \times 60}{7\cdot0 \times 10^6 \times 800 \times 10^{-8}} = 780.$$

The circumference of armature $= \pi d = 47''$, deduct a space of $2''$ for the driving horns, leaving $45''$ for the winding. No. 11 S.W.G. will carry $17\cdot875$ amperes at a density of $1,800$ amperes per square inch : it is $0\cdot116''$ in diameter when bare, and with 16 mils of insulation $= 0\cdot132''$. Therefore, with No. 11 wire the turns per layer will be $\frac{45}{0\cdot132} = 340$. Two layers will $= 680$ turns, *i e.*, say 68 coils of 10 turns each. The resistance when cold will be about $0\cdot37$, and when hot about $0\cdot4$ ohm.

Ampere-turns per coil $= 10 \times 17\cdot875 = 179$ say

Permissible circumflux $= D \, 1000 = 15,000 \quad . \quad . \quad . \quad (9)$

Arranged circumflux $= \dfrac{i_a \, C_a}{2} = \dfrac{35\cdot75 \times 680}{2} = 12,200.$

Two layers of this wire give fewer turns than the value of C_a determined from (8), and *three* layers give too many, besides increasing the air gap unduly and raising the circumflux above the safe limits given by (9). This is a common case in practice; the remedy is to raise the speed, or to increase the armature induction : or to do both,

making $\qquad n = 800 \times \dfrac{780}{680} = 920,$ say ;

or, making $\qquad N_a = 7\cdot0 \times 10^6 \times \dfrac{780}{680} = 8\cdot00 \times 10^6$;

or, by varying both n and N_a.

To illustrate these points the combined method is chosen.

Since B''_a is already $110,000$, in order to increase N_a it is necessary to make the cross-section of the armature larger, and to alter the radial depth of core, r, from $2\cdot5''$ to, say, $2\cdot75''$. Therefore $2\,l\,r$ will $= 82\cdot5$ sq. in., and the effective armature section will be $82\cdot5 \times 0\cdot85 = 70$ sq. in.

The new value of N_a is therefore

$$7 \cdot 0 \times 10^6 \times \frac{2 \cdot 75}{2 \cdot 5} = 7 \cdot 7 \times 10^6.$$

The speed must be raised to give the required conditions. Substitute these new values in (8), and solve for n

$$n = \frac{E\ 60}{C_a\ N_a\ 10^{-8}} = \frac{736 \times 60}{680 \times 7 \cdot 7 \times 10^{-2}} = 835.$$

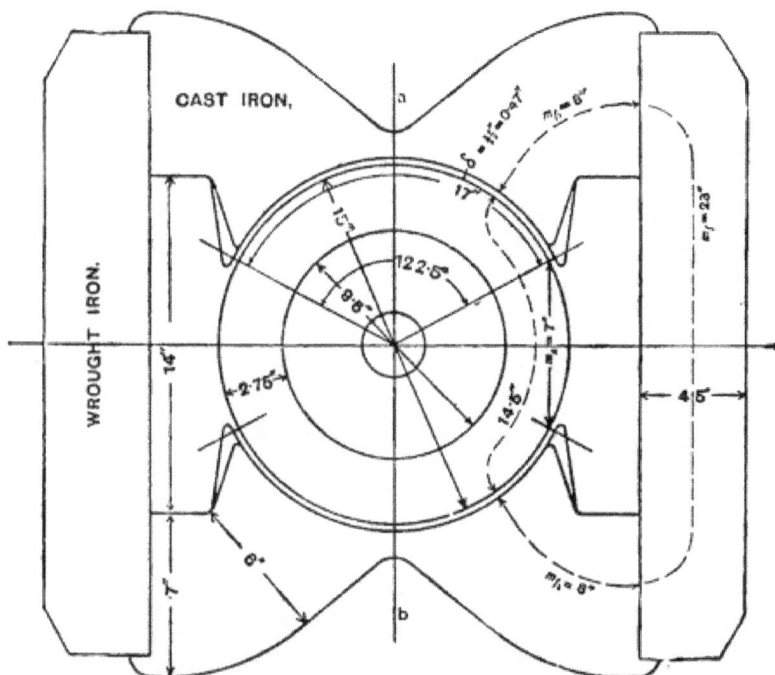

Fig. 14.—Cross Section of Dynamo, showing Method of Design.

The bore of the poles must next be determined. Allowing for two layers of No. 11 S.W.G., suitable insulation, bands, and clearance of, say, 0·0937 (3/32") of an inch all round, the required bore is 15·75, and $\therefore \delta = 0 \cdot 375$. And let the angular width of the pole pieces be 120° at the edges, and 125° in the centre; *i.e.*, let the "horns" be curved in a direction parallel to the winding (*see* Fig. 14). This tends to destroy the abruptness of the magnetic flux, and to distribute it more gradually on the leading horns. The average width of the pole pieces, w, is $122 \cdot 5°$, or, say, $17''$.

The length of the poles should be a little less than that of the armature core, say, in this case, $14 \cdot 5$; therefore the area of each cavity is $17 \times 14 \cdot 5 = 247$ sq. in., and B''_g, the density in the gaps, is found by dividing N_a by 247,

or, $$\frac{7 \cdot 7 \times 10^6}{247} = 31,000.$$

This is a convenient density.

Next, to determine the field induction :
$$N_f = v_1 \, N_a = 1 \cdot 42 \times 7 \cdot 7 \times 10^6 \quad . \quad (v_1 = 1 \cdot 42)$$
$$= 10 \cdot 95 \times 10^6,$$
and half of this flux, $5 \cdot 475 \times 10^6$, must be supplied by each limb of the magnet.

Let the cores be wrought iron, and $B''_f = 85,000$ say, and the area of each core $= \dfrac{5 \cdot 475 \times 10^6}{8 \cdot 5 \times 10^4} = 65$ sq. in. nearly.

Now the length of the pole pieces has been fixed at $14 \cdot 5''$; let the magnet cores be the same, then the width of each core is
$$\frac{65}{14 \cdot 5} = 4 \cdot 5''.$$

The dimensions of the machine are now sufficiently developed to enable a drawing to be made showing a cross-section of the field and armature cores ; let this be to, say, an inch and a-half scale, as in Fig. 14. The length of the magnetic paths can be estimated, and the field winding calculations made.

$m_f = 23''$; $m_{f1} = 16''$; $m_a = 7''$; $A_a = 2 \, l \, r \, 0 \cdot 85 = 70$; $A_f = 14 \cdot 5 \times 4 \cdot 5$ $= 65 \cdot 0$; $A_{f1} = 110$; $B''_a = 110,000$; $B''_f = 85,000$; $B''_{f1} = 50,000$; and $B''_g = 31,000$.

The path of the induction in the armature can only be averaged, since the density is not constant (see Fig. 14A, which has been drawn specially to show the probable distribution of the lines of force). It is noticeable that the maximum induction only takes place in the areas bounded by the lines a a_1, and b b_1. Averaging the path to the density chosen, $B''_a = 110,000$, it is seen that the mean distance between the pole tips is a fair approximation to the value of m_a.*

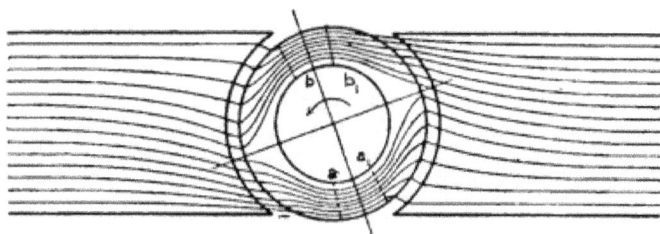

FIG. 14A.

The excitation per limb for full load is from (12)—

ATs for armature	$= m_a f (B''_a) =$	$7 \times$	$280 = 1,960$
„ cores	$= m_t f (B''_t) =$	$23 \times$	$77 = 1,771$
„ pole pieces	$= m_{tl} f (B''_{tl}) =$	$16 \times$	$130 = 2,080$
„ gap	$= 2 \delta f (B''_g) =$	$0.75 \times$	$9,750 = 7,300$

$$13,111$$

The various "functions" for the determined values of the inductions are read direct from the proper curves in Fig. 12 (p. 35). Let the value of the ampere-turns be taken at 13,100.

To find the back ampere-turns : Let the maximum angular lead θ be $27°$, *i.e.*, nearly to the tip of the leading "horn." Then, by (13), the back turns are

$$BTs = C_a \frac{\theta}{180} i_a v_1,$$

$$= 680 \times \frac{27}{180} \times 35.75 \times 1.42,$$

$$= 5,150 \text{ nearly.}$$

(For value of "v_1" *see* Table D, p. 33.)

---

* It is also interesting to mark how the direction of the lines is altered after leaving iron and entering air. This is to be expected from the very different permeabilities of the two media, a subject which is engaging the attention of experts.

And 2,575 ampere-turns must be added to the excitation of
each limb, since (13) estimates the total demagnetising effect.
Therefore the total excitation per limb is

$$\text{ATs} + \text{BTs} = 13,100 + 2,575 = 15,675.$$

With a series machine no calculation can be made for "open
circuit," since it will not excite unless the main circuit be
closed. If the dynamo be compound wound, this calculation is
necessary in order to determine the shunt winding; and for a
simple shunt machine it is usually sufficient to determine the
excitation at full load only.

It now remains to find the gauge of wire for the field windings. This can be done in several ways, according to the
result aimed at.

In a series machine* the current is fixed, and thus the
gauge of wire is given by the heating effect, *see* § 9 (*a*);
and the number of turns must be selected to give the proper
excitation. In the design under consideration the convolutions required

$$= \frac{15675}{37\cdot75} = 440 \text{ per limb,}$$

---

* If the field coils were in shunt, then the potential difference would be
fixed, and the current and turns be variable, the efficiency increasing with
the number of turns for a given value of the excitation. Now there is
one gauge of wire, and one only, which will give a particular excitation, for
the given section of core, at a fixed potential difference. This wire being
used, the number of convolutions affects the efficiency and not the value
of the excitation. Therefore, it is necessary to find the proper area of
cross section of wire, and usually as it is not a stock size the *next size larger*
is chosen. This increases the number of ampere-turns; but an external
resistance is easily put in series with the coils, and hence the potential
difference can be adjusted for any required excitation. The method
followed is to first find the length of a mean turn of the coil, say it $= l_1$
inches; then, the resistance of a mean turn

$$l_1 = \frac{e_1}{\text{ATs}},$$

where e_1 = the potential difference available; and the gauge of wire can
be found by reference to a table of wire resistances.

and No. 5 B.W.G. will carry 38 amperes at a density of 1,000 per square inch of section ; and this will give an ample margin. The resistance when hot is

$$\frac{440 \times l_1 \times 0.00064}{36} = \frac{440 \times 34 \times 0.00064}{36} = 0.258 \text{ ohm,}$$

where l_1 = length of a mean turn in inches, and 0.00064 = resistance of one yard of the wire.

The winding space is 14in. long ; allowing 0.75in. for the two flanges of the " shape," and for insulation, there is left a net length of 13.25in. Now 4.25 turns of No. 5 B.W.G. (with 16 mil insulation) will occupy 1in., and so the convolutions per layer will be 56. The turns per coil are 440 ; hence, 8 layers are required. The depth of the winding is approximately 2in.

The resistance of the two coils is about 0.516 ohm, and the total resistance of the dynamo 0.916 ohm. The pressure lost over this is—

$$35.75 \times 0.916 = 33 \text{ volts nearly.}$$

Now, in fixing the value of E, a fall of 6 per cent., or 41 volts, was allowed. The winding is thus within the required conditions. If desirable, the value for the total E.M.F. can now be readjusted and the calculations amended accordingly. Practically, however, it is sufficient to decrease the speed in the ratio of

$$\frac{695 + 33}{695 + 40} = \frac{728}{735} = 0.99,$$

and speed = $835 \times 0.99 = 830$ revolutions per minute.

The design of the dynamo is now complete. That of the motor is substantially the same ; but special care must be given to the driving horns and general details of the armature. The speed of the motor is proportional to the counter E.M.F., *i.e.*, to

$$E_1 - i_a \left(r_a + r_m \right),$$

F

and the torque is

$$\frac{\text{B.H.P.} \times 33,000}{2\,\pi\,n}, \text{ from (18),}$$

or

$$\frac{C_a\,i_a\,N_a}{852\cdot3 \times 10^6} \text{ from (16).}$$

Now e_1, the volts at motor terminals, $= e - 104 = 591$ volts.

$$i_a = 35\cdot75.$$
$$r_a = 0\cdot4.$$

and let $r_m = 0\cdot3$, say (trial value with six layers on field cores,
$$C_a = 680. \qquad \text{and 660 turns iu all).}$$

And n and N_a are the two unknown variables. If there be any particular speed, then n is also fixed. Let 720 be the required number of revolutions per minute, then $n = 720$, and the torque is given by (18)

$$\frac{25\cdot5 \times 33,000}{2\pi\,720} = 187 \text{ lb. ft.}$$

The peripheral pull on the armature at full working load is given by (19), and

$$= \frac{\text{B.H.P. } 33,000}{2\,\pi\,n} \cdot \frac{1}{g} = 297 \text{ lbs.}$$

where g = radius of armature in feet, $= 0\cdot625$.

The maximum stress at starting may possibly be twice as great, or even more, so it is necessary to allow for a pull of, say, from 800 lbs. to 1000 lbs. Equating to (16), the value of the torque found from (18), and taking out N_a, the required value of the armature induction* is found, thus—

$$N_a = \frac{187 \times 852\cdot3 \times 10^6}{680 \times 35\cdot75}$$
$$= 6\cdot55 \times 10^6,$$

---

* The value of N_a cannot be determined from equation (6) with sufficient accuracy, as the exact value of the counter E.M.F. is not known until the field-winding is settled and its exact resistance, r_m, arrived at.

and,

$$B''_a = \frac{6\cdot55 \times 10^6}{70} = 93,570,$$

$$B''_g = \frac{6\cdot55 \times 10^6}{247} = 26,500,$$

$$N_t = 1\cdot4\ N_a = 9\cdot17 \times 10^6,$$

$$B''_t = \frac{4\cdot585 \times 10^6}{65} = 70,500,$$

$$B''_{t1} = \frac{4\cdot585 \times 10^6}{110} = 41,600.$$

The excitation is by (12) :—

ATs for armature	$= m_a f(B''_a) =$	7×53	$=$	371
„ cores	$= m_t f(B''_t) =$	23×35	$=$	805
„ pole pieces	$= m_{t1} f(B''_{t1}) =$	16×66	$=$	$1,056$
„ gap	$= 2\delta f(B''_g) =$	$0\cdot75 \times 8,280$	$=$	$6,210$
				$8,442$

The back ampere-turns* are practically the same as with the dynamo, and, therefore, the full excitation per limb is

$$ATs + BTs = 8,442 + 2,575 = 11,017.$$

Dividing this by the number of amperes, the number of turns per coil is 310, or six layers, as was assumed in the approximation.

This finishes the calculation for the motor. The figures have been estimated by means of a slide rule, and so are not always quite exact ; but they illustrate the use of the equations, and show the general method of designing dynamos and motors. It is interesting to notice that the motor is rather large for the output, and that the inductions in the field cores and gap are consequently much less than those required in the dynamo. This can be obviated to some extent in practice, by reducing the section of the motor cores, making them, say, 4in., instead of 4·5in., in depth. Such alterations will readily suggest themselves.

---

* Probably the lead at full load will be from 5deg. to 10deg. less than that required with the dynamo. If this be so, the extra turns will be correspondingly less.

§ 13. TABULATED FORM FOR USE IN DESIGNING DYNAMOS AND MOTORS.

In practice, the difficulties attending the designing of machines are very much lessened if the constants and variables are classified under their proper heads and in the order in which they come under the notice of the designer. The author has found the following tables of great utility; they can, of course, be modified to suit special requirements. One of the great advantages possessed by such a regular system is that the details of different machines can be readily compared, and hence fresh designs can be got out with the minimum of trouble and figuring. To derive the greatest possible benefit from this arrangement the printed sheets should be bound together into a volume of suitable size, with an index.

Table E.

DYNAMO or MOTOR.

...... Poles. Output........K.W.	Volts.......... Amps..........
Magnetic Circuits	Revs. per min. ....
Type	

ARMATURE.

Type...........................	Weight of copper ....lbs.
Length ....in.	Weight of iron ....lbs.
Diameter......................in.	Current density ....per sq. in.
Dimensions over iron....in. × ....in.	Length of mean turn ....ft.
Effective area (..*ab*. 0·85)....sq. in.	Resistance of copper (hot) from brush
Size of hole ....in.	to brush ....ohms
No. of coils........., πd ....in.	Bore.............................in.
No. of convolutions per coil... C_a ..	Circumflux ....
Area of bare wire:	Permissible ditto ....
......in. × ......in. = ...sq. in.	
Dimensions over insulation:	
..........in. × ....in.	

COMMUTATOR.

Over-all dimensions	Metal ....
No. of segments	Insulation, mica.......... in thick
Size of ditto	

FIELD MAGNETS.

Cores..........Iron.   Area..........sq. in.   Weight..........lbs.
Yoke ....Iron.   Area..........sq. in.   Weight..........lbs.

AREA OF POLAR CAVITY..............sq. in.   Arc of poles......deg.

LENGTHS OF MAGNETIC PATHS, TAKEN AS FOLLOWS :

m_a ....in. m_f ....in. m_n ....in. δ ....in.

LEAKAGE COEFFICIENT, open circuit, v ....; full load r_1........

OPEN CIRCUIT EXCITATION PER COIL.

$E_a =$ .... $C_a N_a n 10^{-x}$ | ATs for armature ....
∴N_a ...., B''_a ...., | „ gap ....
B''_g ...., | „ field $\begin{cases} m_f \\ m_n \end{cases}$ ....
$N_f = r N_a$ .... |
∴B''_f ...., B_n .... | ∴Shunt ATs ....

FULL LOAD EXCITATION PER COIL.

E_{a1} .... r_1 .... | ATs for armature ....
∴B''_a .... B''_g .... | „ gap..................
$N_f = r_1 N_a$ .... | „ field $\begin{cases} m_a \\ m_{a1} \end{cases}$ ....
∴B''_f .... B''_n .... |
| Total ....

Turns per coil to compensate for fall of pressure over internal resistance
= CTs = ... minus .... = ...

Total back ampere-turns = BTs | Total excitation per coil :
= $C_a i_a r_1 \theta/360$ = ... |
(i_a = armature current, θ = ..., | ATs + CTs + $\dfrac{BTs}{u}$ ....
and is the angle of lead in | (u = number of coils)
degrees ; take 2θ for Grammes) |

SHUNT COIL WINDING.

Mean turn ....feet. | ....layers of ...... turns each.
Res. (Hot) mean turn........ohms. | Depth ....inches.
Res. (Cold) mean turn........ohms. | Weight per coil ....lbs.
Gauge of wire..............S.W.G. | Total weight..................lbs.
Shunt current.................. | Res. per coil ....ohms.
Convolutions .... | Total resistance ....ohms.

MAIN COIL WINDING.

How wound ....

Area of bare wire ....in. × ....in. = ...sq. in.

Dimensions over insulation.................in. × ....in.

Convolutions ....

Weight per coil...............lbs.　Total weight.................lbs.

Resistance per coil ....ohms.　Total resistance ....ohms.

HEAT WASTE IN COPPER.

Watts in armature...........................

„　shunt coils........................

„　main coils ....

Total watts ....

Pressure lost over internal resistance ....volts

Electrical Efficiency ....%

PROBABLE RISE OF TEMPERATURE.

FIELD : area of surface of each coil...............................sq. in.

$$\text{Rise of temp.} = \frac{\text{Watts } 100}{\text{Surface}} = \text{...........................F}^{\circ}$$

ARMATURE : area of surface ....sq. in.

(3,000ft. per min.) } $$\text{Rise of temp.} = \frac{\text{Watts } 64}{\text{Surface}} = \text{...........................F}^{\circ}$$

These Tables show at a glance the most important details of continuous-current dynamos or motors, and can be easily adapted for alternate-current machines. They should be used in conjunction with cross-sectional diagrams, similar to that shown in Fig. 14, page 49. Special requirements for individual designers can be incorporated without difficulty, and small modifications will readily suggest themselves.

§ 14. SERIES, SHUNT, COMPOUND, AND SEPARATE EXCITATION.

It is of the utmost importance to appreciate fully the differences between the various methods of exciting the field-magnets, as all are not equally well adapted for all kinds of work,
some being specially suitable for one class, and worse than
useless for others. It is probable that the connections of the
several windings are well known to most readers; yet it will
serve to fix ideas and to give a firm grip of the various purposes to which each type of machine may be most profitably
applied if the following sets of diagrams are carefully examined.

In each set the first figure shows simply the diagram of connections, irrespective of the arrangement and number of the
field coils and poles.

The second figure refers to a dynamo wound as indicated in
the first figure, and shows the order of the curve connecting the
terminal, or omnibus bar pressure, with the main current.

The third figure refers to a motor wound in the same style,
and indicates roughly the shape of the curve connecting the
number of revolutions per minute with the main or armature
current.

In each case the dynamo speed is assumed to be constant,
and so also is the pressure at which current is supplied to the
motor *except in the case of series winding.* These conditions are
usually essential, and are conformed to, as nearly as possible,
in all central stations.

The diagrams are so explicit as to need little comment,
but a few suggestions may be useful. The pressure in series-
wound dynamos varies with the current, and these machines
are chiefly suitable for working on a fairly constant resistance, such as a definite number of arc lamps or a fixed
number of incandescent ones; they are more especially
adapted for the former work. They present one very apparent advantage in the fact that, since the main current
passes through the field winding, the number of turns of

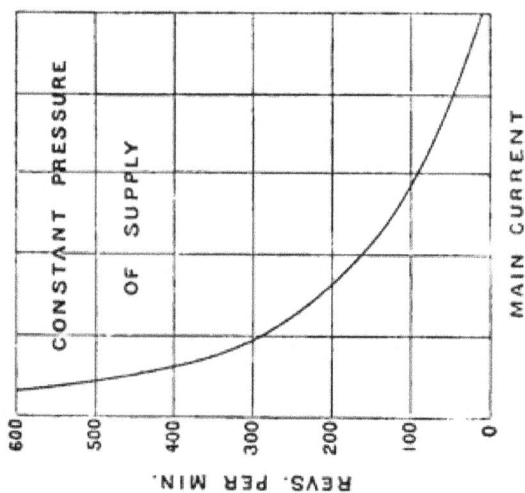

Fig. 15c.—Mechanical Characteristic Curve of Series Motor. Abscissæ Proportional to Torque.

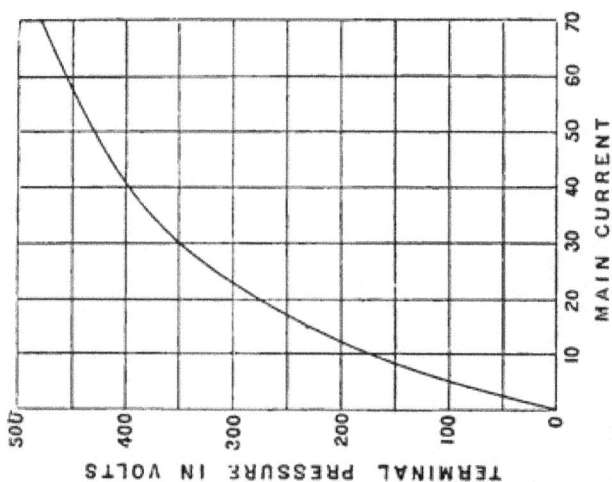

Fig. 15b.—External Characteristic Curve of Series Dynamo.

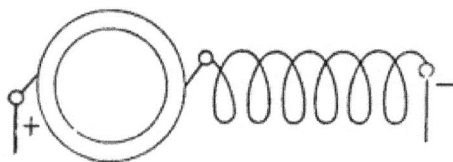

Fig. 15a.
Series Winding.

ARMATURE CURRENT

CONSTANT PRESSURE OF SUPPLY

REVS. PER MIN.

FIG. 16c.—Mechanical Characteristic Curve of Shunt Motor. Abscissæ proportional to Torque.

MAIN CURRENT

OMNIBUS BAR PRESSURE IN VOLTS

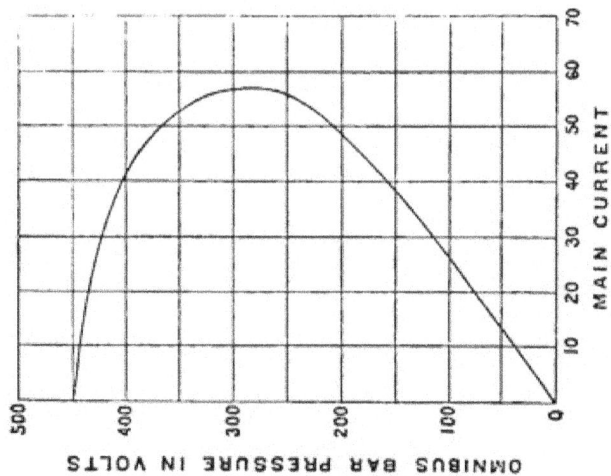

FIG. 16B.—External Characteristic Curve of Shunt Dynamo.

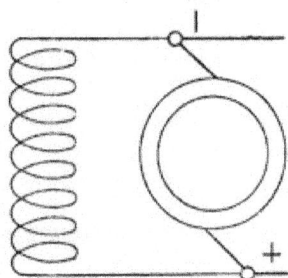

FIG. 16A. Shunt Winding.

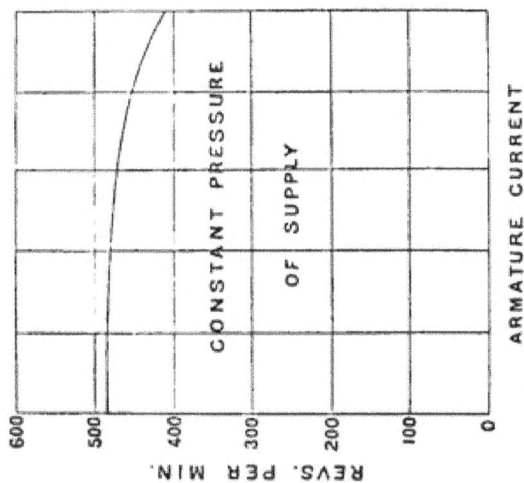

wire on the magnets is comparatively small, and the wire is large, and therefore strong.   High insulation resistance is thus easily obtained without mechanical difficulty or the use of costly material.   For power purposes this winding possesses special value.   Consider Fig. 15c, which refers to a series motor running at constant pressure.   It will be seen that as the current in the motor is increased the speed decreases ; and, since the torque is proportional to a function of the current, it follows that the torque varies inversely with the speed.   Now, look at the series dynamo curve, Fig. 15B.   The pressure increases with the current, which is exactly the condition of supply to make the series motor run at a uniform rate of speed.   Therefore, if a series-wound dynamo be coupled to a series-wound motor, and be driven at a constant speed, it will supply current at a pressure varying directly as the current, and the motor will tend to run at a constant speed.   This is such an important problem in the electric transmission of power that it is referred to again,* and the conditions for success are examined more closely.

Series motors are suitable for a variety of purposes, and are generally used where a large torque is required at starting and a fairly constant load obtains afterwards.   When run off constant-pressure circuits, they are regulated by a variable resistance placed in series with them, which may be altered by hand as occasion requires.

The *feature* of a series motor to be specially remembered is that the torque is greatest at starting, when the current is a maximum : this is so because the main current passing through both field and armature coils then magnetises them to the greatest possible degree.

Shunt-wound dynamos give variable current at fairly constant pressure.   They are, therefore, suitable for running loads, such as banks of lamps, which require varying power ; and also for running *series-wound motors with constant loads*, or *shunt-wound motors with varying loads*.   This is seen to be the case by

---

* *See* § 23, especially pages 121 and 122.

comparing the shunt motor mechanical curve (Fig. 16c) with the shunt dynamo curve (Fig. 16B). The shunt motor when supplied at constant pressure maintains a fairly uniform rate of rotation with varying load : the shunt dynamo *approximately* conforms to the supply conditions.

The shunt motor is not so well adapted for starting against a heavy torque as the series motor. For the shunt field excitation cannot exceed that determined by the supply pressure ; and a sudden rush of current through the armature at starting tends to weaken the field magnetism unless the brushes have a forward lead, when there will be considerable sparking. When starting a shunt motor the field circuit should be first closed, and time be allowed for the magnets to become fully excited before the armature circuit is closed. A starting resistance is always used in the armature circuit of these motors. After the armature has commenced to revolve the resistance may be gradually cut out and the speed raised to its normal value.

The *feature* of a shunt motor to be specially recollected is that it will run at nearly constant speed, with varying load, when supplied with current at constant pressure. This is the supply condition of nearly all central power stations, and hence shunt motors have a special field in this connection.

As will be gathered from Fig. 17A, compound winding is a combination of the shunt and series methods of excitation. The shunt coils may be coupled across the brushes in " short shunt," or across the mains in " long shunt " as in Fig. 17A. The theory is as follows :—The shunt coils give an excitation proportional to the terminal pressure, and the series coils an excitation varying with the armature or the main current, according as the long or short shunt is adopted. It is evident that, by suitably proportioning the two windings, the external characteristic, Fig. 17B, can be made to slope at any angle to the current line ; or briefly, the pressure may be made to increase with the current, and by compensating for the fall in the feeders, maintain a constant pressure at the centre of distribution. The chief difficulty with high-pressure compound dynamos is the

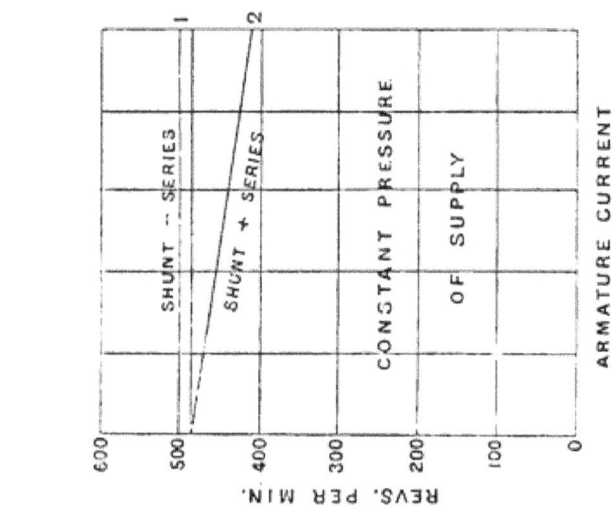

Fig. 17c.—Mechanical Characteristic Curve of Compound Motor. Abscissæ proportional to Torque.

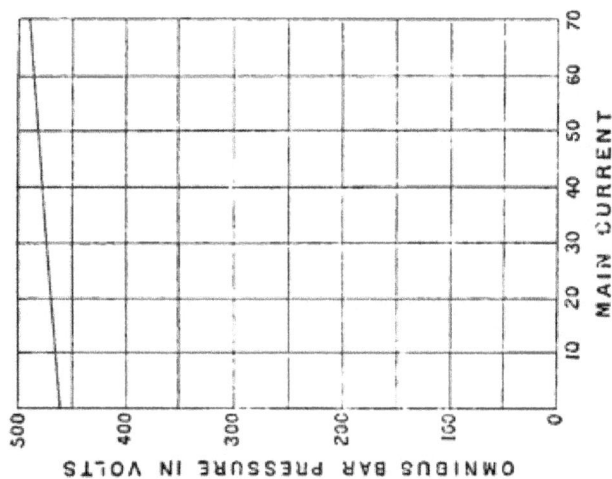

Fig. 17a.—External Characteristic Curve of Over-Compounded Dynamo.

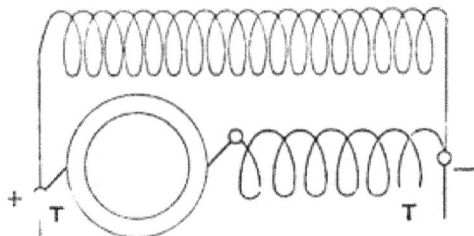

Fig. 17b.—Compound Winding, Long Shunt.

shunt windings, which consist necessarily of many turns of fine wire, and give trouble from a variety of causes, besides being costly.

The mechanical characteristic of the compound-wound motor is shown in Fig. 17c. It will be seen that there are two methods of coupling up the coils, as illustrated in curves 1 and 2. In the top curve the shunt and series excitations are differential, *i.e.*, the series coils demagnetise the field-magnets; in the lower curve they are cumulative, and both act in the same direction. The differential coupling is theoretically the better arrangement, and gives a more constant speed than even a shunt motor, as is apparent from equation $(6)^*$ $E_m = C_a N_a \dfrac{n}{60} 10^{-}$,

where E_m is the motor armature counter-electromotive force, C_a the number of armature conductors, N_a the lines of force, and n the revolutions per minute. Since C_a is constant, and N_a may also be regarded as such when a constant pressure acts on the shunt coils, it is clear that the speed, n, varies directly as the counter electromotive force, E_m. And the armature reactions do not affect the problem if we assume the armature current to have no effect on the value of N_a. Now,

$$E_m = e - i_a r_a,$$

where r_a is the armature resistance, and i_a the current in the armature, and e the pressure of supply which is assumed to be constant. It is clear that E_m varies inversely with i_a, and, therefore, the shunt motor tends to slow as the current increases (for the speed is proportional to E_m). Referring again to equation (6), it is seen that to prevent this falling off in speed it is necessary to decrease the lines of force, N_a, in the same proportion as the fall of counter-electromotive force. This is most readily accomplished by decreasing the excitation, either by intercalating resistance in series with the shunt circuit, or by adding compound series coils wound in an inverse sense to the shunt turns. It should be noticed that the armature reactions always weaken the main magnetic field, and therefore *tend* to make a shunt-motor run at a constant speed. It is thus possible to obtain nearly a straight line speed regulation with a well-designed shunt motor. It is

---

* *See* page 18.

also evident that a motor armature should have a very low internal resistance, the lower the better.

The cumulative method of winding is more frequently used than the differential, for although it obviously does not regulate so well as a simple shunt motor, yet it combines to some extent the starting power of the series winding with the speed regulation of the shunt. Sometimes a few series turns are wound on large shunt motors, and are used only at starting, being short-circuited as soon as the armature is fairly under weigh.

Broadly speaking, differentially-wound compound motors do not seem to be a success, and, except for the few cases where absolute regularity of speed is required, they certainly have not found favour. One great objection to them is a liability to start in the wrong direction owing to the reversed series winding. This difficulty is met to some extent by short-circuiting the series coils at starting. Cumulative compound motors have also a very limited field of usefulness, well-designed shunt machines being more efficient, and giving better speed regulation. Yet there are some few applications where this winding will meet the conditions better than any other.

The *feature* of compound winding to be specially noted is that, by applying it to the generator, the pressure at a distant point can be maintained (within limits) automatically constant with varying current.

The idea of separate excitation is so easily grasped that there is no need to do more than refer to Fig. 18A, which is a diagram of the connections, the top scrolls representing the field-magnet coils. It is obvious that the number of turns on the field-magnets and the gauge of wire can be varied to suit the exciting current so long as the necessary ampere-turns are provided. In Fig. 18B the external characteristic of a separately-excited dynamo is shown at curve 3. It will be noticed that the pressure falls as the output is increased, owing to the volts lost over the internal resistance of the armature. To obviate this the excitation must be proportionately augmented, so as to give curve 2. And if it be desired to compensate for the fall of

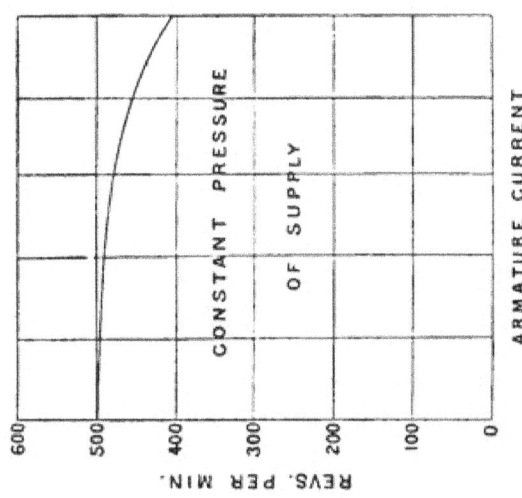

Fig. 18c.—Mechanical Characteristic Curve of Separately Excited Motor. Abscissæ proportional to Torque

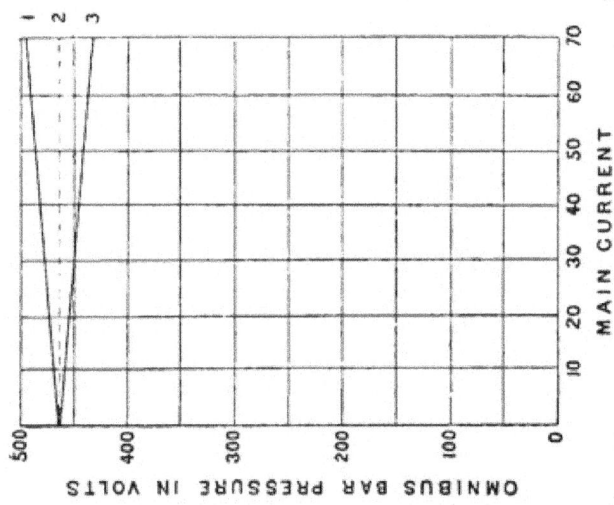

Fig. 18b.—External Characteristic Curve of Separately Excited Dynamo with Constant Excitation.

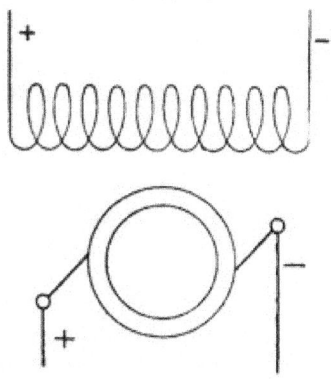

Fig. 18a.—Separate Excitation Winding.

pressure in feeders, then the exciting current must be still further increased until the slope of curve 1 suits the required conditions. The regulation can be accomplished automatically by some mechanical device controlled electrically; but it is usual in central stations, where this type of machine is most frequently used, to alter the excitation by hand.

The separately excited motor is not yet much in use, but it will steadily grow in favour as transmission of power by electricity becomes more general. The mechanical characteristic of this type of motor, supplied at constant pressure, is shown in Fig. 18c. The fall of speed with increase of load can be compensated by *decreasing* the excitation either by hand regulation or reversed series coils.

These motors are specially adapted for large powers, because, since the main current only passes through the armature, a high pressure can be safely used; and moreover, the speed regulation is much superior to that obtainable with the series-winding. The objection is the independent source of power for the exciting current, which will probably have to be supplied in most cases from accumulators. Yet it must be recollected that a small shunt-dynamo, driven by the large motor, will not only charge the cells, but also excite the motor field-magnets after the proper speed is once attained; and, further, this dynamo is available for lighting during the period the motor is at work.

The *advantages* of separate excitation are : firstly, the dynamo fields can be excited at a low-pressure, and hence are free from insulation troubles; secondly, lightning discharges are less likely to affect the magnet coils, since they are not connected with the line circuit (unless reversed series coils be used); and, thirdly, the pressure at any distant point can be easily varied at will. With motors a high-pressure service is permissible, and the speed can be regulated either by the dynamo or motor exciting current.

CHAPTER III.

THE LINE AND THE DISTRIBUTING MAINS.

§ 15. DEFINITION OF THE "LINE."

By the "line" is here understood not only the conductor, but also the posts or culverts, insulators and protective devices. Treated in this way, the subject may be regarded from two distinct points of view : one relating only to the method of erecting and protecting the conductor, and the other having reference to the kind of material, to the area of cross section of the wire, and to the conditions which limit the distance through which a definite quantity of power can be profitably transmitted.

Practice usually precedes theory, and the first attempts at electrical transmission of power were made without any definite conception of the governing laws. An ordinary galvanised iron telegraph wire was the handiest conductor when Marcel Deprez made his memorable experiments between Munich and Miesbach in 1882. And even to-day considerations of convenience or necessity often invite, or compel, the use of material which is not exactly what would be chosen if the engineer had a free choice, and cost were no object.

It will be more in keeping with the fitness of things, however, if the theoretical conditions are first discussed.

§ 16. THE THEORETICAL CONSIDERATIONS AFFECT-ING THE LOSS IN THE CONDUCTOR.

The two conditions absolutely essential in the line circuit are continuity and insulation, *i.e.*, the line must consist of a conducting substance or several conducting materials carefully joined together, and must be so arranged that at least one of the circuits between the dynamo and motor is insulated. The earth, under certain conditions, may be used as part of the circuit if sufficiently good "earth-plates" are provided at each end of the metallic conductor. This is the case with submarine cables and land telegraphs, and is oftentimes so with telephone circuits. An "earth return," as it is generally called, presents serious objections with relatively large currents at high pressures ; and hence, with but few exceptions, existing power plants are designed with insulated return circuits. A compromise between the two methods is found in a continuous metallic *uninsulated* "return," as used with electric street railways, in which the metals are coupled by copper connections, and so form a continuous conductor. In this case, the action of the current is not confined to the iron rails, but considerably affects the adjacent earth ; so that telegraph and telephone lines, using the "earth" in the immediate neighbourhood, may suffer considerable disturbance. A purely practical objection to the use of the earth as part of the circuit, or to an uninsulated wire for any portion of the line, is that the leakage may be largely increased and the difficulties of handling the line and plant intensified. Street railway practice has hitherto been limited to a maximum pressure of 500 volts, and it is probable that 300 volts will be more generally used in the future. At any rate, in Great Britain the Board of Trade regulations restrict the pressure to this limit.

In mining work the author has selected 500 volts as the most convenient pressure for general purposes, as it permits the transmission of considerable power without an excessive outlay in copper, and yet is not dangerous in the event of accidental "shocks." The use of an uninsulated *separate* return is particularly unsafe in mining, owing to the serious risk of fire which is inevitable with such a circuit.* Even

* This remark does not refer to a concentric cable with the outer conductor earthed, as used in the Andrews or Fowler-Waring systems.

with pressures as low as 100 volts, both mains should be carefully insulated.

It may be granted, then, that the practice of the future, excepting in tramways and railways, will demand an insulated lead and return, and that the whole of the circuit, including dynamos, motors and switches, will be insulated in the most careful and thorough manner.

The simplest case of transmitting power, and one that often occurs, is where there are only two machines, a dynamo and a motor. The circuit in this case is obviously a lead and return of the same diameter throughout, since the current is the same at all points (neglecting the small loss from leakage). The line may be an insulated cable buried in a culvert ; or a bare copper wire carried by insulators on poles—a so-called aërial line—or it may combine the two methods, being partly buried, or cased-in, and partly in the open air. Yet since the diameter is the same throughout the circuit, the resistance of the copper per unit length will be constant. Now, for this simple case what is the *best area* of conductor to carry a given current or to transmit a given quantity of power if the pressure be not fixed ? It is oftentimes not easy to give a definite answer to this apparently simple query, as so much depends upon the result desired. Is the prime power at disposal practically unlimited ? If so, the line loss may be large. What is the cost of power at the generator ? If it be cheap, then the line loss may be high ; but if it be dear (although, perhaps, plentiful), the waste in the line must be restricted. But this limitation of loss will mean a corresponding increase in the weight of the copper, and copper is costly. So the answer is practically a compromise, and each case must be settled on its own merits. Lord Kelvin as early as 1881 pointed out that a general solution is obtained by equating the annual cost of the horse-power lost in the cable to the interest on the capital invested in the line and supports, including the cost of erection, plus the annual cost of maintenance. But this condition does not always meet all the practical requirements. Indeed, the size of the conductor is often fixed between narrow limits by the simple conditions of permissible fall of pressure and mechanical considerations. Theoretically, the higher the supply pressure the

smaller the conductor; but prudence, to say nothing of Board of
Trade regulations, limits the pressure, and copper circuits are
not often used smaller than, say, No. 10 or No. 12 S.W.G., or
equivalent cross section in stranded wires, from considerations
of strength and durability. And an excessive outlay in copper
is prevented not only by the price of the cable, but by the
increased expense in the erection of the line. The latter objection has less weight when the conductor is laid in the ground,
for it is practically as cheap to lay a large cable as a small
one; and large mains are invariably run underground.

Prof. G. Forbes, in his Cantor Lectures, 1885,* showed that
Lord Kelvin's law is most accurately expressed in the following
form:—The most economical area of conductor is that for
which the annual value of the energy lost in the copper is
equal to the annual charges on the total cost of the line (including erection, &c.), minus that part of the capital outlay which
is independent of the area of the conductor. Or, in other words,
the annual value of the $i_m^2 R$ loss must be made equal to the
annual interest on that portion of the total capital which may
be considered proportionate to the sectional area or weight of
the copper.

Symbolically, this may be written

$$p \frac{i_m^2 R}{746} = k \left(C_1 - C_3 \right) = C_2 k,$$

where p = the annual cost of one E.H.P. at the dynamo
           terminals,
       R = the resistance of the line,
       i_m = the equivalent or average current in the conductor,
       C_1 = the total cost of line,
       C_2 = part of capital proportionate to area of conductor,
       C_3 = part not proportionate to area,
       k = the rate per cent. of annual interest to cover total
           charges in the several capital sums, C_1, C_2, C_3.

The equivalent current, i_m, must be calculated from the
estimated load curve, and it is by no means easy to predict

* The student should carefully peruse these Lectures, especially the
tables formulated to facilitate the calculations. They appeared in *The
Electrician*, Vols. XV. and XVI., and are published in pamphlet form
by the Society of Arts.

its value. The annual heat waste in the mains may be written $i_m^2 \, R \, T$, where T is the number of hours in a year (8,760 hours, say), and i_m the equivalent current. And suppose it is estimated that in one year a current of value i_1 will be required for, say, t_1 hours, a current of i_2 for t_2 hours, &c.; then the total waste is equal to the summation of the quantity of energy spent severally in the periods t_1, t_2, t_3, &c. Whence

$$i_m^2 \, R \, T = R \, (i_1^2 \, t_1 + i_2^2 \, t_2 + i_3^2 \, t_3 + \ldots \; i_n^2 \, t_n)$$

$$\text{or } i_m = \sqrt{\frac{i_1^2 \, t_1 + i_2^2 \, t_2 + i_3^2 \, t_3 + \ldots + i_n^2 \, t_n}{T}} \quad \ldots \; (20)$$

It is thus evident that there is likely to be a difficulty in finding the value of the equivalent current, and yet on the accurate determination of this quantity depend the estimate of annual heat waste and the selection of the economic ratio between current and area of conductor.

The equation for determining the section of the copper for the predetermined current is usually given in some such form as the following :

$$a = i_m \sqrt{\frac{w \, p \, T}{C_2 \, k}}, \quad \ldots \ldots \ldots \; (21)$$

where a = the area of conductor in square inches, i_m, k, p, T, and C_2 have the values already given to them here, and w is a constant depending on the energy wasted in the conductors.

If the current be constant (or nearly so, as in a pumping plant), the economic area is most readily determined from a modification of (21), which takes into direct consideration the load factor. This equation, (21A), is due to Mr. E. Tremlett Carter. It may be written

$$a = 1 \cdot 168 \; i_m \sqrt{\frac{W \, p \, F}{c \, k}}. \quad \ldots \ldots \; (21\text{A})$$

The new symbols are :—

W = the watts spent in one yard of copper, one square inch in cross section, at a density of one ampere per square inch. This is a constant for commercial samples of copper, and equals $2 \cdot 56 \times 10^{-6}$.

$F =$ the load factor expressed as a rate per cent.

$c =$ the cost of additional copper per square inch per yard length in £ sterling.

The values of c used by the author are given in Table E1.

Table **E1.**— *Values of "c" for different costs of laying one yard of additional copper one square inch in section.*

Copper, including additional cost of laying, at			Corresponding values of c in £ sterling.
d.	£		
6 per lb.	= 56·0	per ton.	0·2900
7 ,,	= 65·3	,,	0·3383
8 ,,	= 74·6	,,	0·3866
9 ,,	= 84·0	,,	0·4349
10 ,,	= 93·3	,,	0·4837
11 ,,	= 102·7	,,	0·5316
12 ,,	= 112·0	,,	0·5800
13 ,,	= 121·3	,,	0·6287
14 ,,	= 130·6	,,	0·6766
15 ,,	= 139·9	,,	0·7249
16 ,,	= 149·2	,,	0·7732
17 ,,	= 158·5	,,	0·8215
18 ,,	= 168·0	,,	0·8700
19 ,,	= 177·3	,,	0·9187
20 ,,	= 196·6	,,	0·9666

It should be noticed that (21A) is applicable to *constant pressure* circuits with *variable* currents if F be taken as 100 and i_m be determined from (20).

When the pressure and current vary approximately in the same ratio, (21A) is applicable, if the equivalent current be determined from (20); and is specially useful when only two machines, both series wound, are used.

There are three separate difficulties to be overcome before the above conditions can be fulfilled, even assuming that practical and financial requirements necessitate no modifications. In the first place, what is the annual cost of an electrical horsepower? This is by no means easy to predetermine, judging from the wide limits of figures recently published in the reports of Central Lighting Stations. Secondly, it is not easy to estimate with certainty how much of the total capital outlay is

independent of the area of the conductor. And, thirdly, the
determination of the equivalent current presents such difficulties
as to make this quantity indeterminate in many cases. Further-
more, the question of fall of pressure in the conductor is not
taken into consideration in the equation.

The cogency of the above reasoning will be more easily
appreciated after studying some of the theoretical considerations
which affect the problem. Transmission of power may usually
be classed under one of the following heads :—

(a) *Definite loss in line and fixed quantity of power at motor
terminals.*—This usually implies that the power at the generating station is either costly or limited, or sometimes that
only a definite horse-power can be profitably spent at the
motor.

(b) *Unrestricted loss in the line and fixed quantity of power
at motor terminals.*—Here the prime energy is cheap and unlimited, and only a portion is required for use.

(c) *Fixed quantity of power at the generating station and as
much at the motor as possible.*—In this case the line loss must
be made a minimum, and the weight of copper as great as the
capital at disposal will permit.

The limiting conditions of pressure, current, weight, cross
section and length of conductor, are determined by relatively
simple laws.

In (a) the loss in the line, $i_m^2 R$, and the energy at the motor
terminals, $i_m e_1$, are fixed. Therefore the sum of these two
quantities—that is, the output of the dynamo—is also fixed.
Let this be $i_m e$.

Now, the resistance of the line, R, will vary directly as the
length and inversely as the area of cross section of the cable ;
and therefore

$$R = k\, l/w, \text{ or } k\, l/d^2,$$

where d is the diameter, w the weight of the copper wire, and
k a constant. And by the assumed conditions of transmission,

$i_m e = i_m e_1 + \dfrac{i_m^2 l k}{d^2} = a$ constant. It is evident that the value of this quantity may be kept the same with considerable variations of $i, e, l,$ and d—that is, with alterations of current and pressure, and with different lengths and gauges of wire.

In (b) the line loss is assumed to be unlimited, and therefore the dynamo output may be made as large as convenient. The size of conductor may be determined simply by the maximum pressure deemed advisable and the condition of a definite quantity of power at the motor terminals.

In (c) the power of the generator is fixed, and that of the motor is to be as large as possible; therefore, the line loss must be as small as consistent with an economic outlay in copper for the conductor. Practically the pressure selected should be as high as the peculiar circumstances permit, and then as much money should be spent in copper as can be spared for this part of the plant.

If the generating station be situated in the middle of the line and power be distributed equally along each branch of the circuit, the weight of copper will be one-fourth what it would be if the station were at one end. With several supply stations, the weight of copper will vary inversely as the square of the number of stations, assuming they are properly placed, i.e., at equal intervals with reference to the load distribution. And the pressure required will vary inversely as the number of stations.

These few general statements of the laws governing power transmission and distribution by electricity show distinctly two facts: firstly, that a high pressure is necessary for economical transmission; and secondly, that for distribution a number of stations is advisable to minimise weight of copper and line loss. But it must also be recollected that every additional power station involves increased outlay for land and buildings and an additional staff for the control and working; therefore it will not be expedient to lay down a new station from considerations which simply refer to copper and line loss. But it is evident that the first station should be placed as near the centre of the district as convenient, and that the subsequent erection of inter-dependent or independent stations will have to be determined by the load factor* and the centre of gravity of the load with reference to the station and mains.

It has already been stated that the most simple case of transmission of power is where the circuit simply joins up a pair of machines. But no practical difficulty, as far as the line is concerned, is involved if a number of motors, in close proximity, be supplied from one distant power station ; for the whole of the current will be carried by the main conductors. Yet, it is evident that there must be one particular gauge of wire which will give the most economic results for the average current ; and this will be the case when Lord Kelvin's rule in a modified form is adhered to.

Various attempts have been made to give a general expression for the current in terms of area of wire, cost of horse-power at generator, &c., and it is noteworthy that the term for the length of the line cancels out of the final equation. This implies that the length of the circuit does not affect the problem—a statement only true when the economic law is adhered to. The length of the line does not affect the economic area of a conductor for the given current, but it obviously affects the power lost in the line. The length of the line is always understood in the terms expressing the resistance of the circuit and the area of the conductor—for, mathematically, the line may be defined, both as regards its length and weight, by the resistance of the circuit and the gauge of the wire. This is done in the usual equations, and hence the difficulty. Profs. Ayrton and Perry, in March, 1886, read an important Paper on " Economy in Electrical Conductors,"† showing that Lord Kelvin's law holds good only when the current is fixed, and is not taken as an independent variable in the equation for determining the area of conductor.

---

* The load factor is the ratio of the average output to the maximum output of the plant, and is usually expressed as a percentage. For central lighting stations the load factor is found to vary from about 10 per cent. to 15 per cent. With power plants it will generally be much higher, and in collieries, &c., may reach as high as from 80 to 90 per cent.

† Jour. Proc. Institution Electrical Engineers, Vol. XV., p. 120.

Mr. Kapp, in his Cantor Lectures,* gives the subject close attention from the transmission of power point of view, and takes as a basis the elementary case of one generator and one motor. (The case of several generators at one power station feeding into a single pair of mains and delivering energy to a group of motors lying near to each other at the end of the line is practically the same.) He shows that a complete solution requires the following conditions to be taken into account :—

> Annual value of brake horse-power at generating station.
> Pressure of supply at generator terminals.
> Brake horse-power required at motor.
> Length of line.
> Capital outlay per ton of copper *erected*.
> Cost per electrical horse-power at dynamo terminals.
> Interest and depreciation of the whole plant.

It is important to note that Mr. Kapp takes into consideration the pressure of supply ; in fact, he fixes the pressure and finds the best current for the given conditions. And he shows that in no case will it be economical to lose more than half the total power in the line, and that for every transmission problem there is one pressure for which the annual cost of a brake horse-power at the motor is a minimum.

This conclusion was foreshadowed in one of the questions asked a few pages back, " What is the best area of conductor to carry a given current if the pressure be not fixed ?" But, as already suggested, the limitations of the Board of Trade or other considerations practically fix the pressure of supply in most cases, and the problem consequently narrows itself to the relative weight of copper for the predetermined current, and the real variable quantity is the amount of energy to be wasted in the line. Under these assumptions, by a few trial-and-error calculations, a gauge of conductor may be selected to permit such an average heat waste that the approximate annual cost of the energy lost shall be as nearly as possible equal to the interest on the capital outlay in the copper (including erection) and maintenance of the line.

* Cantor Lectures delivered before the Society of Arts, February, 1891.

Exception may be taken to this treatment of the subject; yet it accords with practice. Local circumstances have such weight that no hard-and-fast rule can possibly meet the requirements of every case. And the indeterminate character of some of the quantities of the problem, such as the equivalent current and the cost of an electrical horse-power, on the constancy of which the truth of the equations depends, renders a general mathematical solution inferior to a determination based on a practical knowledge of the requirements of the particular case, and a simple application of Kelvin's law to select the gauge of wire and heat waste for a given current. This method does not necessarily give a minimum cost per brake horse-power of the motor; but if the steam (or turbine) plant and the electrical machines be of good make and properly proportioned for the average load, the best results will be very nearly realised.

In concluding this section on the theoretical conditions affecting the line loss, it is noteworthy that the simple condition of fall of pressure may be of paramount importance, and hence may alone determine the gauge of wire and consequent copper loss. This is specially likely to be the case in the distributing network when the current and pressure are fixed. Yet the author wishes to emphasise the importance of the economic law, and to state that whenever circumstances admit of its application its requirements should be carefully satisfied. A competent engineer will attend to this in a thorough manner.

§ 17. THE CONDUCTOR IN PRACTICE—AËRIAL LINES.

Proceeding to practical requirements, the first consideration is the metal for the conductor. In selecting this, the main consideration is all-round cost—i.e., which metal will be the cheapest, taking into account the predetermined heat waste and conditions of erection and insulation. To guide the choice it is necessary, first of all, to know the specific resistance of the metals usually sold in the form of solid or stranded wire. Copper, or its alloys, and iron and steel are practically the only materials at present available for the purpose.

The resistance to the passage of an electric current in a wire of unit length and of unit area of cross section is defined as the *specific resistance* of the substance. The quantity thus designated gives a ready means of comparing the relative *electrical* suitability of the different metals.

Table F.

Metal.	Sp. resistance in microhms.	Specific gravity.	Breaking strain in lbs. per sq. inch.	Price per lb. in pence.
Copper, annealed	1·615	8·8	—	8·5
Copper, hard-drawn ..	1·642	8·9	56,000	8·5
Bronze (tin and copper)	4·42		84,000	10·0
*Silicium bronze	1·7	8·4 to 8·7	65,000	8·5
Phosphor bronze	3·0 to 60		—	to
Aluminium bronze			—	12·0
Iron, ordinary, soft ...	9·827	7·6 to 7·8	47,000 to	1·0
Iron, galvanised, killed.	10·0 to 11·0		85,000	
Steel, ordinary, soft ...	14·5 to 15·0	7·8 to 7·9	135,000	1·5
Steel pianoforte wire...	16·0 to 16·5	7·75	335,000	2·0
Manganese steel	73·0	7·8	110,000	3·0

The above figures are necessarily only approximate. The specific resistance determinations have been made at 32°F.

In Table F. are arranged the copper alloys and various other metals in the order of their specific resistances; the specific gravity, breaking strain in lbs. per square inch, and the approximate price per lb. of the several metals are also given for rough comparison. It will be seen that iron has nearly seven times the specific resistance of copper, and so for a given resistance of line, the area of an iron wire will have to be, say, seven times that of a copper one. The specific weights are in the ratio of $8·9 : 7·8$, and the weight of the iron line, relative to that of the copper line, will be expressed by

$$7 \text{ W } \frac{7·8}{8·9}, \text{ or } 6·18 \text{ W,}$$

where W is the weight of the copper in lbs. And the costs will be respectively $8·5$ W for the copper, and $6·18$ W × 1 for

---

					Per sq. in.
* Class A, 97% conductivity of pure copper, tensile strength ...					65,000lb.
Class B, 80%	,,	,,	,,	,,	... 90,000lb.
Class C, 45%	,,	,,	,,	,,	... 110,000lb.

the iron. A bare copper line will, therefore, cost about 27 per
cent. more than an electrically-equivalent iron wire. From
the simple consideration of first cost of conductor, then, it
appears that iron is superior to copper. In relation to the
total line cost, however, the reverse of this is the case. For
the weight and breaking strain have to be taken into consideration. The breaking strain of hard-drawn copper of 98 per
cent. conductivity is about 56,000lb. per square inch; whereas
that of well-killed galvanised iron is about 85,000lb., so the
spans may be made longer with iron wire in the proportion of
85 to 56; or, say, 50 per cent. Yet, since the iron wire is 6·18
times as heavy as the copper, the supporting devices must be
correspondingly stronger. The figure of merit of a metal *per se*
may be expressed as the product of the reciprocals of the specific
resistance, the specific weight and price multiplied by the breaking strain—the best metal giving the largest figure of merit.
Thus, for copper

$$\frac{1}{\text{Sp. R}} \times \frac{1}{\text{Sp. W}} \times \frac{1}{\text{Price}} \times \text{breaking strain} \quad . \quad . \quad . \quad (22)$$

$$= \frac{1}{1\cdot642} \times \frac{1}{8\cdot9} \times \frac{1}{8\cdot5} \times 56,000 = 450.$$

For iron $\quad = \frac{1}{10\cdot5} \times \frac{1}{7\cdot8} \times \frac{1}{1} \times 85,000 \quad = 1,040.$

From these general considerations, galvanised iron appears
to be, roughly, about twice as well adapted for overhead lines
as hard-drawn copper. But this proportion is not maintained
when all the conditions of the problem are taken into consideration. In the first place, the greater bulk of the iron is objectionable on account of the expensive posts and attachments
necessary to carry the weight; and secondly, iron, although
galvanised,[*] rusts, and has a shorter life than copper; and so the
cost of maintenance is far greater than with copper or its alloys.
Again, old iron wire has practically no marketable value, whereas
copper is always saleable. As regards total prime cost of line
it is not easy to say which metal is the cheaper for all cases ;
but generally for light work, such as telegraph and telephone

* Galvanised iron wire is usually served with linseed oil before erection.

lines, there is very little to choose between the two. With
heavy power lines, however, copper will always work out the
cheaper (unless the relative costs be very different from those
assumed here), for the weight of the iron becomes practically
prohibitive. And even with large copper cables of, say, 19
strands each of No. 16 S.W.G. (and larger) the supports have to
be so frequent as to form a very important item in the cost of
the *whole line*. In such cases it is usual to suspend the copper
conductor from a small steel wire carried on insulators (*see*
Fig. 19); by this means the spans can be materially lengthened,
and the total cost largely decreased.

If, however, the conductor is insulated, copper is by far the
cheaper metal, since the necessary bulk of insulation material is

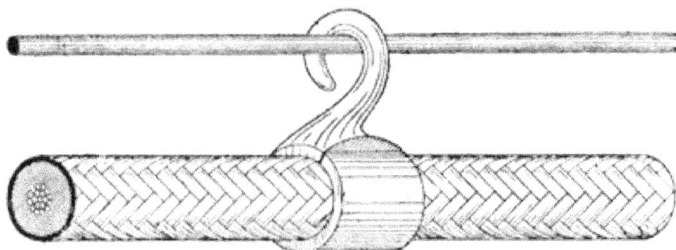

FIG. 19.—Cable Hanger and Bearer Wire.

much less. The areas of the two metals are for equal resistance
as 1 : 7, say, and the diameters will be as $\sqrt{1}$: $\sqrt{7}$, or 1 : 2·63 ;
and for an equal thickness, t, of insulation, the volumes of insulation material will be as $(1 + 2\ t)^2 - 1^2 : (2·63 + 2\ t)^2 - 2·63^2$.
But since to maintain the insulation resistance constant
per unit length of conductor it is necessary that the depth
of the dielectric vary as the diameter of the conductor, the
diameter of the iron cable over the insulation will have to be
$2·63 + 2\ t\ \dfrac{2·63}{1} = 2·63 + 5·26\ t$. To illustrate the meaning of
these ratios let the copper conductor be No. 10 S.W.G., and let
$t = 0·1$in. Then the diameter of copper conductor $= 0·128$in.,
and the diameter over the insulation $= 0·328$in. The diameter
of iron conductor $= 0·336$in., and the diameter over the

insulation $= 0.862$in. Now the insulating materials generally used, such as pure rubber, vulcanised rubber, &c., are costly, and, therefore, the iron conductor is commercially impossible.

These comparisons have been made between iron and copper, because they are the metals most likely to be used for power circuits. The copper alloys, however, present many advantages and combine the high conductivity of copper with the tensile strength of the best iron. Bronze wires are in general use for telephones and telegraphs, and will inevitably come into favour for transmitting power. Silicium bronze has a conductivity about the same as that of hard-drawn copper and a breaking strain of at least 90,000lb. to the square inch. Bearing in mind the relative weights for a given resistance, it is evident that the spans can be made longer with the bronze than with galvanised iron; in practice, it is found that they may be made about 50 per cent. longer, and less bulky and costly insulators are necessary. One reason why the copper alloys have not been used much as yet is a real or fancied difficulty in procuring a wire of a specified resistance and tensile strength, it being found that one of these qualities is frequently gained at the expense of the other. This difficulty is merely a detail of manufacture, and will, no doubt, soon be overcome. The cost, from about 8.5 to 12 pence per lb., is high, but will be reduced as the demand increases.

The figure of merit of silicium bronze (Class A), as previously estimated for hard-drawn copper and iron, is

$$\frac{1}{1.7} \times \frac{1}{8.7} \times \frac{1}{8.5} \times 65,000 = 517.$$

It is better than copper by about 12 per cent.* It should be noticed that the atmosphere has practically no effect upon copper or its alloys, and that the alloys are not so liable to crystallise as hard-drawn copper. These qualities largely reduce the cost of line maintenance, and in conjunction with the other points already discussed render silicium bronze perhaps the best all-round metal for power lines.

---

* For the comparative data of Classes B and C, see foot-note, p. 80.

§ 18.   INSULATED CABLES.

The materials used for insulating electric wires are not very
numerous, being chiefly pure or vulcanised rubber, rubber
compounds, resins, and hydrocarbons, such as the paraffins, bitu-
mens, &c.; vulcanised heavy hydrocarbon oils have also been
tried with some success. The covering on the cables has
usually two functions of entirely different characters. That
part nearest the wire is strictly the electric insulator or
dielectric; while the outside portion simply serves as a
mechanical protection to the real insulator, and is not necessarily a non-conductor of electricity.

It follows then that the insulating properties depend not on
the total thickness of the covering, but only on the nature
and depth of the dielectric. The insulation resistance varies
with the character and quality of the material used, with
the temperature, and even with the pressure at which the test
is made. It is usual to define the insulation resistance of
cables and wires in megohms per mile of cable at 60°F., but
the pressure at which the test is made is not often given.
This, however, is important, and a pressure bearing a suitable
proportion to that at which the cable will be worked should
always be used in the test. Thus, for a cable to be used
in a colliery at 500 volts the testing pressure should be not
less than 1,000 volts, which gives a factor of safety of four,
assuming the insulation resistance to be inversely proportional
to the squares of the two pressures. It is also usual to immerse
the cables in water at 60°F. for 48 hours before making the
test. These precautions are sufficient for the short lengths of
conductors used for most power plants, but if the cable be
many miles long it is necessary to make capacity tests, especially if it be armoured by a continuous metal sheathing of
lead or steel ribbon. This is most important with alternating
currents, since then the condenser effects may cause very
unexpected results. But this will not be a serious question
in general power work, and may be neglected if continuous
currents be used.

It is difficult to place the various dielectrics in order or
absolute merit, but their specific resistances have been determined with close accuracy, and form a rough guide.

The specific resistance of some of the dielectrics in common use are given in the following table, being compared with that of mica :—

Table G.—*Specific Resistances of various Dielectrics compared with that of Mica.*

Insulator.	Specific resistance compared with mica.	Temperature at which the determination was made.
Mica	1·0	63 F.
Callender's cable, bitite	1·8	59 F.
Callender's cable, bitumen......	5·33	59 F.
Gutta-percha	5·33	75 F.
Vulcanised india-rubber........	18·0	59 F.
Ozokerited india-rubber	77·0	49 F.
Fowler-Waring cable	86·0	59 F.
Shellac	107·0	82 F.
India-rubber (untreated)	130·0	75 F.
Siemens' fibre cables	141·0	59 F.
Hooper's submarine cable	178·0	75 F.
India-rubber, Siemens' special..	200·0	59 F.
Ebonite........................	333·0	115 F.
Paraffin	405·0	115 F.

It will be seen that prepared para rubber has the highest resistance (with the exception of ebonite and paraffin), and from its dense, hard texture it is peculiarly adapted for insulating cables. But it is *very* expensive, and becomes hard and cracks when exposed to the atmosphere, if great care has not been taken in the manufacture. Vulcanised rubber, *i.e.*, rubber mixed with a certain proportion of sulphur and subjected to a dry heat of about 360 F. for a certain time, has a lower insulation resistance, but is practically acid-proof, and is not so liable to become hard and brittle as pure rubber. It is specially adapted for use in water, and hence is largely used for underground mains. In applying it to conductors there are usually three servings : the first of pure rubber, the second a separator containing chloride of zinc (to absorb the sulphur and prevent the centre rubber from vulcanising), and the third a mixture of rubber and sulphur; the whole being vulcanised together. In some cases the rubber is vulcanised right through, but there is then a probability of uncombined sulphur damaging the copper. The sulphur blackens the copper, and appears to permeate its substance and to make it brittle. To obviate

H

these difficulties the copper should always be tinned before being coated with rubber : this also assists the soldering of joints.

Paraffin, although having a very high resistance, is not stable and soon shrinks, becoming pierced with small holes. This appears to be the result of the different melting points of the various waxes composing the whole.   Paraffin is largely used for electrical instruments, and the melting point is carefully specified when it is purchased ; a good sample has a melting point of not less than from 180°F. to 200°F., or even higher.

Several forms of resins have been used for insulators, but they all require to be carefully protected from excessive heat or cold, or they will melt or crack.   The Berthoud-Borel lead-covered cable is the best example of this class.   It is not much used, however, now, as various heavy hydrocarbons have an equally high insulation resistance, and are not so liable to crack.

One of the best examples of yarn fibre saturated with heavy hydrocarbon compounds is found in the Fowler-Waring lead-covered cables.   The copper is first covered with a braiding of hemp, or jute, and then thoroughly soaked with the wax at a high temperature, so as to remove all traces of moisture. Lead* is then squirted, under hydraulic pressure, in a fluid state around the core, and so forms a continuous metallic coating.   Since lead is a soft metal and is easily damaged by friction, kinking, or blows, it is usual to cover it with hemp braiding, or servings of tape or yarn.   With tape there should always be two servings, one right-handed and one left-handed, and the whole should then be coated with a preservative compound, such as ozokerite, Stockholm tar, or pitch.

Preparations from bitumen are largely used for insulating cables, but have much lower specific resistances than the previously-mentioned dielectrics.   If used in a solid form without any fibrous material they are practically impervious to water; but being viscous, the conductor may become decentralised

---

* The best Spanish lead, guaranteed to contain less than 3 per cent. of tin or other alloys.

in course of time. To obviate this, Callender's Bitumen, Telegraph and Waterproof Company (Limited), who are the makers of this class of cable, vulcanise the bitumen by adding just enough sulphur to combine with the hydrocarbon. This form of insulation, known as Bitite, is firm, and yet flexible, and appears to be unaffected by water*; but its insulation resistance is very low compared with that of vulcanised rubber, and, hitherto, it has been chiefly used for the distributing and service mains on low-pressure systems. When used for high-pressure mains, the makers recommend laying the cables in solid bitumen in iron or wood troughing (*see* § 20, pp. 94—102).

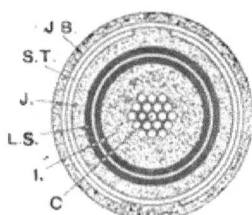

Fig. 20.
Armoured Cable.

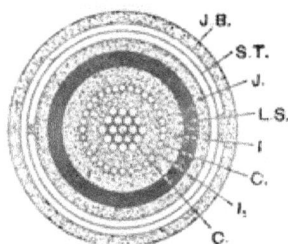

Fig. 21.
Concentric Armoured Cable.

J.B., jute braiding ; S.T., steel tape ; J., jute ; L.S., lead sheathing ; I., insulation ; C., conductor.

The bituminised fibre cables are usually laid in special bitumen culverts, which form a watertight envelope. Both classes of these cables are frequently covered with lead and served with jute ; and when culverts are not used it is customary to armour the outer serving of jute by one or two spiral windings of galvanised iron wire.

Steel tape, braided steel, or spirally-wound galvanised iron-wire is now frequently used to armour all the classes of cable referred to (*see* Fig. 20) ; but it is most generally applied to lead-covered cables, to protect them from mechanical injury, and it is especially useful when they have to be buried in earth the chemical reactions of which are unknown.

---

* It is stated by some that the sulphur works out of the bitumen in course of time.

Armouring is also largely used for concentric feeders (see Fig. 21). The use of a concentric distributing network is open to criticism, but the difficulty of making sound joints is practically overcome by the use of iron service boxes. After the wires have been properly soldered the box is filled with insulating material, which can be vulcanised if desired. Fig. 22 shows a service box adapted for use with single mains; and Figs. 23 and 24 two views of one suitable for concentric cables. These illustrations refer to the Callender Company's system of armoured mains. In all cases, however, the iron or steel armour must be well covered with jute yarn, thoroughly soaked

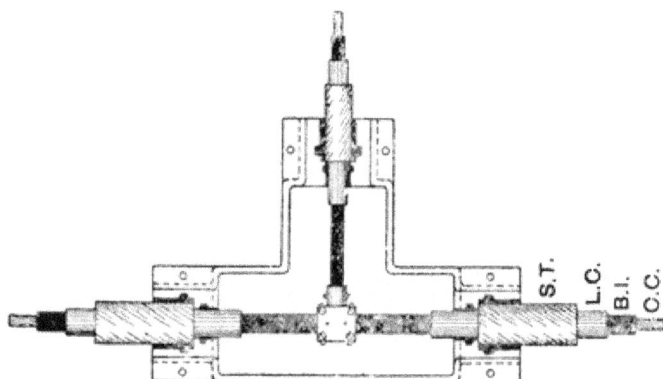

Fig. 22.—Junction Box for Single Cables. Plan.

S.T., Steel Tape, covered with jute: L.C., Lead Coating; B.I., Bitumen Insulation; C.C., Copper Conductor.

in a preservative compound, and the cable should be laid in pitched wood troughing, or in fresh-water sand, or in puddled clay, and thoroughly protected from the atmosphere. Neglect of these precautions is sure to end in breakdowns in course of time.

The chief source of trouble with lead-covered cables seems to be corroding agents, and if they be found in even very small quantities the lead will be attacked at one or more places, and holes will soon be eaten through to the dielectric; the erosion then rapidly increases from the inside as well as the outside, being assisted by leakage of electricity, which is sure to take place at about the same time. Again, lead-covered cables should

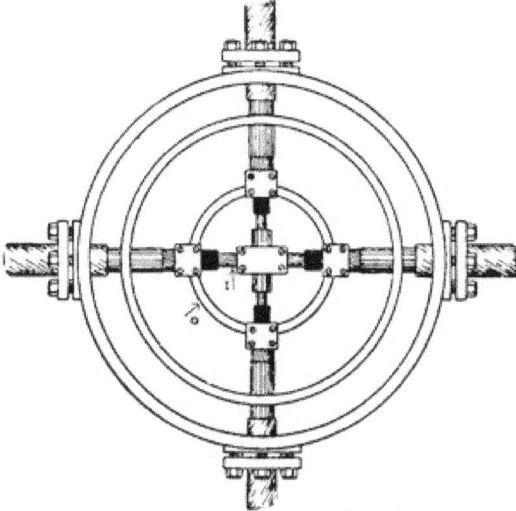

FIG. 23.—Junction Box for Concentric Cables. Plan.
I., Inner Conductor ; O., Outer Conductor.

never be laid in contact with iron pipes, as the slight leakage of electricity, which nearly always exists in damp weather, will

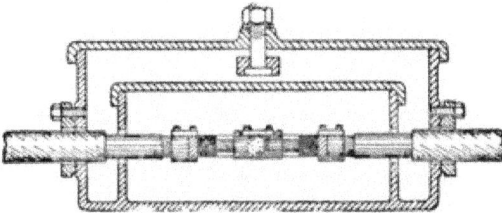

FIG. 24.—Junction Box for Concentric Cables. Sectional Elevation.

rapidly cause the iron to corrode, iron being electro-negative to lead.

§ 19. INSULATORS AND DETAILS OF AËRIAL LINES.

The various considerations affecting the choice of metal for the conductor having now been discussed, it remains to investigate the methods of erecting the line. There are two conditions which cannot be neglected, viz., the mechanical

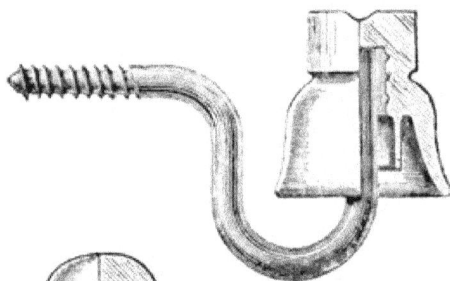

FIG. 25.

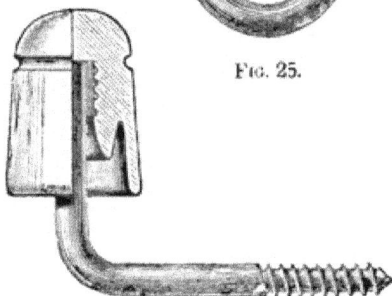

FIG. 27.

FIG. 26.

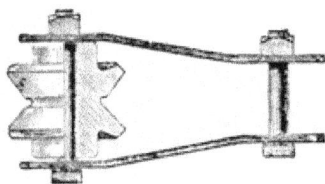

FIG. 28.

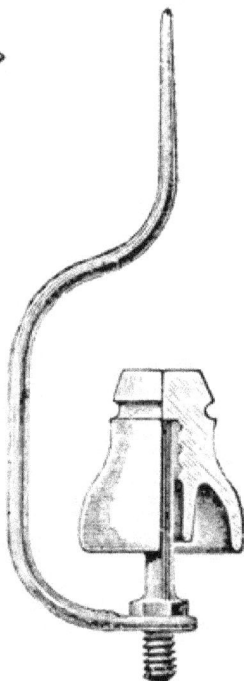

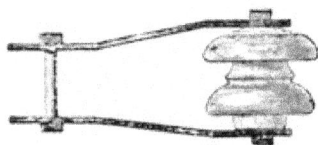

FIG. 29.   FIG. 30.

Iron Stalls, Shackles and Solid Insulators used for Low-Tension
Aërial Lines.

strength of supporting devices and high insulation resistance of the conductor. The general plan of carrying wires on insulators is familiar to us all, by reason of the large number of telegraph and telephone wires in use everywhere, and there is thus no need for a detailed description. The most frequent forms of iron stalks, shackles and solid insulators, suitable for low-tension bare wire lines, or insulated cable circuits run overhead, are shown in Figs. 25 to 30. The uses of the various types of support will be apparent on inspection. The stalks are usually of galvanised iron, either with coach screw threads, or screwed for tapping into metal ; for very heavy power lines, in which copper cables of about 19/16 S.W.G. are used, the stalks are forged or cast steel, sometimes galvanised. The insulators are made of glazed white or brown ware, the latter being the stronger and more durable.

The line wire is laid in the groove at the top or on the sides of the insulators, and is lashed securely in position by binding wire.

If the line pressure be high, and the dielectric covering around the cable of a low quality, it is usual to use an insulator, with a recess moulded in the inside of a bell-shaped bottom, which is filled with a resinous oil of high specific resistance. This plan is illustrated in Figs. 31 to 34, p. 92. Figs. 33 and 34 are adapted for straight lines ; Fig. 31 for curves and straining posts, or for attachments at the end of a line; and Fig. 32 shows a form used for carrying a bare steel wire to which the copper cable is suspended by leather loops (see Fig. 35), or by a metal hook as shown in Fig. 36, p. 93. The function of the oil is, primarily, to prevent surface leakage of current, especially during wet weather, when it necessarily occurs with the solid type of insulator. To maintain the oil in an efficient state, it is important to prevent an accumulation of dust or water on its surface. The arrangement shown in Fig. 33 is the Indian Government pattern, and is, perhaps, superior to any of the other devices ; but the types of insulators shown in Figs. 31 and 34 have given excellent results in most of the plants installed by the author. In very dusty places, or where the insulators are difficult of access, it is better to use the form illustrated in Fig. 33.

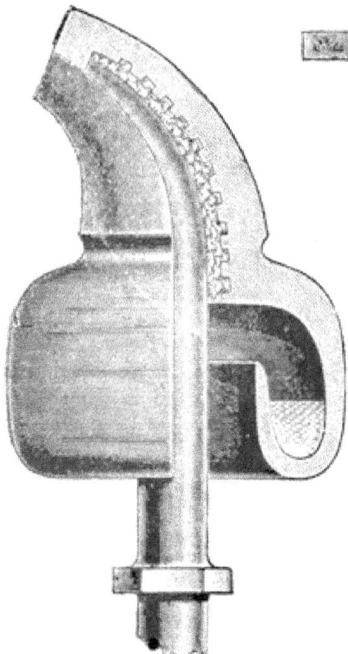

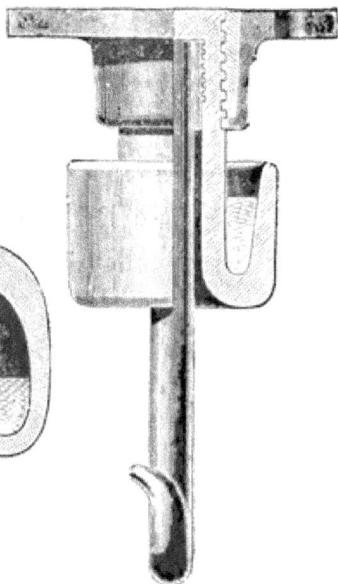

FIG. 32.

FIG. 31.

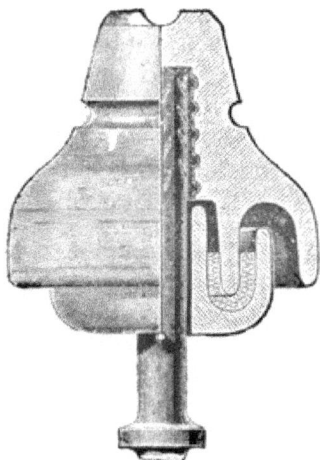

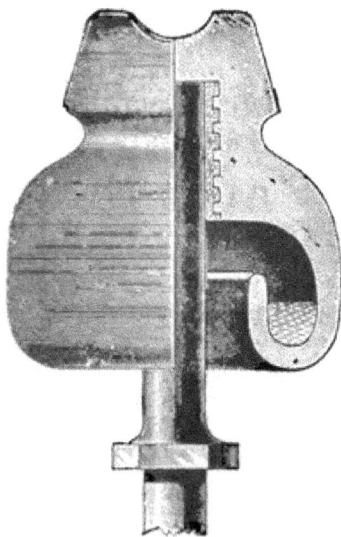

FIG. 33

FIG. 34.

Oil Insulators used for High-Pressure Aërial Lines.

With regard to posts, it is clear that these will only be
employed where there are no buildings or other convenient
objects to which the insulators may be attached. For cross-
country lines wood posts are usually found to answer all the
requirements; but along roadways, and in other special cases
hollow steel posts will be found to be nearly as cheap, and
certainly more durable and ornamental. It is probable, however,
that in the class of work to be carried out in the near future,
wood posts will be more generally preferred. They should be of
well-seasoned larch or pitch pine, and be firmly embedded in the
ground to a depth of from four to six feet, according to their

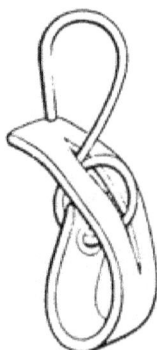 

FIG. 35.   FIG. 36.

Hangers for Suspending Cables from a Bearer Wire.

height and the strain they are designed to carry. It is not easy
to give definite rules for sizes of posts, lengths of spans, and
strain in conductor; but an examination of a few trunk telegraph
lines will exemplify the principles involved. Some idea of the
length of spans and dip of the cable—with and without a
bearer wire—may be gathered from Table H, p. 94, which has
been compiled by the Silvertown Company.

These figures agree fairly well with the author's own experience. In connection with this part of the subject, it should be
noticed that the Board of Trade Regulations fix the maximum
length of span at 200ft. where the direction of the conductor

is straight, and at 150ft. where it is curved, or when the
conductor makes an angle with the support. These regulations
also require a factor of safety of at least 6 in the conductors
and suspending wires, and of at least 12 in the posts, standards,
and fixed attachments, taking the maximum wind pressure to
be 50lb. to the square foot. The accumulation of snow may
be neglected, since the slight heat generated in the conductors
is sufficient to melt the snow in this climate.

Table H.

Copper conductor strands of	Dip.	Span.	
		Without steel bearing wire.	With 7/16 steel bearing wire.
7/20	5 feet	64 feet	238 feet
7/18	5 „	80 „	220 „
7/16	5 „	102 „	230 „
19/18	5 „	118 „	210 „

§ 20. UNDERGROUND LINES, CONDUITS, BUILT-IN AND DRAWING-IN SYSTEMS.

It is becoming generally recognised that aerial lines, whether
of bare copper or insulated cable, though permissible in open
country where they are free from wilful or accidental damage,
are a source of possible danger in towns, &c.   And it is also
apparent that such lines are especially subject to interference
from lightning. (The æsthetic view of the question is irrelevant
to the purely engineering aspect, and need not be discussed.)
It is evident, however, that if a line has to be made permanent,
and be required to work continuously at a minimum cost of
maintenance and depreciation, it must be protected from both
atmospheric influences and accidental injury.   These ends are
best attained by laying the cables in the earth.   Experience
gained with gas and water mains points to two methods : in
one, the cables are simply *buried in the soil* at a safe depth ; in
the other, they are *run in conduits or pipes.*

If the first system be adopted, it is evident that the insulation of the cable must be protected from mechanical injury
during laying, and afterwards from a possible erosive action of

the soil, and, perhaps, from the effects of water. This suggests the use of a metallic sheathing of lead if the Water Company's practice be followed, or iron if the Gas Company's experience be accepted. Both kinds of armouring have been used with success here and failure there. The causes determining the result are often obscure, but electrolytic decomposition, assisted more or less by the peculiar constituents of the soil and water, can always be shown to have taken place. In fact, the problem is complicated by the inevitable leakage of at least one main, usually the negative, and sometimes by weak points on both. The passage of an electric current through water breaks up the chemical combination into its component gases, oxygen and hydrogen. The liberated oxygen immediately attacks any oxidisable material in contact with it—the lead sheathing, for example—and soon changes its nature. Other more obscure chemical changes, resulting in the formation of free chlorine, hydrochloric, and sulphuric acids, and other corrosive agents, may also occur if the soil be impregnated with suitable salts. Hence, wherever there is a leakage of electricity, there is a tendency to destroy the metallic sheathing, the insulation, and, finally, the copper conductor itself. It is, therefore, necessary to protect the metallic armouring by servings of jute soaked in tar, ozokerite, or some preservative compound. And in all cases it is wise to lay the cables in clean soil, clay, or fresh-water gravel, and to protect them by boards from mechanical injury—such as an inadvertent blow from a pickaxe. The armoured concentric mains laid at Bradford for the lighting service are a good example of this class of work. They have been in continual use for about four years, and the author is assured by the engineer that the insulation resistance has increased since they were laid in the ground, and that no trouble is experienced with the coupling-up of consumers through the iron junction boxes already referred to.*

The second method may be considered under two heads:—
(*a*) the "built-in" systems, in which the cables, usually strips of bare copper, are laid in conduits, and covered-in permanently, so that they cannot be reached without taking up the top soil ;

---

* *See* pages 88 and 89.

and (*b*) the "drawing-in" systems, in which continuous lines of waterproof conduits, or pipes, or troughing are laid, with drawing-in boxes at intervals of from 80 to 100 yards in the straight, and at all sharp bends.

A complete discussion of these systems of laying cables is not suitable here, since this properly belongs to a treatise on central

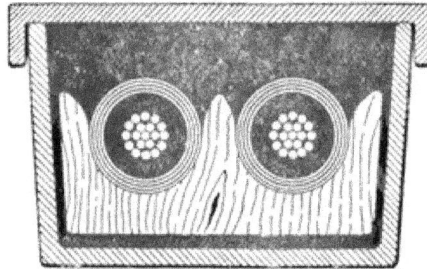

Fig. 37.—Callender Solid Bitumen System.   Cross Section of Trough.

stations for distributing light (and perhaps power) over a large area to a number of points in close contiguity; whereas this book treats, primarily, of the transmission of power between

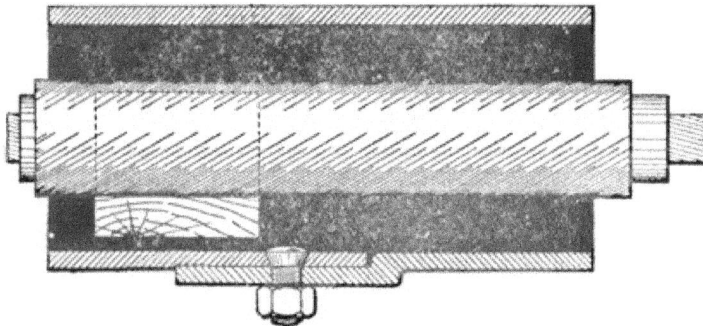

Fig. 38.—Callender Solid Bitumen System   Sectional Elevation of Trough.

places relatively far apart, and, secondarily, of its distribution to a comparatively few points in a limited area.   And should it be deemed advisable to run the cables underground, some modification of the first method (cables simply buried) is sure to be selected from the mere consideration of first cost. If, however, the cables are likely to be frequently disturbed for coupling-up supply mains to new points, or if they are likely to be increased in number within a reasonable time from the first installation, then one of the second systems may prove to be the cheaper in the long run, but this contingency is improb-

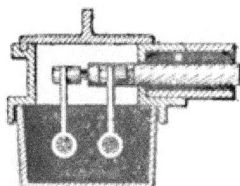

Fig. 39.—Callender Solid Bitumen System.   Cross Section of Service Box.

able on the supposition laid down in the preface and introduction. It will be, therefore, sufficient to illustrate the methods by a few characteristic drawings.

(a) One of the best "built-in" systems is the Callender Solid Bitumen system (illustrated in Figs. 37 and 38). The troughs are made in cast iron, in lengths of 6ft., with sockets at one

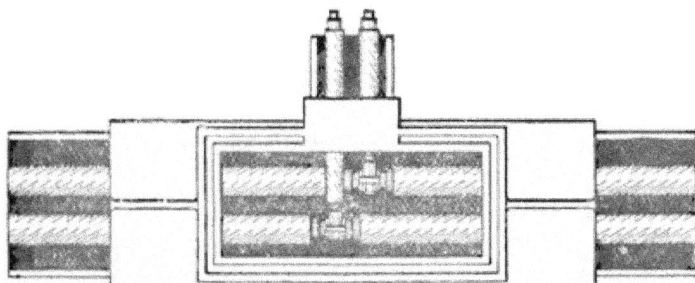

Fig. 40.—Callender Solid Bitumen System.   Plan of Service Box.

end, so that when fitted together there is a free " run " inside. The lengths are coupled together by counter-sunk bolts and nuts. Bends and circular pieces are used to carry the mains round corners and for changing the levels at crossings, but considerable deviation from the straight line is possible with the standard type. The method of laying is as follows :—The

troughs are first connected in position and then a small
quantity of molten bitumen is run in, and before it sets bridges
of bituminised wood are placed in it at intervals of 18in. The
insulated cables are next laid in position and held in place by
these bridges so as to be clear of the sides and bottom of the

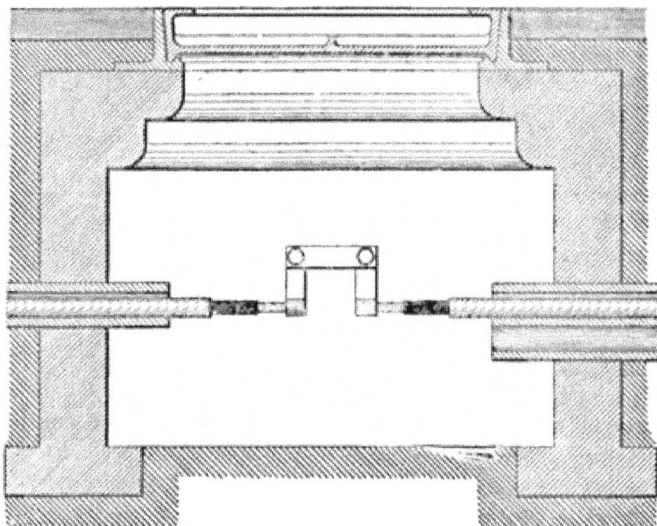

Fig. 41.—Callender-Webber Drawing-in System. Section of Manhole.

trough and of each other. The interior space is then carefully
filled up with bitumen to within half an inch of the top, and
the whole is finished off with a covering of Portland cement or
cast iron lids. It is obvious that cables laid in this manner

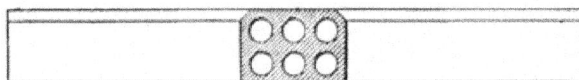

Fig. 42.—Callender-Webber Conduit.

are completely protected from the action of damp, water, or
gases, and yet repairs can be easily and rapidly made. The
method of coupling up the service mains is illustrated in
Figs. 39 and 40, p. 97.

(*b*) The drawing-in systems are illustrated by Fig. 41, which refers to the Callender-Webber system and shows the conduit, a brick pit, and the method of coupling the feeders or service mains without junction boxes. The conduit consists of bitumen concrete formed in blocks 6ft. in length and pierced by "ways" (*see* Fig. 42). The standard sizes of these cases are for 2, 3, 4, and 6 ways of either 1½in., 2in., 2¾in. or 3in. in diameter. To join the lengths, the cases are brought together and an iron mandrel placed in each of the ways. Melted bituminous concrete is then run between the two cases and rammed home. When the bitumen is cold the mandrels are withdrawn, leaving a perfect joint as strong as the main case itself. The "ways" are quite smooth throughout their entire length. It is said that the cables occasionally sink into the bitumen in course

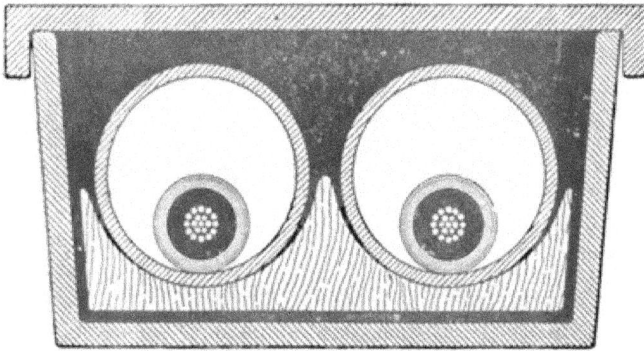

Fig. 43.—Callender-Raworth Drawing-in System. Cross Section of Trough.

of time and cannot, therefore, be withdrawn. The Callender-Raworth system (Fig. 43), however, entirely obviates this difficulty, and combines the mechanical strength of the solid bitumen system with the facilities of a drawing-in system. It consists of cast iron troughs with flanges. The spaced wooden bridges are fixed at short intervals, and on these are laid tubes of specially-made paper impregnated with bitumen. The tubes are joined together by sleeve pieces and are then buried in solid bitumen. The conduit is closed on the top by a cast-iron lid with flanges fitting over the sides of the trough. The high-pressure lighting mains at Huddersfield are thus laid, and are, in the author's opinion, a first-rate example of conduit work.

Messrs. Doulton and Co. make a variety of designs in vitrified stoneware casing, suitable either for built-in or drawing-in systems. They are made both with solid and removable tops, and in every respect are suited to the various requirements of conduit work. The stoneware is acid proof,

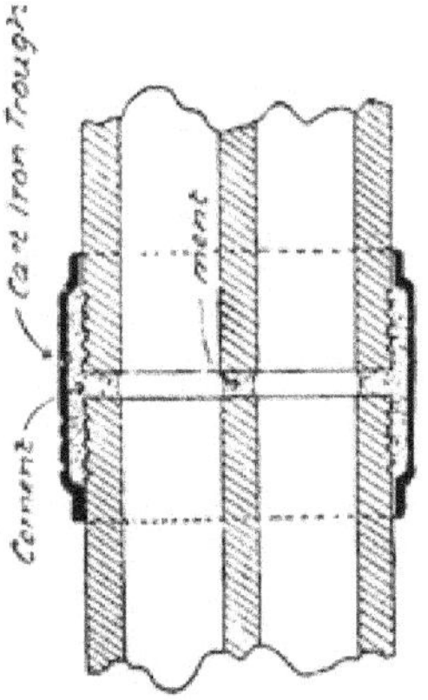

Fig. 44A.—Doulton's Stoneware Casing. Horizontal Section at Joint of 2 way Casing.

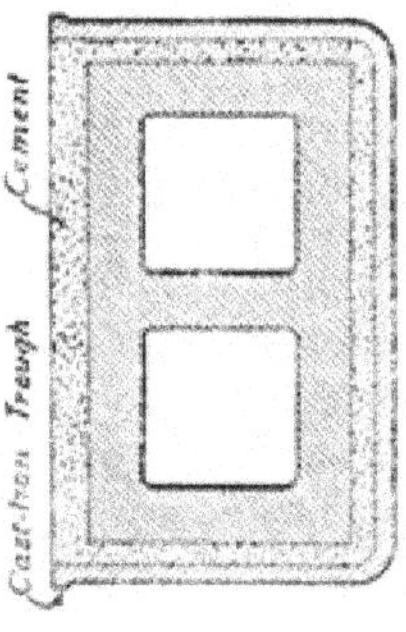

Fig. 44.—Doulton's Stoneware Casing. Vertical Section at Joint.

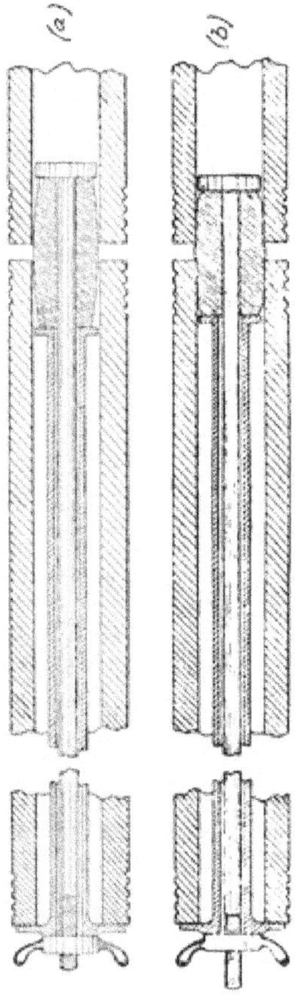

Figs. 45 and 45A.—Doulton's Stoneware Casing. Expanding Mandrel.

impervious to water and gas, and is made in convenient lengths, easily jointed in position by cement or bitumen. The lengths of casing are laid, with a small space between them, in cast-iron or stoneware jointing-collars. Mandrels with india-rubber heads are then introduced into the "ways" and expanded opposite the joints by a screw. Molten cement is next poured in to fill up the space between the lengths. The cement sets in a few minutes and then the mandrel is removed, leaving a smooth joint in the inside, and insulated cables may be safely drawn into the "ways." Solid casing with two "ways" is shown in Figs. 44 and 44A, in vertical and horizontal section at a joint; while the method of using the mandrel is made clear in Figs. 45 and 45A, (*a*) showing the mandrel as introduced into the "way," and (*b*) showing it expanded and closing the duct. This ware can be made in almost any shape, and will probably be largely used in the future instead of cement for conduits with removable tops. In price it competes favourably with most of the methods now in use, and its great strength, combined with its insulating qualities, commend it for a number of purposes incidental to electrical work.

The same firm make round spigot and socket pipes in stoneware with Standford's patent watertight joint. The principle will be at once seen on reference to Fig. 46. The black part

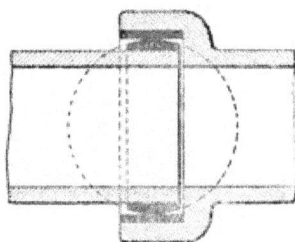

Fig. 46.— Standford's Watertight Joint. Section of Joint as Laid.

in the joint refers to a special preparation of bitumen fitted around the spigot and inside the socket. That in the spigot is curved, while that in the socket tapers gently inwards from the mouth. To joint the pipes it is only necessary to place the ends in line and push them together—a little thick grease being first put on the surface of the bitumen. It is obvious that an exact alignment of the pipes is not necessary, and that the joints will remain watertight with even a considerable settlement or displacement of the line of pipes. This method of jointing will, no doubt, find favour for electrical purposes, as it already has for water works.

I

Iron pipes have not been referred to because their use is sufficiently obvious. It may be mentioned, however, that cast pipes are more durable than wrought ones. They form the cheapest kind of conduit and are most frequently used for high-pressure feeders, which, of course, are heavily insulated. Generally, iron pipes are not so well adapted for the distributors as the Callender or kindred systems.

§ 21.—LIGHTNING DISCHARGES, LIGHTNING CONDUCTORS, AND LIGHTNING ARRESTERS.

It has already been mentioned that aërial lines are specially liable to trouble from lightning. It is, therefore, important to study the conditions under which lightning discharges take place, and also to examine the various methods in vogue for protecting the line and plant from damage. It is well known that trees, tall buildings, and elevated structures in general, are more likely to be damaged by lightning than low-lying parts; and it has also been observed that a certain area around a tall building, such as a church spire, is comparatively safe, and rarely receives damage from this cause. It is therefore customary to erect lightning conductors with barbed platinum points on the summits of tall or isolated buildings, and to connect the rods with the earth by means of suitable earth-plates buried in moist ground. The theory of lightning conductors has been elaborated by Prof. Oliver Lodge, to whose excellent work* the reader is referred for more complete information. It will now serve practical purposes to make a few generalisations with reference to the subject. The function of a lightning conductor is to prevent disruptive discharges by rendering the accumulation of a dangerous charge an impossibility. To attain this end experience shows that the discharging rods should be of stranded wire or ribbon so as to offer a low resistance to the high periodic current of the discharge. The metal may be copper or galvanised iron according to the view taken of the relative values of electro-magnetic induction and ohmic resistance. Prof. Lodge inclines to an iron conductor, but usage certainly points to copper. The conductor should form

* "Lightning Conductors and Lightning Guards," by Prof. Oliver J. Lodge, D.Sc.

a network, completely surrounding the structure of the building, and discharging points should be placed at every prominent place, or semi-detached part. Insulators should be used to carry the rods, in order to lessen the liability of lateral discharges to metallic things *inside* the building, and all external metal-work, such as water-pipes and iron ornaments, should be connected to each other and to earth, but *not* to the lightning conductor. The conductor is usually composed of seven strands each of No. 9 S.W.G., or of a ribbon 1in. wide by $\frac{1}{8}$in. thick. The earth-plate (or plates) should not be of less area than, say, 20 sq. ft., counting both sides, and should be buried about 20 ft. below the surface, so as to ensure contact with moist soil, even during times of drought. Sometimes an iron pipe is used, in which case the area must be reckoned from the outside only.

Lightning conductors as ordinarily understood, however, are not the only protection required for aërial lines, which, as a rule, lie much nearer to the earth than adjacent trees and buildings. Indeed, the difficulties in connection with atmospheric electricity which have to be encountered are of a varied character and demand special precautions. Lightning discharges may be conveniently divided into four classes.

(*a*) The most dangerous of all is a stroke falling on the line from clouds directly overhead, which raises the potential to a dangerously high point.

(*b*) If discharges of electricity occur between clouds in a direction parallel to or approximately parallel to that of the line, they tend to induce charges in the line and cause a disturbance in the line pressure, which may, or may not be compensated for by the effect being in the same direction in the lead and return.

(*c*) If part of the line be earthed at two or more places, disturbances in the surface potential of the earth may cause relatively large currents to pass through the copper line before equilibrium is restored.

And (*d*) The electrostatic induction of banks of clouds may cause rapid and violent changes of pressure in the line. These,

i 2

in a modified degree, are probably of frequent occurrence, but fortunately, the pressure can be rapidly adjusted through lighting arresters.

From an examination of the preceding divisions it may be assumed that *time* is an important factor in the effects of the discharge. Indeed, it seems probable that even a well-designed lightning conductor is insufficient to carry off the enormous discharges which sometimes fall directly on the line from low-lying clouds. Yet the true function of a lightning conductor, in the author's opinion, is not to form a safe path for the surging current of the discharge, but to prevent the accumulation of a dangerous potential by a continuous silent discharge. Viewed in this light, lightning conductors are useful in power work and may be placed at intervals on the posts, as indicated in Fig. 30, p. 90; but to be effective they must be efficiently earthed to plates, preferably of copper, of not less than 6 or 8 sq. ft. in surface on each side. The wire connecting the discharger and the earth-plate may be of galvanised iron wire, say, seven strands of No. 12 S.W.G.—*not solid wire.*

Inductional disturbances produced either by discharges between clouds or by the electrostatic effects of heavily-charged layers of clouds can be generally guarded against by means of devices known as lightning arresters. (These are referred to later.) Or, the line wires may be almost certainly protected from these troubles if a bare grounded wire be run on the tops of the poles, a foot or so higher than the insulated circuit. This is usually of bare stranded iron wires, and is earthed at either end and at intervals along the lines, and is, therefore, at the same pressure as the earth. As soon as there is a tendency to produce a disturbance of pressure in the system, it is annulled by a discharge between the earth and the grounded wire, and there can be no great difference of pressure between the earth and the insulated copper mains from this cause. The safety grounded line is shown in Fig. 47.

The iron wire might with advantage be connected to discharging rods placed on the tops of posts on high ground or exposed positions. Although the use of an earthed wire parallel to the insulated ones is fully appreciated in the telephone and telegraph services, it seems to present several difficulties with high-tension circuits. Since the " live " circuits, *i.e.*, the wires carrying the high-pressure current, are close to the " grounded " wire there is a chance of linesmen receiving shocks, *especially* if an earth return be used as in street railway work. This risk is obviously increased if the " live " wires be bare, and there is also the chance of short-circuiting the " live " and " ground " wires by accidental contact with telegraph and telephone wires. If both lead and return be covered with insulation, these objections have less weight, but

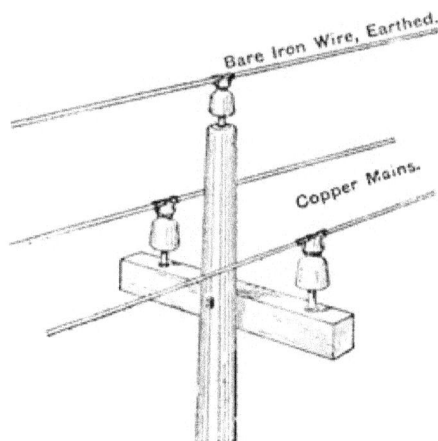

FIG. 47.—Grounded Discharging Wire.

even slight leaks may give serious shocks, when the working pressure is high.

However, this method is so efficient and so generally available that the author strongly recommends it for power lines running across open country, where the previously mentioned difficulties are not of so much importance. But the cheapest and in many respects the most practical way of securing comparative immunity from lightning discharges with an overhead line is to use a concentric cable in which the outer metallic sheath forms the return and is *grounded*. This arrangement only requires one conductor in place of two and the earthed

wire, and as the positive wire is protected by the insulation
between it and the negative wire and also by the grounded
wire itself, it is absolutely impossible for a linesman to get a
shock.   And in the event of a short circuit the worst that can
happen is the blowing of one of the cut-outs.   A concentric
cable used in this manner must not be confounded with a

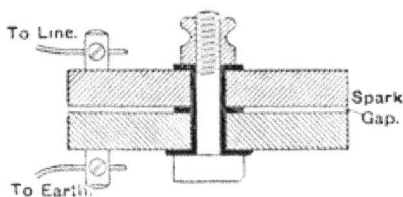

Fig. 48.—Lightning Arrester.

simple earth return or an insulated bare wire return, in both of
which cases there is always the chance of severe shocks to men
working on the line.

Lightning arresters have already been referred to, and two
elementary types are illustrated diagrammatically in Figs. 48
and 49.   The theory underlying the use of these devices is

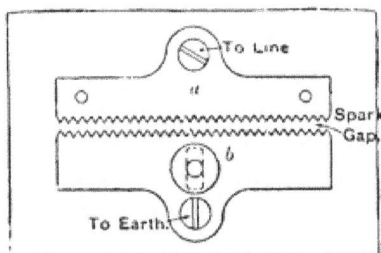

Fig. 49.— Lightning Arrester.

found in the power possessed by high pressure and high frequency currents to leap across air-gaps of high ohmic resistance
rather than traverse the line and dynamo circuits which have
large self-induction but small ohmic resistance.

In Fig. 48 there are two plain-faced discs of metal insulated
from each other by an adjustable air-gap.   The "live" wire

is coupled to one disc, and the other is "grounded." The action is as stated above.

In Fig. 49 another property of an electric charge is utilised to render the device more sensitive and certain in its action. It is found that an electric charge does not distribute itself equally over the surface of a body but tends to accumulate at *points*, from which it is also most readily discharged. This type of arrester has therefore a dual purpose. In the first place it acts as a discharger, and tends to prevent a slow accumulation of electricity; and, secondly, the spark-gap forms a path of low resistance, compared with that of the line, dynamos and motors, in the event of a large disturbance of pressure in the system. Referring to Fig. 49, *a* is a plate connected to the line, *b* a plate of similar size connected to the earth, the two serrated edges being placed opposite to each other. It is evident that *a* will be at the pressure of that part of the line it is in contact with, whilst *b* will be at the same pressure as the earth

The distance of the spark-gap can be regulated to suit the *working* pressure of the line. If this pressure be exceeded by a predetermined amount dependent on the distance between the teeth, a discharge will take place from one plate to the other, thus preventing an excessive difference of pressure. Since the inductive disturbances due to lightning are of a periodic character, it follows that there will not be the same tendency for disruptive discharges all over the area of the system at any moment. In fact, there will be a series of nodes continually shifting their position. To ensure efficient protection, then, the arresters should be placed at intervals, and also at the entrance to the power station, and across the terminals of all dynamos and motors coupled to the circuit. The method of applying them will be clear, after a perusal of the preceding statements. Owing to the fact that a much lower pressure is required to maintain an arc than to initiate it, the simple forms of arrester (shown in Figs. 48 and 49), although affording efficient protection to telegraph and telephone circuits, are not satisfactory on lines which are worked at a high pressure. For there is a risk of short circuiting the dynamo through one of the arresters. In fact, it is necessary to provide some means for extinguishing

the arc as soon as it is formed. This can be accomplished in
various ways, the most apparent being to move the plates apart
sufficiently to break the circuit after each discharge, and then
immediately to return them to the original position. Another
plan is to blow out the arc by means of the heated air resulting
from the energy of the spark. Or the spark-gap may be sub-
divided into a series of very short intervals, which effectually
prevent the formation of an arc by the lower pressure of the
line, but across which the high pressure discharge readily leaps
in a series of small sparks. Or the short-circuiting current

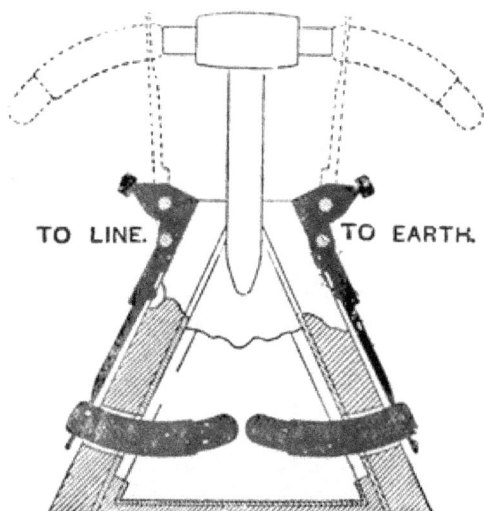

Fig. 50.—Keystone Pattern Lightning Arrester.

may be passed through a few turns of thick wire on a powerful
electro-magnet, and so break the arc by electromagnetic repul-
sion. All of these devices have been tried with more or less
success, and a number of patents embodying the principles
have been taken out at various dates. It is proposed only to
illustrate a few of the most recent or best tried forms, since
these are sufficiently characteristic of the whole.

The air-blast arc-breaking and self-adjusting type of lightning arrester is illustrated in Fig. 50, which shows the so-called

"keystone" pattern, made by the Westinghouse Company. It is largely used in the United States on power circuits. The action will be clear on reference to the diagram. The black lines show the normal position of the discharging points, the dotted lines refer to the movement which takes place immediately after a disruptive discharge, when the heated air has driven the dischargers out of the box. This action takes place on either pole, and so the arc is effectually broken. The swinging arms, after striking against the mallet-shaped stop at the top, rebound again into their normal position, and the arrester is again ready for action. In theory this device is admirable, and practically, perhaps, the only objection is the moving arms which are likely at times to stick. The apparatus as actually

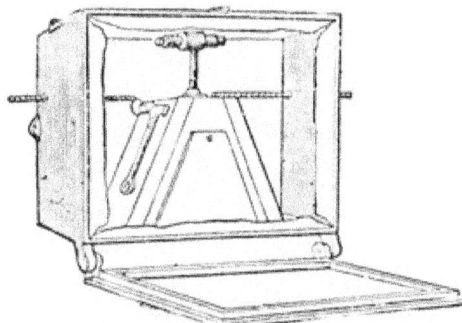

FIG. 51.—The Power Circuit Lightning Arrester.

installed is shown in Fig. 51; the box is opened in order to show the interior arrangements.

The magnetic "blow-out" type is generally not so satisfactory as the "air-blast," it being found that there is a tendency for the spark-gap to fuse across before the magnet has time to act. To obviate this, the strength of the magnet must be carefully adjusted, and in some designs carbon buttons are used on the dischargers. The carbon gradually disintegrates by the heat, and the length of the gap, therefore, requires adjustment from time to time.

Fig. 51A illustrates the Thomson-Houston arc line protector. The spark is formed at the base of the curved metallic horn

plates, but is instantly repelled by the strong magnetic field
produced by the electro magnet in the air space, the lines of
which are perpendicular to the path of the discharge. The arc
is thus pushed along the horns and increasing in length is finally
broken.

One of the most successful of the devices depending on the
bringing into action of new points after each discharge is the

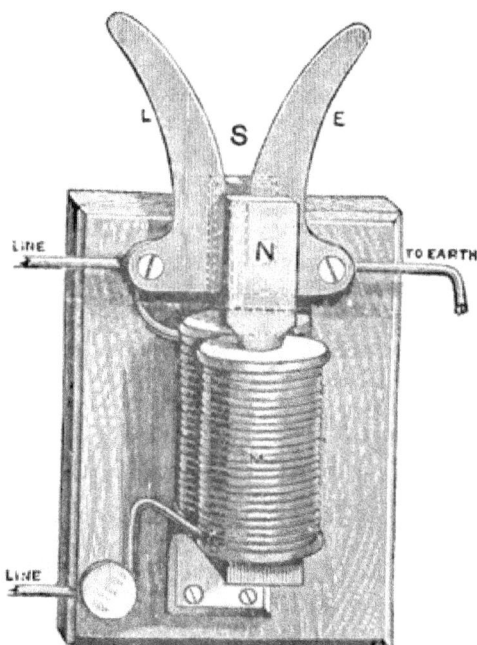

FIG. 51A.—Thomson-Houston Magnetic Blow-Out Lightning Arrester.

Wasson arrester. In this the arc is formed between carbon
buttons, and the current melts a fuse, causing a lever to drop
and thereby bringing into position a fresh couple of carbon
points. The arrangement contains, say, four or five sets of
carbons, and therefore is able to deal with as many discharges,
after which fresh fuses are required, and also a readjustment
of the buttons. It has therefore a more limited scope than
the Westinghouse air-blast type.

The " Wurts" lightning arrester depends on the subdivision of the arc into a number of minute sparks, and practically assumes that an arc is never formed. In its most usual form, and as largely used on tramcars by the Thomson-Houston Company and others, it consists of a number of thin iron washers threaded on an insulated column and separated from each other by thin discs of mica. The top sheet of iron is coupled to the " line," and the bottom to " earth." In order to form a circuit between the two end plates it is necessary for the current to leap from the periphery of one plate to that of the next, and so on through the series. Occasionally, beading takes place, finally resulting in a short circuit ; but generally,

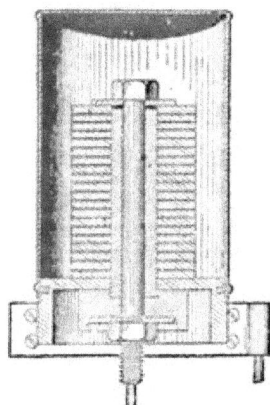

Fig. 52.—The " Wurts " Lightning Arrester for Continuous Current Circuits.

after properly experimenting with the particular line pressure, it is possible to select such a size and number of plates that excellent results are obtained. The portability and automatic action of the arrester commend it specially for tramcar work.

The action seems to be due to the large number of gaps, the length of which is short compared with the breadth, which is equal to the length of the periphery of the plates. The general appearance is given in Fig. 52.

The preceding remarks refer chiefly to direct-current systems ; with alternate currents a much simpler device is sufficient. It has been demonstrated by Mr. Wurts that zinc, antimony, bismuth, cadmium, and mercury have the properties of breaking an alternate-current circuit without sparking, and hence are called by him *non-sparking metals*. This gentleman has availed himself of this peculiar fact to design a very interesting and effective form of lightning arrester which will permit a 1,000-volt alternator to be short-circuited through it as often as desired, without even affecting the lamps on the circuit, so rapidly is the high-pressure arc destroyed. It is

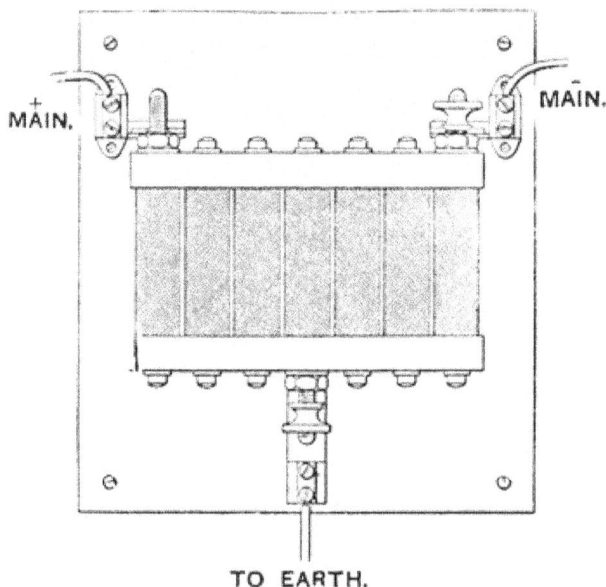

TO EARTH.

Fig. 53.—The " Wurts " Double-Pole Alternate Current Lightning Arrester.

shown in Fig. 53, which is of the double-pole type. The seven pillars, each 3in. long by 1in. in diameter, are solid and cast from an alloy of zinc and copper, and have knurled surfaces. The spark-gaps are each $\frac{1}{64}$ths of an inch in length, and as the earth wire is joined to the middle column, there are three spaces on each side equal to a total gap of $\frac{3}{64}$ths of an inch.

Incidental to non-sparking metals it may be mentioned that the inventor says they only show these remarkable properties where the spark-gaps are very small ; and if the gap be increased to half an inch the short circuit will be maintained with a vicious arc. The author has had no actual experience with these interesting arresters, but Mr. Wurts has given the subject much attention, and his investigations in this field are undoubtedly of importance. Some thousands of these non-arcing arresters are already being used in the United States with success.*

Fortunately, owing to our climate, there is not the same need for lightning arresters in Great Britain as in some other parts of the world. This happy circumstance explains to a large extent why the power circuit and station devices are chiefly of American origin. And due allowance must also be made in the relatively small number of overhead power circuits in the United Kingdom. All telegraph and telephone lines are, of course, protected by arresters, but, as already stated, the overhead grounded line and the devices shown in Figs. 48 and 49 prove amply sufficient.

The author has had, however, in the course of his professional work, to erect a number of overhead circuits for power purposes, and accordingly he designed some years ago the double-pole lightning arrester, shown in Fig. 54, p. 114. The base is of metal, and is grounded ; the faces of the discs are serrated in straight lines, placed at right angles to each other when in the position for action. The two discs are severally coupled to the positive and negative pole mains *outside* the main fuses, so that in the event of a disruptive arc forming across the plates and short-circuiting the dynamos, the cut-outs shall at once open the circuit. Although the author has used a number of these lightning arresters in some of the most open and mountainous parts of Great Britain and the Continent, he has never known a case where an arc has been formed. This does not prove that (under suitable conditions) an arc cannot form, but seems to suggest that the device acts as a silent discharger within the limits of the tests to which it has been subjected. It should also be marked that it is the author's invariable practice to use

* Vide *The Electrician*, December 2, 1892, and Paper read by Mr. Wurts before the American Institute of Electrical Engineers on March 15, 1892, entitled "Lightning Arresters and Non-arcing Metals."

toothed arresters at various parts of the line, mounting them
on the posts in water-tight metal boxes, and also to use discharging rods or lightning conductors at the most exposed parts

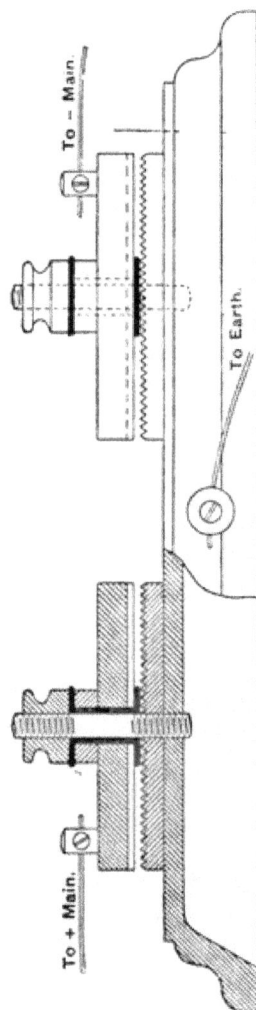

Fig. 54.—Double-Pole Lightning Arrester.

of the circuit.   And he has always used the overhead grounded
line as well as these arresters in mountainous districts, which
are subject to heavy thunderstorms.

CHAPTER IV.

DIRECT CURRENT SYSTEMS OF TRANSMITTING AND DISTRIBUTING ELECTRICAL POWER.

§ 22. INTRODUCTION AND DEFINITIONS.

In the previous chapter the line and *its details* were discussed in theory and practice, but the electrical systems of transmitting power were not differentiated. As it is of the utmost importance to understand the essential differences between them, and to notice the cases to which they are severally applicable, these branches of the subject will be treated in separate chapters.

The electrical systems of transmitting and distributing power may be classified, according to the *direction* of the current, as continuous or alternate ; or, according to its *division*, as series or parallel. The former classification will be adopted here.

A continuous current is one producing electromagnetic effects which are continuous in direction. An alternate current is one producing electromagnetic effects which regularly and rapidly alternate in direction. One complete alternation is called a period, and the number of periods per second the *frequency* of the current. This varies in ordinary practice from about 40 to 130.

A series system is one in which the current is used in a single circuit; a parallel system, one in which the current is divided into two or more circuits.   Both continuous and alternate systems are capable of parallel working, but only the continuous is suitable for series running.

§ 23. SERIES WORKING.

The simplest case of transmission is that in which the current is of the same value at all parts of the circuit, and in which therefore the dynamos and motors are all coupled in series with each other (*see* Fig. 55). The current is usually constant, the system being defined as series or constant current, and its application is limited by considerations of pressure of dynamos and speed regulation of motors.   It is not likely to

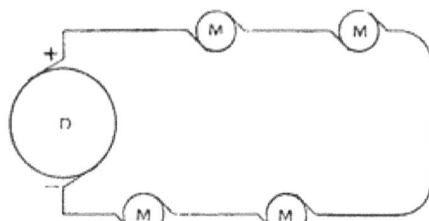

FIG. 55.—Series System.

be largely used, and is worthy of consideration chiefly on account of the particular case in which there are only two machines—one generator and one receiver.   (This, although a true series system, is not necessarily a constant current one, and indeed is generally worked with both pressure and current varying with the load; it is specially discussed at the end of this section.)   The constant-current series system, as its name implies, requires the current to be maintained approximately of the same intensity; and alterations of load are brought about by changes of pressure. Since all electrical work can be expressed by the product of the numbers of volts and amperes in the circuit, or symbolically $e \times i_m$, it is clear that for a given value of i_m, the output of the dynamo (or dynamos) is limited merely by the pressure, e, at which it is found practicable to work.   The insulation of the circuit and machines presents difficulty, and the collection or commutation of the current is influenced by the pressure. It is found that a pressure of about 3,000 volts is as high as can be conveniently dealt with in machines of the open coil type, such as the Thomson-Houston and the Brush arc dynamos; while with closed-coil drum, disc, or Gramme armatures, one of 2,000 volts is, perhaps, all that can be safely handled. (The difficulty will be increased if the current exceed the 6·8 or 10 amperes required for arc circuits.)

The pressure difficulty can be met to some extent by coupling two or more dynamos in series with each other, and this plan has been tried with success. It is a common practice to couple mechanically two dynamos in line with the prime motor— say, a turbine—and to connect the armatures electrically. But in order to maintain the current constant, it is necessary to vary the pressure to suit the work. This may be done in several ways. For instance, the speed of the prime motor, and, consequently, that of the dynamo, may be varied; or a governor may be used to alter the excitation, or to change the position of the brushes, and so cause the armature to have a greater or less demagnetising effect on the field-magnets. The controlling power for either of these governors is the main current which energises the soft iron core of a solenoid, and connection is made by suitable link motion to variable resistance coils, or to the brush rocking frame. The effect of altering the excitation is sufficiently obvious; but that of shifting the brushes is not so clear. It may be roughly stated thus:—When the external resistance of the circuit is lowered, by cutting out one of the motors or arc lamps, or by a reduction of the motor speed, more current tends to flow in the circuit. This increase of current energises the solenoid, causing it to shift the brushes to a fresh position where the pressure is lower, and thereby the current is decreased. When the current falls below the designed value, the governor also shifts the brushes, but in the opposite sense, and increases the pressure so as to tend to raise the current to its normal value. However, in the nature of things, with such a method of control, it is clear that the current cannot be maintained constant, for the governing depends on a departure

K

from the designed value. In fact, it is found that the current
varies between certain limits for full load and no load, the full-
load current being less than the no-load current by an appreciable amount. The result of this variation is, that there is always
a tendency to stop any motor which is temporarily loaded
beyond the torque corresponding to the particular current.

The constant-current dynamo has hitherto been chiefly
used for arc lighting, and as many as 60 lamps, each of
2,000 c.p., and requiring about 50 volts and 10 amperes, have
been run in series off a single machine. As a rule, it is not
necessary to vary the number of lamps on an arc circuit—either
all lamps or no lamps being required to be in use—and, therefore, the fluctuation of the current is not of much moment,
and, in any case, would only increase or diminish the illuminat
ing power. But with *motors* it is a very different matter.
The torque of a given motor [*see* equation (16), $T = \dfrac{C_a\, i_a\, N_a}{K}$,
p. 43], depends simply on the current in the armature, if the
field excitation be kept constant. In considering the action
of motors in series on a constant current circuit, first let it
be assumed that there is only one motor in the circuit, and
that its field excitation is constant. The motor will run at a
stable speed, determined by the relative values of the torque,
excitation, and current. Next suppose that the main current
is decreased by the interpolation in the circuit of another
motor, also with a constant field excitation. As soon as the
current passes through the second machine its armature will
start into motion, if the current be large enough to overcome
the static resistance, and will steadily increase in speed until
the counter pressure is just sufficient to permit the particular
value of current to flow under the forward pressure ; or when

$$i_m = \frac{e - E}{r},$$

where i_m = the main current,

       e = the forward pressure,

      E = counter pressure of the motor, and

      r = the resistance of the motor.

But if the torque be too large for the particular current and the excitation, the armature cannot move ; and the introduction of the second motor will increase the resistance of the circuit, and so lower the current, the immediate effect of which will be to stop the first motor, unless the dynamo is governed so as to keep the current constant.

These effects are intensified by the fact that a constant current dynamo or motor is usually series wound, and that consequently the field coils are traversed by the main current ; therefore, a variation of current affects the torque not only by virtue of its effect on the armature, but also by altering the excitation. Thus the constant current motor (as well as the dynamo) requires a governor, controlled by the *current* or the *speed*, which will strengthen the excitation as the speed diminishes, and *vice versâ*. The problem of running each motor independently of all others in series with it has been attempted by means of centrifugal governors, which, essentially, move a short-circuiting piece so as to cut in or out of circuit more or less of the winding, or else vary the resistance of a circuit placed as a shunt to the main exciting coils. The same end might be attained by means of a solenoid actuated by the main current, if only one motor were used ; but there do not appear to be any published accounts of practical devices embodying this idea. The author has built a number of motors with centrifugal governors, which have worked satisfactorily enough on power circuits ; and in the United States similar motors are used on arc lamp circuits, but probably the number is not large. The power plant at Genoa is, at present, the most interesting illustration of constant-current working. It is noteworthy that the designers have found it convenient to control the pressure by varying the speed of some of the generators, and by altering the excitation of others. But the motors are all controlled by governors which vary the field excitation. This plant is most successful.

Evident objections to constant current working are the heavy field-magnets necessary to prevent stoppages with temporary overloading, and the difficulties of maintaining the current constant. In cases where the load is constant, such as in driving ventilating-fans, for example, no governor is required, and series motors are more convenient than parallel ones, since no starting resistance is necessary. And in cases where the excitation can be controlled by hand, as on tramcars, there is an *apparent* field for series motors on constant current circuits. In practice, however, it is found that if more than three or four cars are worked in series the one with the heaviest load is likely to stop, and cannot be re-started until the load on the others is decreased. In fact, the possible output of the dynamo is fixed by its speed and the arrangement of the brush-rocking governors, and there is only a limited number of volts *free* at any particular moment to deal with sudden temporary demands. As is well known, the load diagrams of tramcars show large variations occurring far too suddenly for the steam governors to control ; and therefore, just when the dynamo should increase rather than decrease its speed, the engine is checked and an increase of pressure is impossible for the moment.

The series traction system was tried between Gravesend and Northfleet from 1888 to 1891. There were two trams with series motors regulated by hand, and a self-regulating constant current dynamo. No special difficulty was met with from the reasons just discussed ; but if there had been five or six trams the experience would have been similar to that of the Short Company and others in America. At the present moment there seems to be no probability of series traction competing with parallel.

Some stress has been laid on constant current transmission because the system presents such favourable conditions for the conductor. Since the current is only fixed by the limits of pressure deemed advisable for the given work, it follows that in cases where the distances are considerable a current of few amperes can be chosen, and the copper wire be as small as mechanical considerations will permit. And further, assuming the governors to work properly, two or more motors far apart from each other may be run in series without variations of load affecting the speed of either through changes in the fall of pressure in the line. The constancy of loss in the line, $i_m^2 R$, whilst advantageous as regards speed regulation, is objectionable from economic considerations, for the same quantity of energy is lost under all conditions of load. With parallel working, the line loss is roughly proportional to the load; but the speed of the motors falls off as the load increases, owing to the loss or drop of pressure. Some means of compensating for this drop

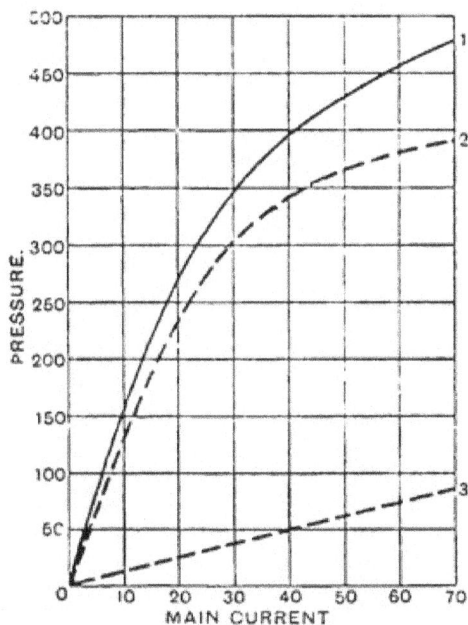

Fig. 56.

Curve 1 is the internal characteristic of a series dynamo running at a constant speed. Curve 2 is the internal characteristic of a series motor designed to be driven by the dynamo. Curve 3 represents geometrically the summation of the resistances of dynamo, motor, and cables. Note that with series machines the field and main currents are the same.

is necessary. The usual methods are referred to in § 26, p. 142. A compromise between these two systems is found in a particular case of the series system in which high pressure, small current, light conductor, line loss proportionate to load, and good speed regulation, can be obtained ; but it is only applicable to one dynamo and one motor. The necessary conditions are :—

Both machines must be series wound and designed for the same maximum current ; the dynamo must be run at a constant speed ; and the characteristics of the two machines must be such that the difference in pressure between the total E.M.F., E_d, of the dynamo and the counter E.M.F. of the motor, E_m (at least on the working part of the curve), is always equal to the pressure lost over the entire circuit, viz., line and internal resistance of the two machines. The idea will be readily grasped on reference to Fig. 56, in which the top curve connects the total dynamo pressure with the current, and the dotted curve shows the motor counter-pressure for corresponding values of the same current. The difference between the heights of the ordinates is suggested by the dotted line at an angle to the abscissæ, the slope of which is proportional to the total resistance of the circuit.

It is not easy to fulfil exactly the conditions necessary to give a perfectly constant speed through a long range of the curve ; but there is no difficulty in a close approximation through, say, a current variation of about 60 per cent., and this is really all that is required in most cases, and no attempt is made to get a better regulation. When the load is taken off the motor the speed will increase by perhaps 15 per cent., but this is of little importance in general work.

Fortunately, many power problems are capable of solution by this method, and no doubt its simplicity has done much to commend the use of electrical transmission of power. Further reference is made to it in the chapter describing actual plants, in which some figures taken from practice illustrate the usual limits of speed regulation.

§ 24. PARALLEL WORKING.

The term "parallel" is applicable to a number of systems of distribution, in all of which the current is divided into branch circuits, and the power is supplied to the distributing mains at approximately constant pressure. The conducting circuit may be complex, and as it will be necessary to make frequent reference to its various parts, it will simplify matters to define the terms used.

The feeding mains, or simply the *feeders*, are the large conductors which carry (high or low pressure) current to a central point of distribution, or to the nearest motor, in a simple parallel case. They may be long or short, according to circumstances, but since their function is that of carriers only, they relate to the transmission part of the problem, and are distinct from the distributing network of mains. Feeders, therefore, are never tapped by branch connections.

The distributing mains, or briefly the *distributors*, are designed to carry the current from a centre of distribution, *i.e.*, the end of a feeder, to the service mains which supply each consumer. They vary in section and length according to the area to be served, and may form simple radial lines from the distributing station, or in some cases a complicated network looped at the extremities by ring mains. Distributors may be tapped at any point to supply current to consumers.

The *Service mains* are the pairs of conductors connecting a motor or bank of motors (or lamps) to a distributor.

The parallel systems suitable for power work are :—

(*a*) Simple parallel, reverse parallel.
(*b*) Feeder system.
(*c*) Three-wire system.
(*d*) Multiple-wire feeder system with regulators.

Inspecting Figs. 57 to 64,* which illustrate diagrammatically the several arrangements, it will be seen that the condition of constant pressure between the distributors is common to all, and that since the energy at any point is equal to the product $e\, i_m$, it follows that variations of load have to be made by changing the current of the particular motor (or lamps).

---

* Pages 124 to 136.

This is exactly the converse of the series system of supply. The immediate effect of the variable current is to cause the loss in the mains (feeders and distributors) to vary with the load. This is an advantage, since the line loss is then roughly proportionate to the useful effect; but it also causes the pressure at different parts of the line to vary inversely as the load, which is exactly the opposite of what is required.

The line loss is expressed in watts by :—

$$i^2 R + i_1^2 R_1 + i_2^2 R_2 \ldots + i_n^2 R_n,$$

where i, i_1, $i_2 \ldots i_n$ are the several currents in different parts of the circuit, of which the respective resistances are R, R_1, $R_2 \ldots R_n$. And the fall of pressure in volts on the several sections is given by $i R$, $i_1 R_1$, $i_2 R_2 \ldots i_n R_n$.

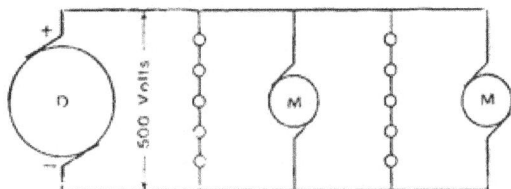

FIG. 57.

Incandescent Lamps in Series on High-Pressure Motor Circuit.

The aim, then, in a parallel system is to compensate for the drop in the feeders, and to keep the pressure constant at the centre of distribution. There will remain a slight fall of pressure uncompensated for between this centre and the points of supply, i.e., the motor terminals. But this is practically unavoidable, and in most cases can be made so small as to give no trouble if sufficient copper be put into the service mains; for then the product $i R$ becomes so small, even at full load, as to be insignificant. The weak point in simple parallel working is that the pressure is limited by the voltage of the smallest motor, or that of the lamps run in parallel with the motors. In lighting circuits, where incandescent lamps requiring about 100 volts are mainly used, the pressure at the centre of distribution is necessarily about 100 volts, although at the station it must be higher by an amount determined by the resistance of the conductors. In power circuits where lighting is of secondary importance the pressure on the distributors may be, say, from 300 to 500 volts ; and lamps may be coupled in series to make up the required voltage. Thus three incandescent lamps of 100 volts each may be coupled in series across a pair of mains having a pressure between them of 300 volts ; and, similarly, five 100-volt lamps may be placed across a 500-volt circuit (*see* Fig. 57).

Raising the pressure of distribution permits a reduction in the weight of the copper, but at the same time introduces complications which require careful attention.

In the first place, consider a constant length of conductor, with constant power at supply station, and let the pressure of supply be varied. As before :—

a = area of conductor.

R = resistance of conductor = $\dfrac{l}{a} k$, where l = the length of conductor and k is a constant.

e = pressure of distribution.

Then we have

$i^2 R$ = line loss ; $\dfrac{i}{a}$ = the current density.

$i R$ = drop of pressure in volts.

$e i$ = the power transmitted = a constant.

Now, assume that the pressure is doubled, so that it may be represented by $2 e$, then the current will be $\dfrac{i}{2}$, since the power is to be the same as before.

The line loss will now be $\left(\dfrac{i}{2}\right)^2 R = \dfrac{i^2 R}{4}$, *i.e.*, only one-fourth as great ; and the drop of pressure will be $\dfrac{i}{2} R$, or one-half. The section of the conductor may, therefore, be safely reduced.

There are two cases of importance in practice.

Firstly, when it is essential to keep the *line loss*, i^2 R, constant. In this case the new conductor need have only one-fourth of the area of the old one, but will have four times the resistance; for the line loss will be $\left(\dfrac{i}{2}\right)^2 4$ R $= i^2$ R, the same as before; and the line drop will be $\dfrac{i}{2} 4$ R $= 2\,i$ R, or the same *percentage* drop as before, which is equivalent to saying that *the line loss remains constant when the percentage of the line drop is kept the same for variations of supply pressure*. It must be noticed that the current density has been doubled, for it is now equal to $\dfrac{i}{2} \div \dfrac{a}{4} = 2\dfrac{i}{a}$.

Secondly, when it is necessary to keep the *line drop*, i R, constant. This can be accomplished by using a conductor of half the area and twice the resistance, for then the drop $= \dfrac{i}{2} 2$ R $= i$ R. And the line loss will be $\left(\dfrac{i}{2}\right)^2 2$ R $= \dfrac{i^2 \text{ R}}{2}$, or only one-half as great as before. The economic current ratio, $\dfrac{i}{a}$, is constant, for both current and area of conductors are halved. Or, generally, *when the line drop is kept constant for variations of supply pressure, the current density is also constant*.

The latter of these two cases is, perhaps, the more important with respect to the feeders, for in relation to them the economic current density is the chief consideration. And the ratio $\dfrac{i}{a}$ should have the same value in both long and short feeders, the drop of pressure in each being compensated for at the supply station. Strictly speaking, on the score of economy, the same considerations should determine the gauges of the distributors, but frequently the fall of pressure will be the only criterion, and then the conditions suggested in the first case will obtain, viz., a constant percentage of drop.

If the current density, $\dfrac{i}{a}$, be settled, as it should be, in accordance with the economic law, the current varies as the area of the conductor, and the drop, $i\,R$, is simply proportional to the length of the conductor. For each pressure of supply and current density, then, there will be a definite distance through which the power can be transmitted for a given drop. This distance includes the feeders, the distributors, and the service mains, if the drop be measured from the dynamo to the motor terminals; but in most cases it will be convenient to consider the three drops separately.

The fall of pressure for any section may be determined in the following manner if the current density, $\dfrac{i}{a}$, be fixed :—

$$ i\,R = l\,\frac{i}{a}\,k. \qquad \ldots \ldots \quad (23) $$

And if l be expressed in yards and a in square inches (the most convenient and practical units for these calculations), then k will be the numerical value of the resistance of one yard of copper of one square inch in cross section, and will be approximately equal to $\dfrac{25\cdot6}{10^{5}}$ for ordinary commercial samples. Therefore

$$ i\,R = l\,\frac{i}{a}\,.\,\frac{25\cdot6}{10^{5}}. \qquad \ldots \ldots \quad (24) $$

From this equation, if we know the drop, we can find the possible length of conductor; and if l be given, then the corresponding drop can be estimated.

It will sometimes be more convenient to estimate the drop in terms of the length of the section (measured along the conductors) rather than in terms of the length of conductor. To do this it is only necessary to put l equal to the length of the section, and multiply the right hand member by 2; thus

$$ i\,R = 2\,l_{1}\,\frac{i}{a}\,.\,\frac{25\cdot6}{10^{5}}. \qquad \ldots \ldots \quad (25) $$

If $i\,R = 1$, then l_{1} will give the number of yards per volt drop for the particular value of the economic ratio $\dfrac{i}{a}$.

The following table shows the relation between drop and distance for various pressures and current densities :—

Table I.—*Two Wire Mains. Drop of 1 per cent.*

Current density in amperes per square inch.	Distance in yards from the power station for a drop of 1 per cent. at the following pressures of supply.			
	100 v.	200 v.	500 v.	1,000 v.
1,000	19·5	39·0	97·5	195·0
900	21·75	43·5	108·75	218·0
800	24·5	49·0	122·5	245·0
700	27·75	55·5	138·75	278·0
600	32·5	65·0	162·5	325·0
500	39·0	78·0	175·0	390·0
4 0	49·0	98·0	245·0	490·0
300	65·25	130·5	316·0	653·0

To find the distance for a drop of *n* per cent., multiply the distance in the columns against the given pressure and current density by *n*.

In the above table the percentage of drop in the line and the current density are kept constant for pressure variations, both current and area of the copper varying inversely as the pressure of supply. The line loss, $i_{in}^2 R$, will remain constant under these conditions.

It is thus sufficiently clear that a parallel system requires conductors of larger section than is necessary for a series one ; and a compromise involving a series transmission and a parallel distribution is at once suggested.

Let us examine the various methods a little more in detail.

The simple parallel system is shown in Fig. 58. In this the whole of the current necessary for motors *b*, *c*, and *d* has to traverse the two conductors included in the section of the line *a b*, and that required for motors *c* and *d* has to pass through section *b c*, while *c d* only carries the current for motor *d*. Now, the areas of the conductors in the several sections of the line will be suited to the maximum, or more probably the average, current required to pass through them. There will, therefore, be a different pressure of supply at each of the points *b, c,* and *d.* This in itself would not matter if the actual values remained constant, but since the pressure of the line varies inversely as the current, it follows that the pressure will be lowest when the three motors are each absorbing their greatest current, and that it will vary at the several points of supply according to the current at each. Now, the speed of a motor is *proportional to the pressure of supply* for given currents, and therefore the speed will decrease as the load increases, from this cause, as well as from the drop over the internal resistance of the motor, which is necessarily greater with large than with small loads.

This difficulty is enhanced if the distance from the dynamo to the first motor, from thence to the second motor, and from thence to the third motor, &c., be considerable. Indeed, in this

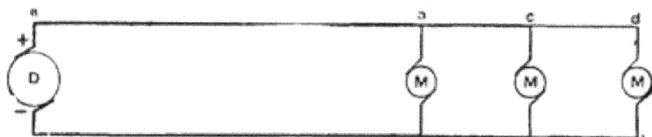

FIG. 58.—Simple Parallel System.

case a limit is very soon reached where the system hopelessly breaks down from the difficulties of regulation already referred to, and from the cost of the copper required to keep the line drop and loss within practicable bounds. To fix ideas, suppose the maximum current to be carried in the feeders *ab* to be 100 amperes, and the current density in them to be selected such that one square inch of area of cross section of copper is allowed to every 500 amperes. Then, since the maximum current is 100 amperes, the area of the conducting wire will be $\frac{100}{500} \times 1 = 0.2$ sq. in. The resistance of this wire is 0·22 of an ohm per mile, and if we assume the feeders *a b* to be one mile—a very fair assumption for a transmission of power problem—the fall in pressure from the dynamo to motor *b* will be $100 \times 0.44 = 44$ volts. Now, if the pressure at *b* be 100 volts, it is clear that that at the dynamo must be 144, and the line efficiency, assuming all the power to be absorbed at point b, is only $\dfrac{100\,i}{144\,i}$, or 70 per cent. It is needless to say that a loss in the line of 30 per cent. is prohibitive in most cases. If the points c and d be severally far removed from b and from each other the line loss is still further increased, and ultimately may become so large as to absorb the greater part of the total power.

The apparent remedy is to raise the pressure. Let it be doubled (say, 200 volts), and the current consequently halved (50 amperes), and let the same section of conductor be used. The current being halved, the line loss will be one quarter as great, the fall of pressure will be halved, and the line efficiency becomes $\dfrac{200}{222}$ or 91 per cent. at the point b, and is correspondingly raised at points c and d. But the current

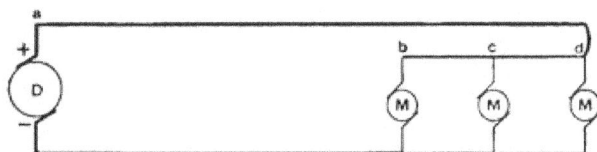

Fig. 59.—Reverse Parallel System.

density in the conductors will be at the rate of only 250 amperes to the square inch, for the current is now only 50 amperes. This implies an extremely costly conductor for the current. Practically, a current density of from 500 to 800 amperes will usually be found to be about correct.

If it be required to cause the pressure at different parts of the circuit to vary approximately equally, the arrangement shown in Fig. 59, which is sometimes useful in lighting circuits, may be adopted ; but the weight of copper in one of the feeders is evidently increased, and if the distance $a\,b$ be at all considerable compared with $b\,c$ and $c\,d$, the variation of pressure in the feeders may still be so great as to practically outweigh the better regulation gained over the distributors by the looped wire. The scheme at best is but an approximation, and only likely to be used in cases where the motors are not grouped but lie conveniently in a line. In practice, they are much more likely to lie in a group, and some point fairly equidistant from the majority of them will be selected as the centre of distribution. The pressure there will be maintained as nearly constant as possible; and each motor or group of motors will be run on separate distributors, the areas of which will be so calculated with reference to their length and to the maximum or average current, that the maximum or average fall of pressure will be fixed between reasonable limits.

This brings us to the feeder system (*b*) shown in Fig. 60. As before, D represents the dynamo, and *a* the power station if there be a number of dynamos in parallel; *a b* represents the

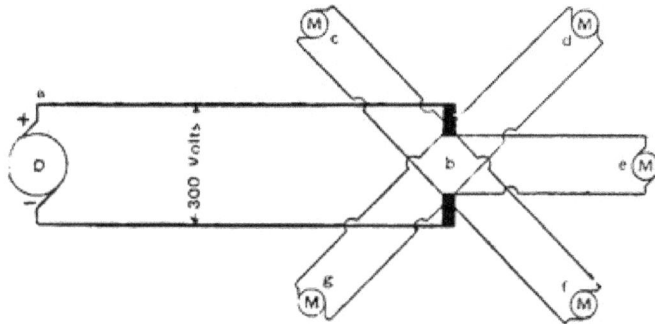

FIG. 60.—Feeder System.

transmission mains or feeders, *b* the centre of distribution and the point at which the pressure is to be maintained constant; and *c, d, e, f* and *g* severally represent motors at the ends of distributors *b c, b d, b e, b f*, and *b g*. Let the pressure of distribution be fixed at 300 volts at sub-station *b*, and let the feeder efficiency at full load be 90 per cent., a drop of, say, 30 volts being permitted along *a b*. The pressure at the power station will be 330 volts at full load, and will fall as the load is decreased until the smallest output is reached, when the pressures at the station *a* and the distributing centre *b* will be approximately the same. The fall of pressure in each of the distributors may be regulated according to the class of work and regularity of speed required—an average drop of about

five per cent. will usually be permissible for power work.
The efficiency of the entire line will be, say, 85 per cent.
under these assumptions; but it must be remembered that the
conditions selected are entirely arbitrary, and the economic
limitations discussed in Chapter III. have not been taken into
consideration in any way whatever.

The next practical point bears on the limiting distances
through which power can be efficiently transmitted. The two-
wire feeder and distributor system (Fig. 60) has already been
examined, and it is clear that the distance of transmission is
practically fixed by the station pressure. Now, the pressure
of the distributors cannot be greater than that permitted by
the conditions of supply to the consumers—500 volts will

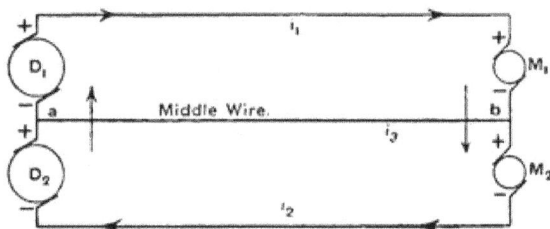

FIG. 61.—Three-wire System.

probably be the maximum.   But there is no restriction to
the pressure between the feeders, other than that of convenience.   The *three*-wire feeder system, shown in its simplest
form in Fig. 61, meets this difficulty to some extent, for by its
use the feeder-pressure may be doubled without raising that of
the distributors.   In this, essentially, two dynamos are coupled
in series across the feeders, each giving, say, 500 volts and 50
amperes, and thus making a line pressure of 1,000 volts.

At the end of the feeders two suitable motors are also coupled
in series, and a third wire is coupled to the junction of the
dynamos and also of the motors as is indicated in the diagram.
The path of the current, when the motors are equally loaded,
is shown by the arrow heads, and it is seen to be entirely confined to the machines and the two outer conductors, the middle

or third wire carrying no current. But if one motor absorbs more current than the other, the difference between the two currents will pass through the middle wire. Let i_1 be the current in the circuit $D_1 M_1$; i_2 that in the circuit $D_2 M_2$; and i_3 that in the middle wire. Now when $i_1 = i_2$, then $i_3 = 0$; and also when $i_1 > i_2$, then $i_1 - i_2 = i_3$. When i_2 is zero, *i.e.*, when M_2 is not running, then $i_1 = i_3$, and the middle wire has to carry the whole current in the circuit $D_1 M_1$. In this case it is not advisable to make the middle wire of less section than the outer ones; but if M_1 and M_2 each represent a group of motors, it is reasonable to assume that full load on one circuit and no load on the other will be unlikely to occur at the same time, and, therefore, the middle wire will have to carry only the difference between the two currents. In practice, with a load of lamps, it is found to be sufficient if the area of the middle wire be half that of one of the outer wires. And it should be recollected that if the distributing centre be a large one it will be provided with suitable switching arrangements for transferring part of the load on the distributors from one circuit to the other, as may be necessary to secure a proper balance, and to keep the current in the middle wire as small as possible.

Assuming the load to be so arranged that it will be generally divided nearly equally between the two circuits, it will be safe to follow the practice usual in lighting and to make the middle wire of one-half the area of the outer ones. This will permit a sufficiently wide difference of load in the two circuits to meet the requirements of practice. Now it is seen that for the same output from the power station the three-wire system requires twice the pressure and half the current necessary in the simple parallel systems shown in Figs. 58 to 60. And, if the same gauge of wire be used in the outer mains, the weight of copper in the three feeders will be 25 per cent. greater than that in the two, being in the ratio of 5 : 4. But since the middle wire is nominally idle, and the outer feeders carry a current which is equal to only half of that in the two-wire mains, it is clear that the power lost with the three wires will be only one-quarter of that with the two wires, and that the fall

of pressure will be only half as great. For *equal conductors*
and the same drop of pressure, the three-wire may be therefore twice as long as the two-wire system ; and for equal line
loss or feeder efficiency and the same percentage of drop, four
times as long. And if the relative weight of copper for *equal
lengths and line drop* be considered, since the area of each
outside conductor is one-half, and of the middle conductor one-
fourth, of the area of either conductor in the two-wire system,
the ratio of total weights will be as 5 : 8, showing a net
saving of $37\frac{1}{2}$ per cent. in favour of the three-wire system.

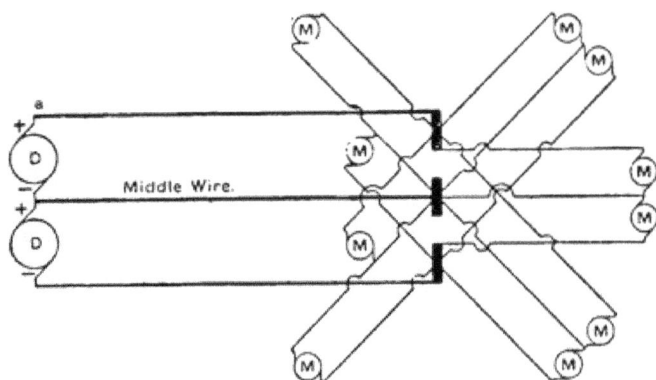

Fig. 62.—Three-wire System in Feeders and Distributors.

This system is evidently capable of extension. For if four
wires be used and coupled to three dynamos and three motors,
the pressure may be 1,500 volts over the outer feeders and the
current only 33·3 amperes ; and if the outer mains have the
same section as before, the distance through which the power
may be transmitted with the same drop is again increased by
50 per cent., or is three times as great as that with the parallel
system. And the weights of copper for equal lengths and line
drop are as 5 : 12, when compared with the two-wire system. At
Manchester a five-wire system has been erected for lighting, to
the specifications of Dr. J. Hopkinson. The outer mains are
at 420 volts, approximately, and between each pair there is a
pressure of 105 volts.

The three-wire system is applicable to the distributors as well as to the feeders, and is an important means of limiting the fall of pressure in the distributing network (*see* Fig. 62, which represents the disposition of feeders, distributors, and motors). It is apparent, however, that the three-wire feeder system, as well as the two-wire, ultimately reaches a limit beyond which it does not pay to transmit power; for though the pressure may be raised almost indefinitely as regards the feeders, it is not so with the distributors—500 volts being probably a maximum for the supply service even in power work.

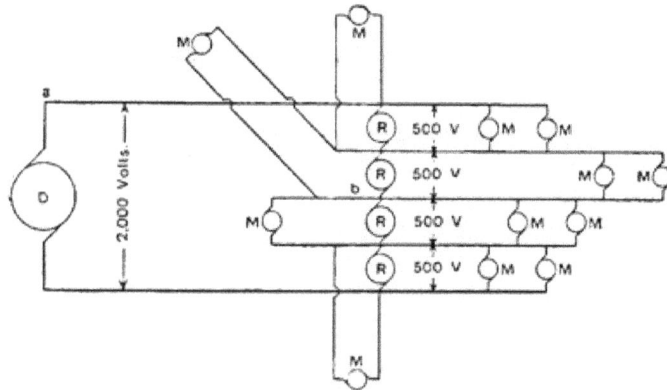

Fig. 63.—Feeder System with Regulator.

A convenient combination of the feeder and multiple-wire systems consists of a pair of high-pressure feeders coupled to a *regulator*, which may be a motor having as many armatures as there are sub-circuits, and feeding each pair of distributors at the desired pressure. The general idea is represented in Fig. 63. In this diagram the armatures are sketched as if they were independent of each other. But in practice they are rigidly coupled or built on one shaft, and essentially run in magnetic fields of the same strength and direction or in a common field. The armatures are of identical construction, and their internal resistance is made as low as possible. The field excitation is preferably supplied from an independent source, such as accumulators, which may be

L 2

charged in parallel across one of the sub-circuits at suitable times. Normally, when the load is equally divided between the sub-circuits, the four armatures absorb just sufficient current to overcome friction. But when one of the circuits takes more current than the others the armature in shunt to it will tend to lag. And the armatures in parallel with the other circuits will resist this alteration, and being coupled to the lagging armature will spend power in keeping the speed constant. Thus the lagging armature will be driven as a generator and supply the increase of current to *its* circuit, and the necessary power for the increase of torque will be supplied from the generating station. The degree of regulation attainable with this device depends on the internal resistance of the armatures; and it can only be efficient when

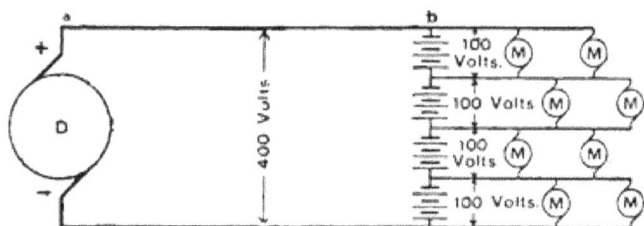

Fig. 64.—Feeder System with Accumulators.

the pressure lost over the armatures is negligible. The regulator principle is capable of considerable modification, and is likely to be used, in various forms, for lighting in cases where the feeders are of great length.

Secondary batteries have already been used with undoubted success in connection with feeder systems with sub-stations, by Mr. Crompton and others. The method is illustrated in Fig. 64. It will be apparent that the secondary battery acts as an accumulator as well as a regulator, and can be arranged to store up, during periods of light load, energy which will be available for temporary outputs exceeding the rated engine and dynamo power. In many cases also the accumulators can deal with the small loads during parts of the early morning, and the steam plant may be shut down for that period.

It is probable that secondary batteries will be found profitable for simple power work in some cases; and for combined light and power they certainly are always worth serious consideration. The author believes strongly in the future of the storage battery, and insists upon its importance in all direct current installations when the output is intermittent, or the load factor small.

§ 25. CONTINUOUS CURRENT TRANSFORMERS OR DYNAMOTORS.

The continuous current transformer, motor generator, or dynamotor (Fig. 65, p. 138) presents some points of interest for long-distance transmissions; for, when used in parallel, each dynamotor forms an independent sub-station.

The principle of the dynamotor has already been mentioned when considering regulators. The essential difference between the two machines is, that in the *dynamotor* the primary and secondary circuits are separate, whereas in the *regulator* they are connected. In its simplest form the dynamotor consists of a motor coupled to a dynamo, the motor being connected to the primary circuit, usually a high-pressure one, and the dynamo feeding the secondary circuit with a low-pressure current. Of various feasible arrangements the most common consists of two armatures mounted on one shaft and run in common field magnets *excited in shunt* from the low-pressure secondary mains. The magnets have a few series turns which are temporarily coupled to the primary circuit at starting. Or, the motor and dynamo windings may be placed on the same armature, and the commutators arranged one at each end, the field magnet details being as before (*see* Fig. 65, p. 138). These machines are adapted for parallel working, and have been used by the Electric Construction Company at the Crystal Palace and Oxford Central Stations and elsewhere. They are placed at the distributor end of high-pressure feeders, and feed the distributing network with current at a pressure of about 100 volts. Each dynamotor forms a sub-station. The general arrangement is illustrated in Fig. 66. It will be seen that the transformers feed into different parts of a network of

FIG. 65.—Dynamotor with Primary and Secondary Windings interplaced on one Core.

distributing mains. The proportion of the total current given out by any one transformer depends simply upon its pressure, and since voltmeter wires are taken from each point to the central station it is easy to alter the pressure of the feeders, and consequently that of the distributors. By this simple means the load can be divided as desired between the dynamotors, the number of which in use at any time can also be readily controlled. Thus supposing it is necessary to start a fresh dynamotor at, say,

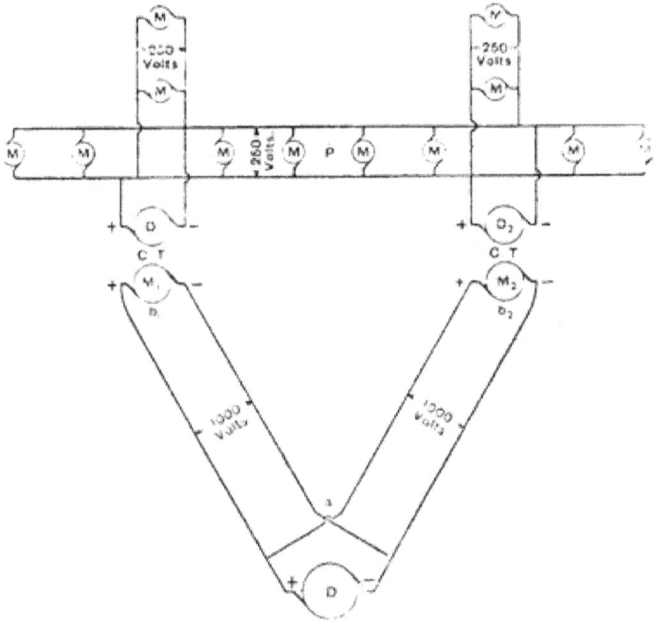

Fig. 66.—Dynamotor Parallel Sub-Station System.

sub-station b_1, the field of the machine is first excited, then the armature is started and the pressure adjusted approximately to that between the distributors to which it is to be coupled. The circuit between the dynamotor and the distributors is next closed, and finally the pressure is raised at the station end of the feeder until the dynamotor has its proper share of the load as gauged by the feeder current. A dynamotor is removed from the circuit in a similar way, the feeder pressure

being lowered until the feeder ammeter shows only the exciting
and friction current; the distributor circuit can then be opened
without sparking or disturbing the pressure of the distributing
network. The whole of the necessary switching and regulating
can be done from the power station.

It may, however, fairly be questioned whether parallel
dynamotors present any real advantages for power work, or
for installations in which power work is of more importance
than lighting. For it is evident that the feeders can be
brought close to the motors in most cases by means of one of
the multiple-wire and feeder systems already described. Yet
the author has found dynamotors to be of use as a detail in
certain cases, especially where the pressure of the main distributors (500 volts) is too high for the small motors, or where
from any other cause it is deemed to be unsafe to carry the
high-pressure mains into certain parts of the area of supply.
A dynamotor may then be driven off a pair of distributors
to deliver power at a suitable pressure, as at the Greenside
Lead Mines.   The author drew attention to this field for
continuous-current transformers as far back as 1889, and
he has already installed several for running rock drills, and
also for working electric locomotives inside mines with bare
overhead conductors.   Dynamotors with the primary circuits
in series and fed with a constant current are not, in the
author's opinion, likely to prove of much use in practice ; for
the arrangement requires the secondary circuits also to be in
series, and governors would probably be required to control the
field excitation. But there are a few cases, perhaps, where
series dynamotors might prove commercially profitable. The
general scheme for series coupling is illustrated in Fig. 67,
which is sufficiently explanatory to show the chief features.
It will be seen that the system has all the drawbacks of a
simple series transmission, and although some economy may
be obtained in the weight of the conductors, yet, since the
loss in both primary and secondary conductors is necessarily
constant and independent of the useful load, the system cannot be economical unless the load factor is very high. The
difficulties attending its use are further increased by the
fact that the dynamotors are necessarily designed for the

maximum power of the motors, while the average output is not likely, except in special cases, to exceed, say, 50 per cent. of this.

If it were possible to run the primary circuits in series at constant current, and to feed the various secondary circuits at constant pressure, dynamotors would combine the advantages of the series system as regards the transmission mains and that

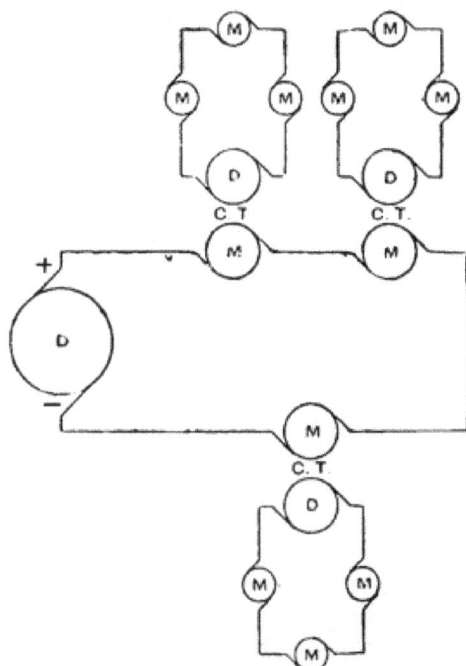

Fig. 67.—Dynamotor Series Sub-Station System.

of the parallel as regards proportionality of line loss to load in the distributing circuits. But the solution is too complicated for practical consideration. And, moreover, the author is strongly of opinion that, assuming such a scheme to be practicable, it would not be so economical, all things being considered, as a two-wire feeder and multiple-wire distributor system, with or without regulators, unless the load factor were very high—a rare thing indeed, even in power work.

§ 26. COMPENSATING FOR FALL OF PRESSURE ON FEEDERS AND MAINS.

In the preceding description of continuous current systems of transmitting power, it has been assumed that the pressure at the distributing centre—the sub-station or sub-stations—can be maintained practically constant. This is evidently a necessary condition of satisfactory distribution either of power or light. In order to secure it the pressure of supply at the

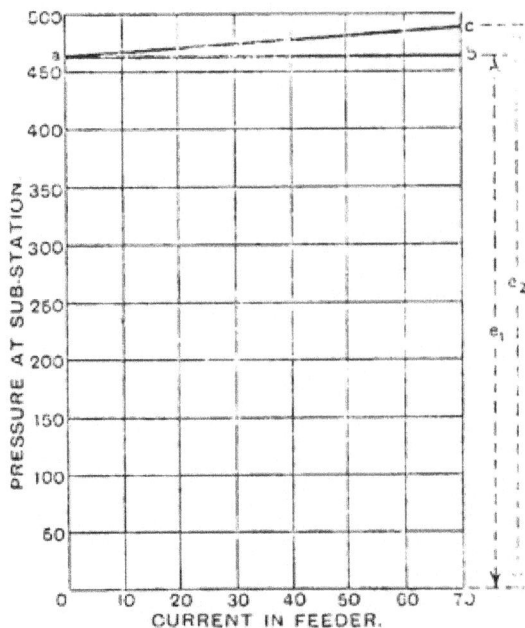

FIG. 68.

Rising Characteristic for Compensating for Fall of Pressure in Feeders.

power station must vary in some definite ratio to the output. (It has already been shown that the rise of station pressure is simply proportional to the resistance of the transmission feeders and to the current in them.) The easiest way of accomplishing this end in a small installation with one dynamo is to over-compound the dynamo so that it gives a rising characteristic. See Fig. 68, where ab shows the characteristic for constant pressure, and ac that required to compensate for the loss in the feeders. The number of volts lost over the feeder resistance is given by the heights of the ordinates between ac and ab; e_1 being the pressure at the sub-station, and e_2 that at the power station.

Shunt excitation is only practicable with moderate pressures, and, speaking broadly, is not advisable with terminal pressures of more than about 350 volts, although the author has found it convenient to run compound dynamos at even 600 volts. In such cases it is necessary to subdivide the shunt winding as much as possible, and to use extra precautions in insulating

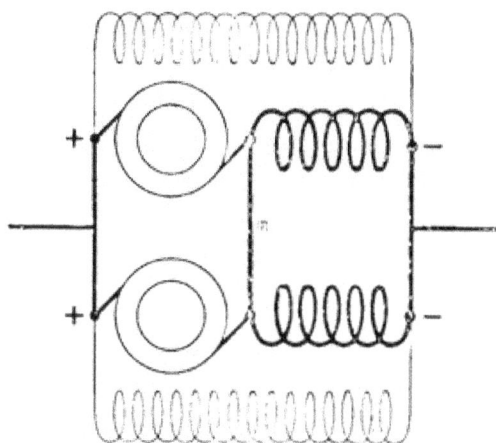

FIG. 69.—Compound Dynamos Coupled in Parallel.

the coils; and further, to guard against lightning discharges, which are a frequent source of injury to shunt wound dynamos coupled to power circuits (*see* § 21, page 102).

If two or more compound dynamos be run in parallel, it will be necessary to connect the machines at *three* points, *i.e.*, at the positive and negative brushes, and also at the extremities of the series coils, as shown in Fig. 69. This coupling will tend to divide the load between the dynamos according to their designed output, and help to ensure that all the machines give the same pressure at any load. Yet, since compounded dynamos, even if of similar design, fail to give characteristics of exactly the same slope, it follows that the output may not be properly divided between them.

To obviate these difficulties central station dynamos are generally separately excited by small dynamos specially arranged for the work. The plan which is the most flexible and allows most scope for regulation of load and pressure is to excite all the dynamos in parallel through separate variable resistances for each set of field coils, as represented in Fig. 70. By these means the pressure of any dynamo can be varied, and its output altered, at will.

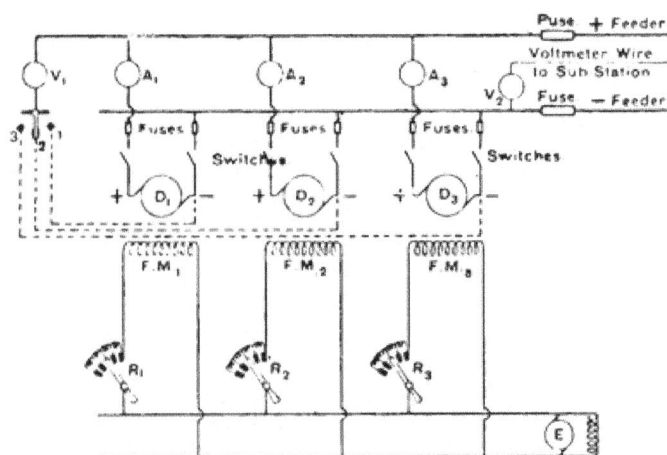

FIG. 70.

FIG. 70.—Diagram showing three separately excited dynamos coupled in parallel and supplying current to one pair of feeders. D_1, D_2, D_3, dynamos. F.M.$_1$, F.M.$_2$, F.M.$_3$, Field Magnets. E, Exciting dynamo. R_1, R_2, R_3, Regulating resistances in series with field magnets. A_1, A_2, A_3, Ammeters for reading current given out by each dynamo. V_1, Voltmeter for reading the separate pressure of the dynamos. V_2, Voltmeter for reading pressure at sub-station.

If the load varies slowly the regulation for fall of pressure on the feeders can be adjusted by hand, as occasion requires, by simply altering the resistances. But if the work necessitates rapid fluctuations of load, and consequent changes of pressure, then a few series turns may be wound round each of the

field magnets and the *main* current of the particular dynamo
caused to pass through it. By this simple means the pressure
can be regulated sufficiently well for most purposes.

Another method rapidly coming into favour involves the use
of a series dynamo, both the field and armature of which are
placed in series with the feeder whose pressure is to be regulated,
the dynamo being driven at a constant speed. Since a series
dynamo running at uniform speed gives a pressure proportionate
to the current, it follows that when it is to be used as a regulator it must be designed so that with the minimum current in
the feeder (and consequently through its windings) it will
generate pressure just sufficient to drive the current. Then as
the feeder current increases it must give a terminal pressure
equal to the drop due to the feeder resistance. These conditions
require a straight line characteristic, and therefore the weight-
output efficiency is small. Machines used for regulating may
be placed either at the dynamo station or at the distributing
end of the feeder ; and they are frequently driven by shunt-
wound motors.

Mr. W. B. Sayers' self-exciting dynamos are specially suited
for this purpose, as in them the field excitation is provided by
the armature reactions; and the difficulty of controlling sparking
with the weak field is entirely obviated by the use of commutator coils. One of these regulators is in use at the Bradford
Station Hotel, and is found to compensate for the drop in the
mains with great exactitude. The relation between the current
and the terminal pressure of the regulator, in this case, is
shown in Fig. 71, p. 146, which is taken from a Paper by
Mr. W. B. Sayers, entitled " The Prevention and Control of
Sparking and Self Exciting Dynamos."*

It will be observed that at 10 amperes the pressure is about
0·4 volt, which is just sufficient to drive this current through
the armature coils. And at the maximum load of 350 amperes
the regulator adds 13 volts to the pressure of the mains.

A modification of this idea has been successfully applied to
a three-wire system. In the case in point two armatures are

coupled on the same shaft and fitted with separate keepers or
fields, the current in one of each of the outer mains passing
through one of the armatures, and that in the middle conductor
through a few turns of wire wound around the keepers. By
this means the pressure between the middle conductor and each
of the outer mains is kept practically constant. An objection
to the dynamotor regulator is the initial cost, which will be
sufficient in many cases to prohibit its use.

Automatic regulators, depending on variations of pressure or current, have been tried from time to time, but

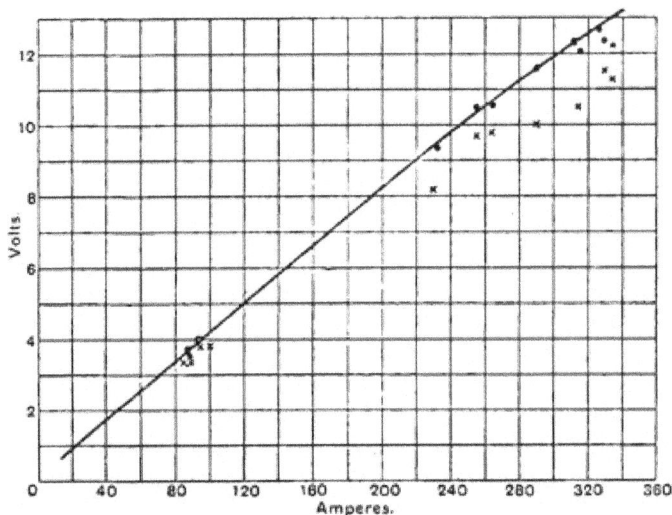

Fig. 71.—Curve showing Relation between Volts and Amperes with a
Sayers' Regulator.

their use has not been attended with much success, and in
most of the present central lighting stations hand regulation
is found to be preferable. The objection to mechanical devices
which depend on minute changes of pressure or current to
work them is the delicate construction of the mechanism.
Since they work by reason of the very error which they are
designed to correct it is obvious that makers will arrange things
so that the smallest possible variation will cause a maximum

effect on the moving parts, which are necessarily light and delicate. From very slight causes, such as dust, thickening of oil in bearings, &c., a large increase of power may be required to move the governor, and the pressure or current may vary by a dangerous quantity before it comes into action. The author is in favour of separately excited dynamos with hand regulation for most transmission of power cases; and of compensating series excitation by the main current when the load variation is too irregular for hand adjustment. Motor regulators will generally be found to be unnecessary and too costly for transmission of power plants.

CHAPTER V.

SINGLE-PHASE ALTERNATORS AND ALTERNATE CURRENT MOTORS.

§ 27. GENERAL FEATURES ; CLASSIFICATION OF TYPES OF MACHINES.

In Chapter II. the general principles of dynamo design were discussed with special reference to continuous currents, because these are at present in more general use for power transmission than alternate currents. The reasons for this preference are made clear in the sequel, and practical deductions drawn from experience are used to suggest the probable systems of the future. Now, it is necessary to examine the alternator, both as a dynamo and as a motor. This chapter is confined to the consideration of single current machines, because these have already established a justly wide reputation for lighting, and also for certain cases of power transmission, while the multiple current systems are not so well known, although

M

they possess many features that render them especially suitable for the distribution of power.

An alternator may be defined as a machine giving periodic electric currents which are reversed in direction many times per second, the number of complete changes per second being called the frequency of the current. Most types of the modern alternator have a series of salient pole pieces arranged symmetrically around the axis of the shaft and excited by a continuous current derived from an independent source, or else by a redressed current. The armatures have also a series of coils, with or without iron cores, the number of which is either equal to, half as great, or twice as great, as that of the pairs of field poles. The current induced in them is of course alternating in direction, with a period depending simply on the number of pairs of poles (*i.e.*, pairs of N and S poles) in the field coils and the number of revolutions per second. Thus the frequency, usually expressed by the symbol $\sim$, is equal to $p\,\dfrac{n}{60}$, where p is the number of pairs of field poles and n is the number of revolutions per minute. This cyclic change in the armature currents is produced by causing the magnetic induction through the coils either to alternate rapidly in direction, or else alternately to increase and decrease in density. These distinctions divide alternators into two classes—

(*a*) Alternators in which the magnetism changes in direction, and consequently passes through all values of density between a positive and a negative maximum ;

(*b*) Alternators in which the magnetism is of constant direction, but alternately increases and decreases in density, varying between the maximum and a lower limit depending on the design, and which is always nearly zero.

The first class is the older, having been devised by Wilde in 1867.

Frequently the field-magnets consist of two opposing crowns of poles of alternate sign, between which the armature coils revolve, as in the old Siemens alternator and the modern Ferranti machines; or the cores are spaced radially around the *inside* of a cast-iron ring, forming the yoke of the field-magnets, with the faces of the poles pointing to the axis of the shaft; and the armature coils are laid flat on the periphery of a drum-shaped core, and so face the field-poles, as in the Westinghouse and Gülcher machines.

In the two examples just referred to the field-magnets are stationary, but in some designs they revolve inside the armature coils, which are then fixed radially so as to form an outer ring, with the faces of the coils opposite to the field-poles. A

Fig. 72.—Field Magnet of Mordey Alternator.

recent development of this type of field-magnet is C. E. L. Brown's multiphase alternator, in which the armature is stationary and the field-magnets are excited by a single coil. The Lowrie-Hall alternator also has revolving field-magnets, but each of the poles has a separate exciting coil.

The second class (*b*) is of comparatively modern development, and is most familiar to us in the Brush alternator, the field-magnet of which is shown in Fig. 72 and the armature in

M 2

Fig. 73. In this case the armature is stationary. This arrangement, although not typical of the particular class, is very common with the alternators of many makers, as it offers facilities for coupling up the armature coils in different groups

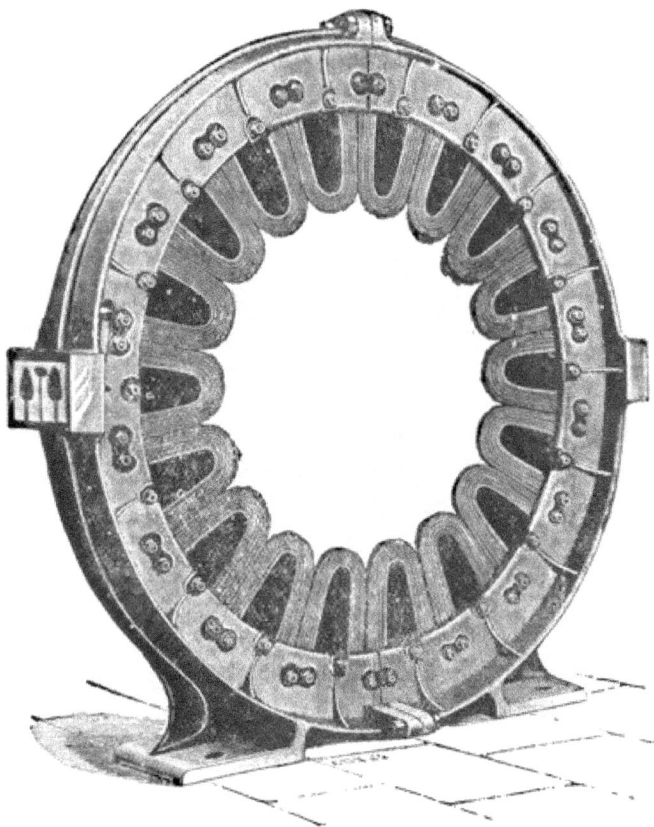

Fig. 73.—Armature of Mordey Alternator.

without the complication of brushes. The field-magnets consist of two crowns of pole-pieces, but all the poles on one side are N poles, and on the other all are S poles, not alternately N and S, as in type (a).

§ 28. MAXIMUM, INSTANTANEOUS, AVERAGE, AND
EFFECTIVE VALUES OF HARMONIC FUNC-
TIONS ; RATIO BETWEEN EFFECTIVE AND
AVERAGE VALUES ; EQUATIONS FOR ELEC-
TROMOTIVE FORCE ; PITCH OF POLES AND
COILS; DIAGRAMS OF TYPICAL ALTERNATORS;
SELF-INDUCTION OF ARMATURE COILS; EQUA-
TION FOR SELF-INDUCTION.

It will be apparent that the laws which govern the determination of the excitation for a given magnetic flux in
continuous-current machines are equally applicable to the
alternate-current designs, it being necessary, however, to modify
the equation for the total electromotive force of the armature,
and to arrange suitable current densities in the copper and
magnetic densities in the iron. These modifications are of
importance, and require elucidation. The total armature impressed electromotive force, V_{imp}, may be expressed in terms of
the number of turns, the number of lines of force per pair of
poles, and the number of revolutions per minute, as in the continuous-current machine in equation (6), p. 18. But since
the pressure is periodic, and therefore changes in value through
each cycle, there will be a maximum pressure considerably
greater than either the *effective* or *average* value. It is evident
that both the pressure and current have all values between
zero and a maximum (one positive and one negative) twice in
every period, and it is, therefore, necessary to define what is
understood by the terms alternating pressure and current.

In practice we require to know the *effective** values, and not
the maximum or instantaneous values. The readiest means
of comparing the *effective* values of an alternating and direct
current is afforded by the several heating effects produced in a
wire of known resistance having no self-induction. When the
heat caused by the current is the same in either case, then the
effective pressure of the alternate current is equivalent to the
continuous pressure.

---

* Prof. Silvanus P. Thompson prefers to use the older term, "*virtual*"
value.

Instruments which depend on the heat waste, $i^2 R$, are calibrated by continuous currents, the effective values of which are proportional to the squares of the uniform currents, and therefore the readings of the instruments are proportional to the square root of the mean square of the currents (usually written $\sqrt{\text{mean square}}$). For example, the heat produced by a continuous current of 10 amperes is proportional to 10^2, and the readings of an ammeter of the dynamometer type placed in the circuit are proportional to $\sqrt{10^2}$. If the same instrument were placed in an alternate-current circuit and gave the same deflection, the $\sqrt{\text{mean square}}$ value of the current would be as before, 10 amperes, and this is called the *effective* strength of the alternate current. This $\sqrt{\text{mean square}}$ value is independent of the shape of the curve, which, however, determines its ratio to the maximum value.

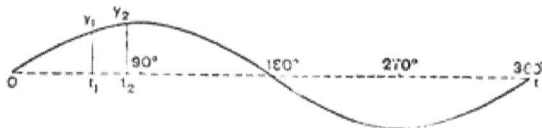

FIG. 74 —Sine Curve of Pressure or Current.

If it be assumed that the pressure and current are harmonic functions of the periodic time, the curves of either will be sinusoidal, and may be expressed by an equation of the order—

$$y = a \sin t,$$

where y is the instantaneous value of the pressure or current, t is the time of one period (it is most conveniently reckoned in angular measure), and a is a constant numerically equal to the maximum value of y, which occurs when

$$t = \frac{\pi}{2}, \text{ for } \sin \frac{\pi}{2} = 1.$$

Fig. 74 shows a sine curve of wave length t. The ordinates y_1 and y_2 may represent either values of pressure or current, at the corresponding instants t_1 and t_2.

The arithmetical mean value of the ordinates y_1, y_2, y_3, &c., for the first half of the curve in Fig. 74 is $\frac{2}{\pi}$ times that of the maximum ordinate; or the average value $= 0.637$ times the maximum value.

But the total heat generated per second in the wire is a numerical measure of half the square of the maximum ordinate of a simple periodic current; for the mean of the squares of the sine, either for one quadrant or a whole circle, is one-half of the square of the maximum value, or $\frac{i^2_{max}}{2}$. And therefore instruments depending on the estimation of the heat waste ($i^2 R$) will measure $\frac{1}{\sqrt{2}} = 0.707$ of the maximum value of a periodic current or pressure.* And the continuous current which will give the same deflection on the instrument, that is, the same heating effect, is equal to $\frac{1}{\sqrt{2}}$ times the maximum value of the alternate current.

Therefore, an alternating current, if it follow the sine law, will cause a heating effect greater than that of a continuous current of the same *average* strength by the ratio of 0.707 to 0.637, that is, about 1.1 times greater. In these pages when an alternate current or pressure is referred to, unless otherwise stated, it is assumed that effective values are understood, and that pressure measurements are made by hot wire or electrostatic voltmeters, and currents are read by dynamometers. The *effective* value of the total armature electromotive force may be written

$$V_e = k\, N_a\, C_a\, p\, \frac{n}{60}\, 10^{-8} \quad . \quad . \quad . \quad . \quad (26)$$

if all the coils are in two parallels, or

$$V_e = 2\, k\, N_a\, C_a\, p\, \frac{n}{60}\, 10^{-8} \quad . \quad . \quad . \quad . \quad (27)$$

if they are in one series.

---

* For a more complete explanation of the interesting phenomena of alternate currents reference should be made to "The Alternate Current Transformer, in Theory and Practice," by Dr. J. A. Fleming, F.R.S.

The symbol V_e = the effective pressure in volts,

p = the number of pairs of poles,

n = the number of revolutions per minute,

N_a = the magnetic flux from one pair of poles,

$C_a = 2\,p\,w$, where w = the number of active conductors on one side of one coil; or, briefly, C_a = the number of turns counted all round.

The symbol k requires special explanation.

The quantity expressed by k is a constant for the particular arrangement of pole-pieces and armature coils; it is the ratio between the effective and the average pressures. If the pressure curve obeys a sine law (it does very nearly in many types of alternators), $k = 1\cdot1$, and this value may be taken in most rough approximations. The equations for the electromotive forces of alternators, assuming a *sine law* to be followed, may be written thus :—

Two parallels : average total volts $= N_a C_a\,p\,\dfrac{n}{60}\,10^{-8}$ . . (28)

One series : „   „   „ $= 2\,N_a C_a\,p\,\dfrac{n}{60}\,10^{-8}$. (29)

Two parallels : maximum „   „ $= \dfrac{\pi}{2}\,N_a C_a\,p\,\dfrac{n}{60}\,10^{-8}$. . (30)

One series : „   „   „ $= \pi\,N_a C_a\,p\,\dfrac{n}{60}\,10^{-8}$. . (31)

Two parallels : effective „   „ $= \dfrac{\pi}{2\sqrt{2}}\,N_a C_a\,p\,\dfrac{n}{60}\,10^{-8}$ (32)

One series : „   „   „ $= \dfrac{\pi}{\sqrt{2}}\,N_a C_a\,p\,\dfrac{n}{60}\,10^{-8}$ (33)

The numerical value of the constant in equation (32) is approximately $1\cdot1$. In equation (33) for series coupling the coefficient must be twice as great as that in equation (32), if the same value of C_a be retained, and it is therefore equal to $2 \times 1\cdot1 = 2\cdot2$. In practice it is found to vary between $2\cdot2$ and $2\cdot3$ when the coils are all in series.

The pitch of the pole-pieces and the width of the armature cores are of great importance in designing alternators, for they affect the shape of the pressure curve, and consequently the effective pressure and the output. Four of the more frequent

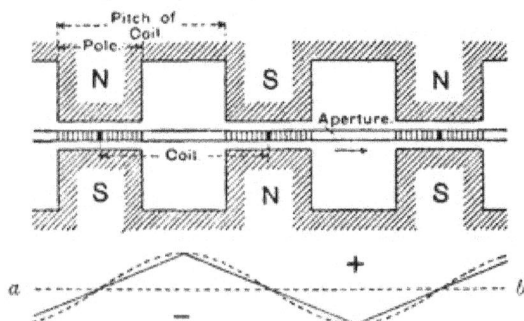

FIG. 75.—Diagram of Typical Alternator giving a Triangular Pressure Curve.

arrangements are illustrated in Figs. 75, 76, 77 and 78, which it will be necessary to examine carefully before proceeding with the calculation of the output. In these diagrams the field poles are

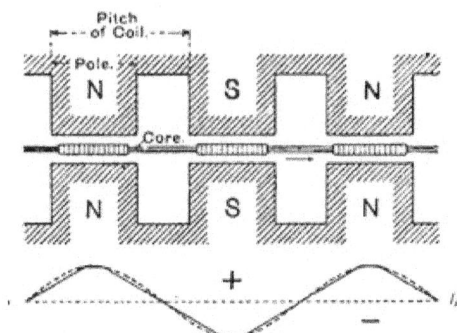

FIG. 76.—Diagram of Typical Alternator giving a Flat-Topped Triangular Pressure Curve.

developed horizontally, and the polarity is indicated by capital letters. The winding and bare spaces on the armatures and the apertures of the coils are clearly marked. The pitch of the field poles is equal to half a wave length, and, therefore, in

Figs. 75, 76 and 78, it is equal to the angular width of one pole and one space; whilst in Fig. 77 (which refers to the Brush alternator) it is equal to that of one pole only, for there the spaces, strictly speaking, act as if they were poles of opposite sign to those hatched in the diagram. It will be noticed in this case that the armature coils are twice as numerous as the pairs of field poles.

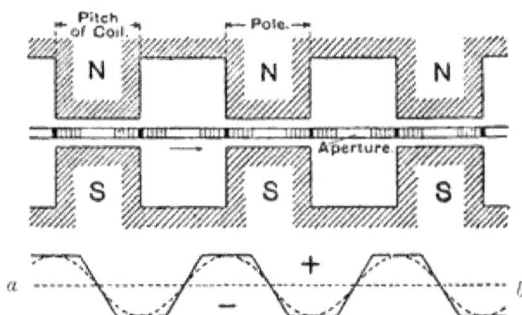

Fig. 77.—Diagram of Typical Alternator giving a nearly Rectangular Pressure Curve.

Below each figure is the theoretical shape of the pressure curve in full lines, determined on the assumption that there

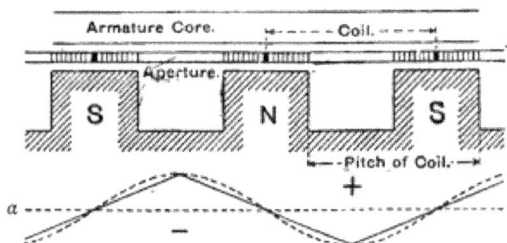

Fig. 78.—Diagram of Typical Alternator giving a Triangular Pressure Curve.

is no magnetic leakage between the poles, and, therefore, no magnetic fringe at the corners of the pole-pieces. This is not the case in practice, and the actual curve will therefore be less angular, and approximate more or less closely to sine curves, as shown in dotted lines. This is easily seen to be the case with Figs. 75 and 76, and practically it is found that the design in Fig. 77 gives an almost perfect sine curve.

In tracing the pressure curve for any arrangement of pitch and angular width of poles and coils the zero points may be first determined and marked off on the line *a b*, which shows the angular position of the coils. Having found these points the wave-length is known, and the positions of maximum pressure should lie midway between the zero points. If the *winding spaces* are exactly equal to the *width of the poles*, then, assuming no magnetic fringe, the intermediate values lie in straight lines, joining the maximum and zero points as in Figs. 75 and 78. The height of the maximum ordinate depends on the scale selected, but its numerical value can be found from the known magnetic flux and the linear speed of the coils. In these diagrams an arbitrary height has been selected. If the coils be in one series, and a simple zigzag curve be assumed, then

$$V_{max} = 4\, C_a\, N_a\, p\, \frac{n}{60}\, 10^{-8} \text{ volts,} \quad . \quad . \quad . \quad (34)$$

where the maximum pressure is assumed to be twice the average pressure. If the winding space be shorter than the polar width, then the maximum pressure will be maintained for that period of time during which the coil as a whole is passing under the pole piece. Thus, if the polar arc be 10in. in width and the winding space only 8in., then for the fraction of time required for the coil to move through the angular distance of $10 - 8 = 2$in., the pressure will have the same value, and hence the top of the crest will be flat, as in Fig. 76, which refers approximately to the design of the Kapp alternator.

In some designs the coils are made to exceed the poles in width by a small percentage, the arrangement being found to give a very close approximation to a sine curve. Prof. Elihu Thomson is in favour of such proportions.

The Gülcher alternator, designed by Mr. W. B. Esson, also embodies this idea, but the armature coils are only half as

numerous as the field poles, and are equal to the width of
three and a-half poles. There is a distance piece between the

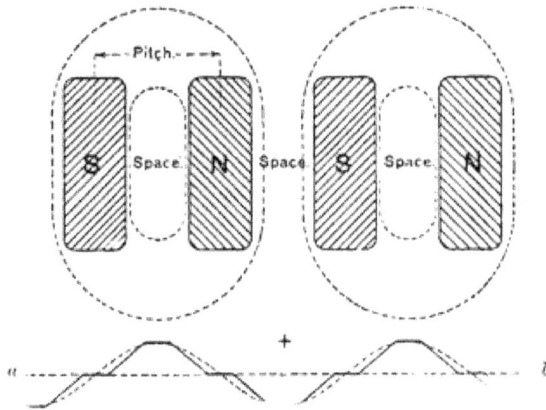

FIG. 79.—Diagram of Typical Alternator showing a Stepped Pressure
Curve.

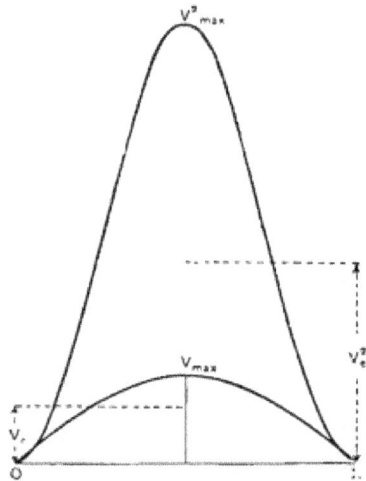

FIG. 80.—Diagram showing Effective Pressure—Sine Curve.

coils equal to half a pole, the pole being of the same width
as the space (see Fig. 79). The theoretical curve of pressure is

here shown in full lines, and a sine curve is indicated by dotted lines. It is evident that this arrangement approximates very closely indeed to the necessary conditions for obtaining a sine function.

The curve of pressure for any arrangement of pole pieces and armature coils can be traced approximately, and then it is easy

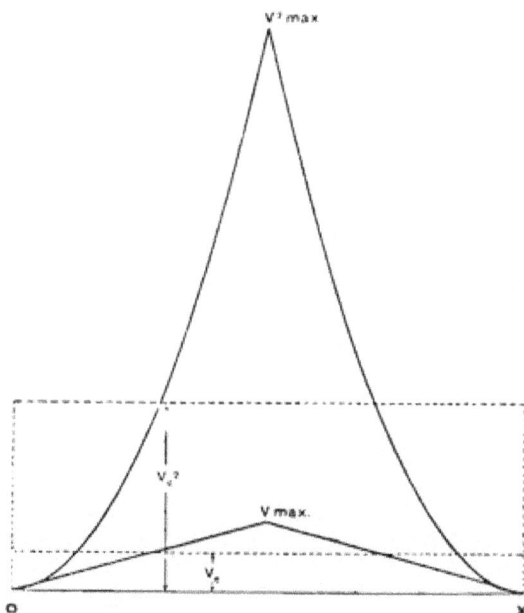

Fig. 81.—Diagram show n ; Effective Pressure—Triangular Curve.

to determine the effective pressure with sufficient accuracy for practical purposes. Consider the sine curve in Fig. 80. The half wave length is equal to, say, OX, and V_{max} equals the maximum ordinate. Now, bearing in mind that the instruments used for measuring alternate currents give readings proportional to the $\sqrt{}$mean square, it is necessary to square the ordinates of OVX and plot a new curve on the same base, and then to find the mean ordinate, say V_e^2, of the squares. The square root V_e will be the effective pressure. This can be readily found for most symmetrical figures by integration ; but many engineers prefer the graphical method, which is as follows :—Plot the curve to any convenient scale on squared paper and divide the base line O X into any number of equal parts—the more the better. Measure the heights of the ordinates at the middle of each and divide the sum of their values by the number of parts : the result is the numerical value of the mean square, V_e^2, in the selected units. Next

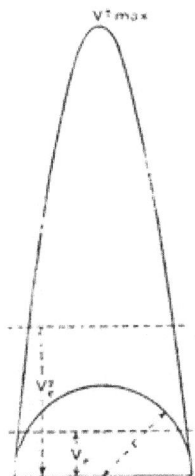

Fig. 82.—Diagram showing Effective Pressure—Semicircular Curve.

extract the square root, and then the effective pressure, V_e, is known. It will be seen that for a sine curve

$$V_e = 0.707 \; V_{max}.$$

The same operation has been performed in Fig. 81, for the triangular pressure curve O V_{max} X, and the value of V_e is found to be 0.58 of the maximum pressure ; or

$$V_e = 0.58 \; V_{max}.$$

Thus equation (34), which gives the value of the maximum pressure with a simple zigzag curve, must be multiplied by

0·58 to equal the effective pressure. Therefore, if the coils be in one series, since

$$V_{max} = 4 \, C_a \, N_a \, p \, \frac{n}{60} \, 10^{-8},$$

$$V_c = 0·58 \times 4 \, C_a \, p \, N_a \, \frac{n}{60} \, 10^{-8}$$

$$= 2·32 \, C_a \, N_a \, p \, \frac{n}{60} \, 10^{-8}. \quad . \quad . \quad . \quad (35)$$

And if the coils be coupled in two parallels, then the effective pressure is half as great as the value given by (35).

In Fig. 82 the same operation has been performed for a semi-circular curve, and it is found that

$$V_c = 0·835 \, V_{max}.$$

The relation between the effective and average values, for the more common curves, and the theoretical values of k are shown in Table J :—

Table J.

Order of curve.	Average value as a decimal of the max. ordinate.	Effective value as a decimal of the max. ordinate.	k = Ratio of effective to average value.
Sine	0·637	0·707	1·1
Semicircle	0·7854	0·835	1·063
Triangle	0·5	0·58	1·16
Rectangle	1·0	1·0	1·0

The above values will not all be realised in practice, since the magnetic fringe will cause the angles to be rounded. The slope of the curves will also be largely affected by the pitch ratio of the poles and coils. The shape of the coils is also important in determining the pressure curve. For example, a rectangular coil will tend to give a very rapid increase and decrease of pressure, because it will become instantly active or inactive; while a coil of rounded form will cut the lines of force in steadily increasing or diminishing numbers.

There is some difficulty in finding the actual value of k in practice, since the exact magnetic density B''_a is not easily found without special experimenting. The designer, of course, knows the calculated value of N_a, and consequently B''_a, and then from the measured pressure may calculate k with a fair degree of approximation to the truth. But if a small test coil be arranged so as to explore the gap at various places, then the average value of B''_a may be readily found ; this, however, is not an easy thing to do in an ordinary shop with workshop instruments.

Mr. Kapp has made a special study of the coefficient k, and the following table is based on his researches. The windings are assumed to be in one series, and the width of the poles and coils is expressed in decimals of the pitch.

Table K.

Proportions of Field-Poles and Armature Coils in Terms of the Pitch, and Consequent Values of k.

Ref. No.	Pitch.	Pole.	Total breadth of copper in each coil.	$2k$.
1	1	1	1 (covering whole surface)	1·16
2	1	1	0·5 (covering half surface)	1·635
3	1	0·5	1 (covering whole surface)	1·635
4	1	0·62	0·5 (covering half surface)	2·06
5	1	0·5	0·5 (covering half surface)	2·30
6	1	0·33	0·33 (covering third of surface)	2·832
7			Sine Function	2·22

Nos. 4 and 5 approach most nearly to designs found in practical machines.

Having thus determined the total effective pressure of the alternator, it is easy to find the terminal effective pressure.

This will be given as for continuous currents by equation (5), page 18. Thus, if v_e be the effective terminal pressure, then

$$v_e = V_e \pm i r,$$

where i is the current in the armature and r the resistance of the coils.

It remains to examine more closely the armature pressure and to consider the relation of its components. Essentially there are three active electromotive forces in every alternate-current device, the capacity of which is either zero or negligible, whether it receives or generates power; they are the impressed pressure V_{imp}, which represents the total electromotive force of the circuit, v_s the electromotive force of self induction, and v_r the resultant or *current-driving* electromotive force.

The self-induction pressure is always in advance of the resultant pressure by 90deg. The pressures may be represented by the diagram in Fig. 83, and numerically their maximum values are related by the equation

$$V^2_{imp} = v^2_r + v^2_s. \qquad . \qquad . \qquad . \quad (36)$$

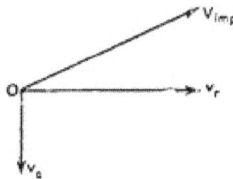

Fig. 83 — Diagram of Phases of Pressure.

These three pressures have different amplitudes, but the same wave length, and it is necessary to distinguish between their maximum and effective values; but equation (36) is equally true for both. In practice, however, the effective pressure is of the greater importance, and is generally understood when the word pressure is used without any qualification.

The self-induction electromotive force requires further explanation. Whenever an electric current is started in a conductor there is a tendency to stop the passage of the current, and when the circuit is opened so as to break the path of the current there is a tendency to continue the current, which is evidenced by a spark at the gap. These tendencies are clearly of a different nature to the ohmic resistance, which absorbs pressure ($i R$), and simply regulates the quantity of current. Since it

requires an electromotive force to start a current, an electromotive force will also check a current; and hence in alternate-current working, as the current is rapidly alternated, *i.e.*,
first started in one direction, stopped, and then restarted in the
other direction many times in each second, it is clear that this
apparent *inertia* of the electrical current may at times assume
considerable proportions.

The phenomenon is said to be due to self-induction, which is
expressed as a quantity whose magnitude is measured in units
called *henries*. The self-induction of a circuit is proportional to
the square of the number of turns of wire. The number of lines
of force embraced by the circuit is proportional to the product
of the self-induction and the current. The electromotive force
generated by the variation of the lines of force is proportional
to the frequency of the current, and also to the total flux of
lines, and may therefore be expressed in volts by an equation
of the form

$$V_b \max = 2\pi \sim i\, C^2\, N\; 1{\cdot}41, \quad . \quad . \quad . \quad (37)$$

where $\sim$ is the frequency, or number of complete cycles per
second, C is the number of turns of wire, N is the number of
lines of force due to one turn of the winding, and i is the
current in one coil in amperes.

The quantity expressed by the product $N\,C^2$ is called the
coefficient of self-induction of the circuit, and is usually sym-
bolised by the letter L.

$$\therefore\; V_s \max = 2\pi \sim i\, L\; 1{\cdot}41, \quad . \quad . \quad (38)$$

and the effective pressure

$$V_{s\,e} = \frac{V_s \max}{1{\cdot}41} = 2\pi \sim i\, L. \quad . \quad . \quad (39)$$

In considering the self-induction of an alternate-current
armature, N is the number of lines of force due to one turn of
the winding where one ampere flows through the coil, and C
is the number of turns in series. The self-induction can
therefore be made small by keeping the quantity $C^2 N$ small.
But C is obviously proportional to the required pressure, and

therefore practically fixed between narrow limits for a given design; and the magnitude of N will depend on the permeability of the field. If the magnetic density be sufficiently great to practically saturate the iron of the cores (field and armature) the self-induced field will be small compared with the main field, and will have little effect. In designing alternators, then, it is advisable to make the field induction strong in order that this induction per turn of armature may be as large as possible. In armatures with air cores it is possible to work at a high magnetic density, and therefore the self-induction can be made low; but with iron-cored armatures at the density usually permissible the self-induction is considerable. It must also be noted that the electromotive force of self-induction of an armature depends on the frequency (*i.e.*, the rate of change of induction), and also on the armature current if the permeability of the magnetic path be constant.

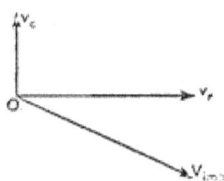

Fig. 84. —Diagram of Phases of Current.

It is found in a practical alternator that L has a smaller value at full load than on open circuit. This is the case, because the permeability of the field is then less. The electromotive force of self-induction in most cases will be greatest at light loads, since, though i is then small, L, on the contrary, attains its greatest value.

If a circuit have capacity, such as is caused by a condenser, an armoured cable, or a concentric cable of considerable length, then there will be an electromotive force lagging 90deg. behind the resultant pressure, and therefore in opposition to the self-induction pressure if the circuit be an inductive one. The relative position of the phases and magnitudes of the pressures are shown in Fig. 84, where V_{imp} is the impressed pressure, v_r the resultant pressure, and v_c the capacity pressure.

§ 29. IMPRESSED, DYNAMIC OR WORKING, AND
CONDENSER OR WATTLESS CURRENTS; RE-
LATION BETWEEN PHASES AND MAGNITUDES
OF PRESSURES AND CURRENTS; POWER
MEASUREMENTS; EFFECTS OF CURRENT LAG.

It has been shown that the resultant pressure can be divided
into two components, the impressed pressure in phase with the
dynamic current, and the self-induction pressure at right

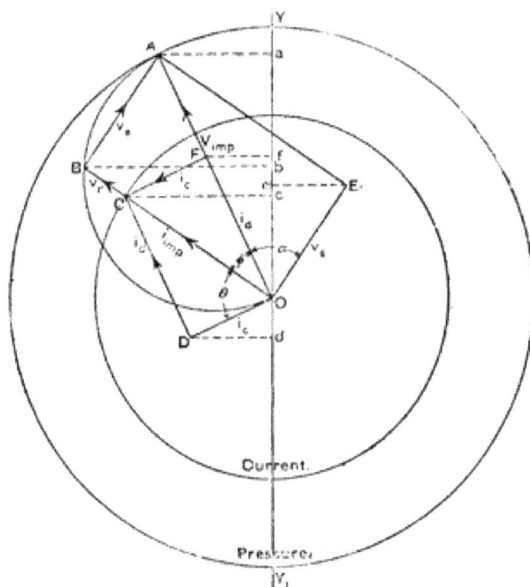

FIG. 85.—Clock Diagram.

angles to it. A similar relationship is true of the impressed
current, i.e., the current measured by a dynamometer, which
can be resolved into a dynamic or work-producing current in
phase with the impressed electromotive force, and a wattless or
condenser current lagging 90deg. behind it. Let the three be
distinguished by the several symbols i_{imp}, i_d, and i_c. Then

$$i^2_{imp} = i^2_d + i^2_c. \qquad \ldots \ldots \qquad (40)$$

The relation between the pressures and currents is illustrated in Fig. 85, which is known as a clock diagram. The radius O A is supposed to revolve clockwise, and to represent the impressed electromotive force, one complete revolution being equal to one period. It is evident that the projections of O A on any straight line, say, Y Y₁, passing through the origin O, will represent the fluctuations of pressure from instant to instant, assuming the curve to follow a sine law. The instantaneous values of the several pressures may therefore be read on the axis Y Y₁.

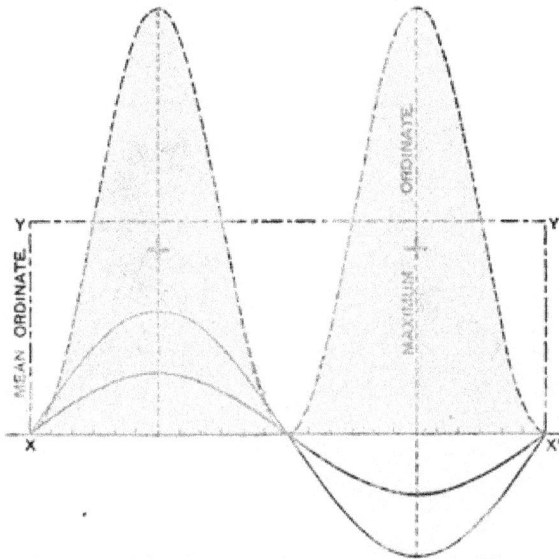

Fig. 86.—Work Diagram—Current and Pressure in Phase.

On A O describe the semi-circle A B O. Let any convenient unit length be chosen to represent a volt and an ampere. Through O draw O B, touching A O B at B, proportional to the resultant pressure and lagging behind O A by an angle ϕ, and join A and B. Then the line A B will be equal to the self-induction pressure expressed in the same units. Draw O E parallel to B A, and complete the parallelogram B A E O. Then O E shows the relative phase of the self-induction electromotive force, V_{sl}, which is seen to lead V_{imp} by an angle

x . Now the impressed current is due to the resultant pressure
and therefore is in phase with it.   Hence on O B (prolonged if
necessary) mark off a part O C equal to the impressed current.
From C draw C F perpendicular to O A, draw O D parallel to
C F, and join C and D.   Then the resolved parts of O C, viz., O F
and O D, will severally be proportional to the dynamic and the
wattless currents, for they are the components of O C resolved
at right angles.   It is seen that the impressed current lags
behind the impressed pressure by an angle ϕ—arbitrarily chosen
in this case—and the dynamic current leads the impressed

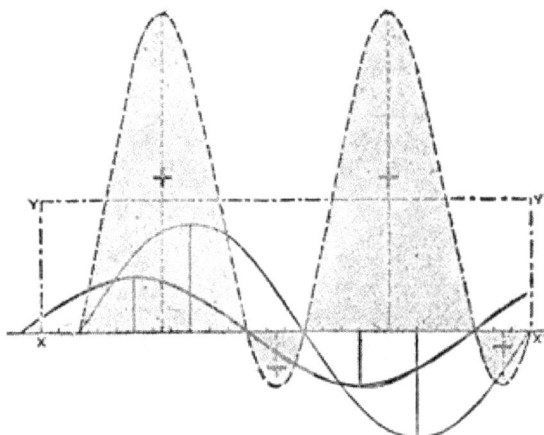

FIG. 87.—Work Diagram—Current lagging 45-deg. in Phase behind
Pressure.

current by the same angle ϕ, while the wattless current lags
behind it by the angle θ.   It is also apparent that the greater
the difference in phase between the impressed pressure and
impressed current the greater the magnitude of the wattless
current.   The import of this will be more readily appreciated by
reference to Figs 86 and 87, which show respectively curves
of instantaneous values of pressure and current coinciding in
phase in Fig. 86, and the same curves differing by 45deg. in
Fig. 87.   The work performed in each cycle, corresponding to
products of V i, is shown in the shaded areas bounded by dotted
lines, the ordinates of which are simply proportional to the pro-

duct of the corresponding ordinates of V and i, with due regard to the algebraic signs. The curves beneath the time line are considered as of negative value; and, since the product of two negative quantities is algebraically positive, the work areas are chiefly positive or useful. In Fig. 87 the small negative areas denoted by negative signs must be subtracted from the positive areas when estimating the total or mean power.

We are now in a position to deduce the equation for the mean power. It has just been shown that the mean power is simply the product of the impressed effective pressure and current, as measured by a hot wire voltmeter and a dynamometer, *if there be no difference of phase.* If there be a phase difference, then the expression must be multiplied by some function of the angle of phase. Referring to Fig. 85, p. 168, it is clear that the mean power is proportional to $AO.FO$, or to $BO.CO$; or to the product of $V_{imp} i_{d}$, or $i_{imp} V_{d}$, all being effective values.

Now
$$AO.FO = AO.CO \cos AOB.$$
$$= V_{imp} i_{imp} \cos \phi,$$
and
$$BO.CO = AO.CO \cos AOB.$$
$$= V_{imp} i_{imp} \cos \phi.$$

And if we consider the maximum values, since
$$V_{max} = \sqrt{2} V_e;$$
$$\therefore \text{ mean power} = \frac{V_{max\ imp}}{\sqrt{2}} \frac{i_{max\ imp}}{\sqrt{2}} \cos \phi.$$
$$= \frac{1}{2} V_{max\ imp} i_{max\ imp} \cos \phi.$$

The effective values are practically the only directly measurable ones, and therefore it may be stated finally that
$$\text{Mean power} = V_{imp} i_{imp} \cos \phi. \quad . \quad . \quad (41)$$

It is evident that if the angle ϕ be large the power may be relatively small, although the product of impressed pressure and current be considerable. And it is also seen that a large angle of lag is detrimental in two ways:

Firstly, it decreases the output for a given plant, *i.e.*, lowers the plant efficiency, and

Secondly, it increases the heat waste in the copper of mains and machines.

§ 30. EFFECT OF IRON CORES; POWER WASTED BY HYSTERESIS; EFFECT OF FREQUENCY.

There is considerable difference of opinion as to whether an alternate-current armature should have iron cores or not. In the thin disc type there is no doubt that iron is inadmissible, as the armature would inevitably be drawn out of flat and jammed against the pole-pieces; but in drum and Gramme-wound machines there is no mechanical reason why the magnetic resistance should not be decreased by a liberal use of thin iron plates. But the magnetic density must be kept low or the losses from hysteresis will be serious: B_a must not exceed from 40,000 to 65,000. And iron cores are said to have the effect of reducing the loss from eddy currents in the copper. Regarded from a purely mechanical standing, the author is of opinion that the iron-cored drum-wound type of alternators, such as those built by the Westinghouse and the Gülcher companies, are the better adapted for power purposes. The Brush and Ferranti armatures, compared with these, are fragile in make, and appear to be much more likely to give trouble. The experience of Continental and American engineers seems to confirm this, and the gradual development of multiphase alternators with iron cores undoubtedly points in the same direction.

From a purely electrical point of view the Mordey alternator is almost perfect, and certainly possesses the important feature of a high efficiency of conversion at light loads—a most important consideration with some kinds of work. On the other hand, it is an undoubted fact that the light-load efficiency of many of the iron core alternators is extremely low. In designing alternators with iron in the armatures it is always important to estimate the power which will be wasted in simply magnetising and demagnetising the cores, as this will be sufficiently large to be comparable with the copper loss. Table L, which is calculated for 50, 83, 100, and 130 periods per second, gives the data necessary for most practical cases.

Table L.—*Power Wasted by Hysteresis.*

B C.G.S.	B" per square inch.	Watts wasted per cubic inch. $\curvearrowright = 50$.	Watts wasted per cubic inch. $\curvearrowright = 83$.	Watts wasted per cubic inch. $\curvearrowright = 100$.	Watts wasted per cubic inch. $\curvearrowright = 130$.
4,000	25,800	0·116	0·192	0·233	0·310
5,000	32,230	0·1665	0·277	0·333	0·434
6,000	38,700	0·2165	0·360	0·433	0·564
7,000	45,180	0·2675	0·444	0·535	0·695
8,000	51,600	0·3 05	0·532	0·641	0 832
10,000	64,500	0·450	0·744	0·900	1·170
12,000	77,400	0·597	0·965	1·194	1·550
14,000	90,300	0·7 0	1·260	1·520	1·970
16,000	103,200	0·910	1·510	1·820	2·360
17,000	109,650	1·1375	1·830	2·275	2·950
18,000	116,100	1·4075	2·340	2·815	3·660

It will be noticed that for a fixed induction the waste of power increases in direct proportion to the frequency; and that for a given frequency the loss varies as the induction raised to the 1·6 power.

A further loss may also arise from eddy currents unless the iron plates are very thin, but with the usual gauges the waste of power is unimportant.* (*See also* § 37, p. 207.)

The effect of the frequency on the size of an alternator for a given output demands careful attention. It was shown in § 27

---

* Mr. Steinmetz's law gives the expression for the loss in core thus :— Watts $= a \curvearrowright B^{1·6} + b \curvearrowright^2 B^2$, the first term referring to hysteresis and the second to eddy currents. The constants a and b have to be determined for each sample. The second term decreases in value as the temperature of the core is raised. Generally, eddy current loss will not exceed about five per cent. of that due to hysteresis. See *The Electrician*, February 12, 19, and 26, 1892; also *Journal of the American Institute of Electrical Engineers*, January 19, 1892.

that the frequency is proportional to $p\frac{n}{60}$; that is, it varies
directly as the number of pairs of poles in the field magnets and
the number of revolutions. If the speed be fixed, the frequency
determines the number of poles, and, consequently, to a large
extent, the size and cost of the machine. The bearing of this
point in the design of alternators will be readily appreciated on
inspecting Table M, which shows the necessary revolutions per
minute at the frequencies customary in this country, with
various numbers of pairs of field poles.

It is seen that for a given number of pairs of poles the revolutions vary directly as the frequency, which on this account
should be low. But the output of a given alternator, other
things being equal, is proportional to the frequency, and therefore a low frequency means a heavy machine for the output.
This disadvantage is, however, counterbalanced to a large
extent by the higher induction which is permissible as the
frequency is diminished.

Table M.*

Number of pairs of field poles.†	$\sim = 50$ revs. per min. =	$\sim = 80$ revs. per min. =	$\sim = 100$ revs. per min. =	$\sim = 130$ revs. per min. =
2	1,500	2,400	3,000	3,900
4	750	1,200	1,500	1,950
6	500	800	1,000	1,300
8	375	600	750	975
10	300	480	600	780
12	250	400	500	650
14	215	343	432	557
16	187	300	375	487
18	166	266	332	433
20	150	240	300	390

The ordinary values of B''_a are given in Table N, and are
practically the same as those found advisable in the design of
transformers.

---

* *See also* Table T, p. 184.

† The number of pole pieces in a machine is equal to $2p$.

Table N.

Frequencies $\sim$ =	Induction B''_0 = from
50	32,000 to 40,000
60	30,000 to 32,000
70	28,000 to 30,000
80	26,000 to 28,000
90	24,000 to 26,000
100	22,500 to 24,000
110	21,000 to 22,500
120	19,500 to 21,000
130	18,000 to 19,500

If the magnetic field vary in strength the induction must be the $\sqrt{\text{mean square}}$ value.

§ 31. DESIGN OF ALTERNATORS ; MAGNETIC DENSITY ; EFFECT OF SHAPE OF PRESSURE AND CURRENT CURVES ; PERIPHERAL SPEED ; EXCITATION.

The preceding investigation of alternate-current working will be sufficient to show the modifications necessary to apply the method of designing dynamos and motors described in Chapter II. But special attention must be given to the magnetic induction, and the current density must be chosen according to the ventilation provided. For example, in an iron-cored armature having a large heat waste from hysteresis, the loss in the copper must be kept relatively small, whereas in a coreless armature of the Brush or Ferranti type, the current density may be as high as 4,000 amperes per square inch, or even more. The effective pressure and current will in all cases be of chief importance, since the output, heat waste, and average torque are proportional to these values. But the maximum pressure will determine the limiting stress on the insulation, and must be considered accordingly. From this point of view the Kapp and Gülcher alternators appear to be excellent, for they give a flat top to the theoretical curve of pressure (*see* Figs. 76 and 79, pp. 157 and 159). Also the maximum torque will be proportional to the maximum value of the current, and the stresses must be calculated with reference to this value. In cases likely to be met with in practice, both pressure and currents closely approximate to a sine law in the outer circuit, and it is probably safe to assume that the armature pressure curve also does so; at any rate, judging from the illustrations given in Figs. 75, 76, 77, 78 and 79, the probable error is not sufficiently large to seriously affect the results.

The peripheral velocity of alternators is generally greater than that of continuous-current machines, and is a question of mechanical construction. Generally, with non-magnetic core armatures of the Ferranti type, the moving parts are so light that the centrifugal stresses are easily dealt with at even from 5,000 to 8,000ft. per minute. And in revolving field machines of the Brush and Brown type it is clear that the speed is only restricted by the tensile strength of soft cast iron. In revolving fields, of the Electric Construction Company's type, since the magnets are mainly built up of wrought iron, the chief factor is the ultimate strength of the bolts and screws holding the parts together, &c.

§ 32. EXCITATION; PARALLEL WORKING; CRITICAL FIELD CURRENT OF ALTERNATORS AND ALTERNATE-CURRENT MOTORS.

The estimate of the excitation presents some difficulty, but since the current is almost invariably provided by a separate direct-current dynamo, often coupled to the shaft of the alter. nator, a limited adjustment can be made by a variable resistance placed in series with the coils. However, it is important to notice that for every driving power there is one value of exciting current which will make the output, and, consequently, the efficiency, a maximum (*see* Fig. 88). In fact, the output does not vary directly as the field strength, as is the case with continuous-current machines; and it follows that a proper regulation cannot be made by the exciting current alone, but requires as well an adjustment of the driving power for each load. This is of special importance when alternators are coupled in parallel. If the exciting current be either above or below this critical value, the efficiency will be lowered and the output reduced (for the given driving power). In the first case, the armature current will be increased, and the terminal pressure also; but the efficiency will be diminished, owing to the increased loss over the armature resistance. In the second, the current will also be increased, and the terminal pressure will be lowered.

This effect is not very marked with many alternators, and in most cases the best excitation is nearly the same through the working range of output. Perhaps it is of more importance in the case of alternate-current motors, for then the increase of

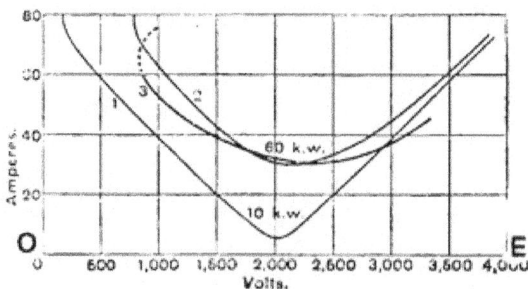

Fig. 88.

* Curves connecting Terminal Volts and Amperes in Alternators supplied with Constant Power, the excitation being increased continuously from zero at **O** to a maximum at **E**. Curve 1 refers to a 10 k.w. alternator; curve 2 refers to a 60 k.w. alternator; and curve 3 refers to the larger machine when the armature reactions are taken into account. The Volts and Amperes are in phase at the apex of each curve.

main current may cause a serious loss in the line, as well as in the armature circuits. With motors it is found that for every load there is *one* excitation which will give a minimum armature (main) current.

This important fact was apparently first discovered in 1890, at a lighting plant in Cassel, where the generating station, comprising turbines and alternators, lies several miles from the town. The electric power is transmitted through a concentric cable to two sub-stations, each of which contains a large alternate-current motor driving dynamos. The

*See " Dynamos, Alternators and Transformers," by G. Kapp, page 409.

men in charge of these machines found that the main
current could be varied for the same load by altering the
motor excitation.   A similar effect is illustrated in Fig. 88A,
which refers to experiments made by Mr. Mordey with two
Brush alternate motors running light.   The terminal pressure
was maintained constant at 2,000 volts and 100 ∿, the speed
also being constant throughout the range of observation.   The

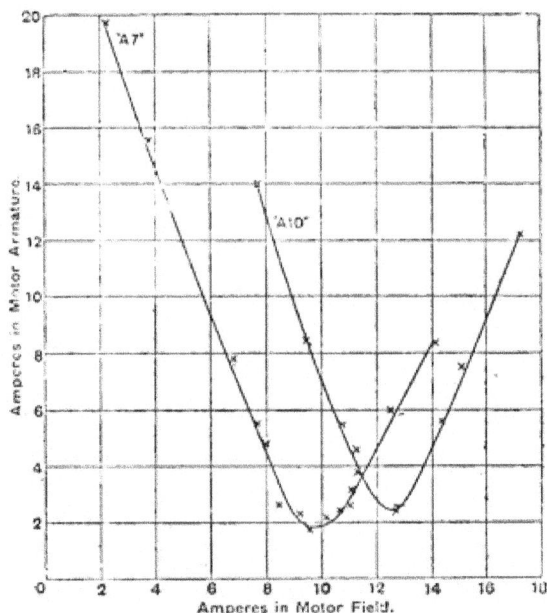

FIG. 83A.—Curves Connecting Exciting and Armature Currents.

tests were continued with various loads, and it was found that
for every output there was one value of exciting current
corresponding to minimum armature current, and that the
critical excitation was approximately the same for all loads.
Professor S. P. Thompson has shown that minimum current in
a motor armature occurs when the pressure and current are in
phase, that under-excitation causes the current to lag, and over-
excitation causes it to lead.

Dr. Sumpner, Mr. G. Kapp, Mr. Swinburne, Mr. Mordey, and others have investigated the theory of exciting alternate-current machines. And it appears that the *output* of an alternator depends on the steam admission, or simply the driving power ; while the *lag*, and, consequently, the wattless current, is controlled by the excitation. And for a given brake horse-power of an alternate-current motor, the excitation determines the lag and wattless current.

The problem is complicated by armature reactions, self-induction, and capacity effects in the circuit.[*] Armature self-induction tends to weaken the field, and therefore to necessitate more than the calculated excitation, while capacity in the armature tends to strengthen the field, and hence to reduce the value of the necessary excitation. The actual number of ampere-turns for any load will, therefore, depend on the balancing of these two effects, and no general rule can be stated. When alternators are running in parallel, with separate engines, the steam admission should be controlled by a common single governor, and not by separate governors to each engine. And the real output of each dynamo, after the throttle valve is properly set for the full output of the machine, should be measured by a wattmeter, and controlled (in accordance with its readings) through the field excitation. An ammeter is of no use for this purpose, as it gives no indication of the lag between current and pressure ; and the omnibus pressure, being the same for all the machines, is no criterion of the actual total electromotive force of any of the alternators. It may therefore happen that two alternators in parallel may be giving pressures differing as widely as 2,500 and 1,500 volts, and yet the omnibus pressure will show 2,000 volts, the proper station pressure, say.

§ 33. SINGLE-PHASE ASYNCHRONOUS MOTORS—SELF-EXCITING AND SELF-STARTING.

This type of motor is not yet in general use in England, but it is steadily gaining favour on the Continent, and is worth some attention in connection with the present subject.

---

[*] For further information and experiments with synchronous motors on long-distance lines, see *Proceedings of the Institution of Electrical Engineers*, Vol. XXIII.

Messrs. Hutin and Leblanc, Mr. C. E. L. Brown and others, have devised various forms which run with single-phase currents, asynchronously and with good efficiency. By the combination of an additional rotary magnetic field, they can be made to start against a considerable torque, and the auxiliary field can be stopped after the proper speed is attained. Various devices to secure such a field are suggested by Mr. C. E. L. Brown, the most feasible being two parallel windings on the field magnets—one with small self-induction (or even capacity), and the other with larger self-induction. These two circuits when supplied with a single-phase current will produce a rotary field, and so cause the armature to revolve.*

The construction and general appearance of these motors is similar to that of multiphase machines (see § 58, page 281). It may be noted now, however, that the armatures have no collectors or brushes, and that the windings consist simply of straight conductors laid in grooves or tunnels, and connected at either end to solid metal rings (see Figs. 169 to 171, and page 292); or are laid on the periphery of the core so as to look not unlike a squirrel cage.

The general principle will be readily grasped from diagram Fig. 89, which refers to the ordinary method of coupling the Brown motors.

The winding of the motor is in two parallels, each of as many coils as there are poles in the field magnets, the coils of each series being wound in the spaces between the coils of the other. The starting set is in series with a voltameter or condenser and serves to lead the phase of the current passing through it, while the main coil, having considerable self-induction, retards the phase of its current.

To start the motor, the paths a b and c d are closed by their respective switches. Current from the transformer T then passes from the terminal m to the field magnet coils of the motor M. On dividing between the two coils, one portion of

---

* Other forms of this ingenious motor have been suggested by Mr. Brown, for which rference should be made to the technical journals of January and March, 1893.

the current passes through *a b* and through the regulating handle J and the inductive resistance R, and returns to the other terminal *n* of the transformer through the ammeter A. The second current, in parallel with the first, goes through the voltameter V, and passes direct to the other terminal of the transformer. Thus it will be seen that the phase of the current passing through the main coils and the resistance R will lag, while the other will be advanced by means of the voltameter V. By a suitable adjustment of the condensers to the

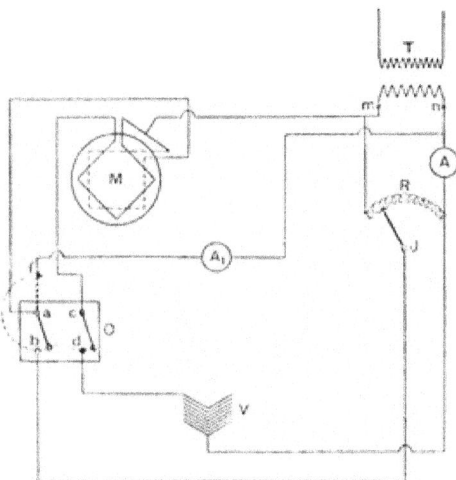

Fig. 89.—Diagram of Connections of C. E. L. Brown's Single-Phase Asynchronous Motor.

periodicity, a phase difference of a quarter of a period is obtained between the two currents. A steady deflection of the ammeter A shows the motor has arrived at synchronism. Then the connection *a b* is broken by the commutator switch, and that through *a f* is made, the motor then running by means of the main coils only. By a suitable coupling of the switches, the making of the contact *a f* and the breaking of *c d* can be accomplished at one operation.

The resistance R is always in circuit across the transformer terminals, absorbing a current which, however, is very small owing to its large self-induction. This arrangement is used to eliminate the induced current on opening and closing the circuit, and is the subject of a special patent by Mr. Brown. The condenser is built up of layers of V-shaped iron plates, insulated from one another, and immersed in a solution of caustic soda contained in an iron jar. It is to be noted that the ammeter, the resistance and the condenser, may be made to serve all the motors of a given installation.

For very small motors of from $\frac{1}{10}$ to $\frac{1}{2}$ H.P. a few spirals of German silver or iron wire put in the starting circuit are sufficient to give the requisite lag. This resistance is conveniently arranged on a hand frame, and can be used consecutively for a number of machines.

By the courtesy of the Directors of the Gülcher Company, the author was enabled, in November, 1893, to make, in conjunction with Mr. W. B. Esson, a series of tests with Brown motors of various sizes. The results obtained are given in Tables O, P, Q, and R, and may be of interest as showing the commercial efficiency and general performance of this type of motor.

Table O.—*100 volts, 80 ～, 4 poles, ¼ H.P.*

Amperes.	Watts actual.	Watts apparent.	Power factor.	Revolutions.	Watts given out.	Actual efficiency %	Apparent efficiency %
3	128	300	0·43	2,330	Empty.	0	0
5·5	343	550	0·62	2,280	245	71·4	44·6
36·5	2,188	3,285	0·666	0	0	At starting.	

Table P.—*100 volts, 80 ～, 6 poles, 1 H.P.*

Amperes.	Watts actual.	Watts apparent.	Power factor.	Revolutions.	Watts given out.	Actual efficiency %	Apparent efficiency %
8·48	176·4	848	0·21	1,560	Empty.	0	0
10·0	364·0	1,000	0·364	1,510	213	58·5	21·3
12·45	706	1,245	0·567	1,520	500	70·8	40·18
16·7	1,109	1,670	0·663	1,490	800	72·1	48·0
22·0	1,125	2,200	0·511	0	0	At starting.	

Table Q.—*100 volts, 80 $\sim$, 6 poles, 2 H.P.*

	Absorbed.			Revolu- tions.	Watts given out.	Actual efficiency %	Apparent efficiency %
Amperes.	Watts actual.	Watts apparent.	Power factor.				
12·7	252	1,270	0·2	1,570	Empty.	0	0
14·5	520	1,450	0·38	1,550	290	55·8	20·0
18·0	1,040	1,800	0·557	1,520	637	66·0	38·1
24·5	1,630	2,450	0·665	1,500	1,100	67·5	44·9
30·5	2,050	3,050	0·672	1,490	1,400	68·3	45·9
31·2	2,150	3,120	0·69	1,480	1,520	70·7	48·6
30·0	1,848	3,000	0·616	0	0	At starting.	

Table R.—*100 volts, 60 $\sim$, 6 poles, 2 H.P.*

	Absorbed.			Revolu- tions.	Watts given out.	Actual efficiency %	Apparent efficiency %
Amperes.	Watts actual.	Watts apparent.	Power factor.				
16·5	320	1,650	0·194	1,185	Empty.	0	0
17·5	500	1,750	0·236	1,170	210	42·0	12·0
19·6	1,040	1,960	0·53	1,160	690	66·3	35·7
23·4	1,320	2,340	0·564	1,150	980	74·2	42·0
27·0	1,710	2,700	0·631	1,140	1,270	74·3	47·0
44·5	2,850	4,450	0·64	0	0	At starting.	

Mr. Banti has made some tests with larger Brown machines of this class, the figures of which are given in Table S. It will be seen that the efficiency is high at from half to full load.

Table S.—*106·3 volts, 45 $\sim$, 6 poles, 5 H.P.*

	Absorbed.			Revolu- tions.	Watts given out.	Actual efficiency %	Apparent efficiency %
Amperes.	Watts actual.	Watts apparent.	Power factor.				
15·0	343·0	1594·5	0·215	902	Empty.	0	0
17·5	967·7	1860·25	0·521	890	588·8	61·0	31·6
21·0	1421·0	2232·0	0·638	870	1015·0	72·0	46·2
24·0	1911·0	2551·0	0·751	884	1479·0	77·4	57·1
27·5	2401·0	2925·0	0·820	872	1869·0	77·9	64·0
32·3	2976·7	3433·0	0·859	872	2325·0	78·9	67·6
37·0	3503·5	3933·0	0·894	860	2699·0	77·0	68·2
43·0	4189·5	4570·0	0·917	868	3083·0	73·6	67·4
48·5	4806·9	5155·0	0·934	848	3383·0	70·3	65·6
55·0	5439·9	5846·0	0·934	850	3731·0	68·0	67·0

Table T.

Oerlikon Company's Asynchronous Alternate-Current Motors.

Horse-Power	1/16	¼	½	1	1½	2	3	6	9	12
42 and 50 ~										
Actual Watts	164	353	910	1,132	1,580	1,960	2,860	5,500	8,000	10,100
Amperes	2·3	4	10·5	15	20	25	36	75	110	145
Efficiency per cent	40	52	60	65	70	75	77	80	83	86
Revolutions per minute at 42 ~	2,400	1,200	1,200	1,200	1,200	1,200	1,200	1,200	1,200	1,200
50 ~	2,850	1,440	1,440	1,440	1,440	1,440	1,440	1,440	1,440	1,440
Weight in lbs.	26·4	110	165	220	275	330	396	550	926	1,276
65 ~										
Actual Watts	164	353	…	1,132	1,580	1,960	2,860	5,500	8,000	10,100
Amperes	2·3	4	…	15	20	25	36	75	110	145
Efficiency per cent	45	52	…	65	70	75	77	80	83	86
Revolutions per minute	3,700	1,875	…	1,875	1,875	1,875	1,875	1,250	1,250	1,250
Weight in lbs.	26·4	66	…	165	220	275	330	550	926	1,276
85 and 100 ~										
Actual Watts	164	353	…	1,132	1,580	1,960	2,860	5,500	8,000	…
Amperes at 85 ~	2·5	4	…	15	22	27	39	80	120	…
100 ~	2·5	4·5	…	16·5	22	27	39	85	125	…
Efficiency per cent	45	52	…	65	70	75	77	80	83	…
Revolutions per minute 85 ~	2,400	2,400	…	2,400	1,600	1,600	1,600	1,600	1,600	…
100 ~	2,850	1,900	…	1,900	1,900	1,900	1,900	1,440	1,440	…
Weight in lbs.	23	66	…	145	187	264	320	540	990	…

All at 110 volts.

Mr. Banti conducted his tests in the following manner :— The motor was started by sending the single-phase current from the transformer through the auxiliary and main windings. As soon as synchronism was attained, the ammeter inserted in the principal circuit assumed a definite deflection. The condenser circuit was then opened and the belt applied to the motor. Readings were then taken of the watts and amperes supplied, the speed, and the torque on the motor shaft ; the volts being kept at constant pressure. The watts were measured by a Ganz wattmeter, a Cardew voltmeter was used for observing the pressure, whilst a previously standardised tachometer indicated the speed. The output was measured by an absorption brake.

The effect of frequency on the outputs, weight, and speed is of great importance, and has been made the subject of investigation by some of the leading firms. It is illustrated in Table T, which is compiled from a series of tests made by the Oerlikon Company with their stock machines, designed for frequencies varying from 42 to 100 $\sim$ per second, at a uniform pressure of 110 volts. The figures are only approximate.

The weights bear no particular ratio, to the outputs, because the same patterns have to be used for machines working under different conditions. It will be clearly seen, however, that the weight for a given speed and output and the same number of poles varies, roughly, inversely as the frequency. The actual watts and the efficiency are practically the same at each frequency within the limits of the Tables. In connection with this, reference should be made to Table M, page 174, which shows the alteration of speed caused by varying the number of poles for various frequencies.

CHAPTER VI.

TRANSFORMERS.

§ 34. INTRODUCTION; TYPES OF TRANSFORMERS.

REFERENCE has already been made to the function of the continuous-current transformer or dynamotor (§ 25, p. 138). That of the alternate-current transformer or converter, generally denoted simply by transformer, is essentially the same; it converts power, $i_1 V_1$, at one pressure into a corresponding power at another pressure and current, $i_2 V_2$ (say). Assuming no loss in the transforming,

$$i_1 V_1 = i_2 V_2.$$

The left-hand product represents the power in the primary circuit at the high-pressure terminals of the transformer; i_1 and V_1 severally expressing the primary current and pressure. The right-hand product gives the power in the secondary circuit at the terminals of the consumer's circuit, V_2 being the pressure of supply. It is clear that the primary pressure will be selected as high as possible in order to keep the current small, and thereby to economise in copper. A usual value for V_1 is 1,000 volts. Let the horse-power be 100 ($i.e. = 74,600$ watts) and the secondary pressure, V_2, be 100 volts, then

Primary circuit, $i_1 V_1 = 74 \cdot 6 \times 1,000$,
Secondary circuit, $i_2 V_2 = 746 \times 100$,

or the currents are in inverse ratio to the pressures.

In practice, owing to the loss in conversion, the secondary output will be less than the primary by an appreciable quantity, varying with the design, size and load of the transformer It will fix ideas to examine the following results obtained by Dr. J. A. Fleming from some of the best known transformers. Table U shows that with an output of one tenth of the full load a well-designed transformer has an efficiency of rather more than 86 per cent., and with one-twentieth about 76 per cent.

It will be seen also that the efficiency is highest at from three-quarters to full load, and that the efficiency decreases as the load is diminished. It is thus of the highest importance to secure a high-load diagram for each transformer, and to ensure, if possible, that the all-day load shall not be less than 50 per cent. of the maximum. Then the efficiency will average about 95 per cent. This is a very important consideration in distributing power, and will be referred to again when discussing power problems. Another point is the *power-factor*, or ratio of true watts (effective volts × effective amperes × cos ϕ) to apparent watts (effective volts × amperes). This is a variable quantity, being from 0·77 to 0·80 at no load, and attaining unity at about one-tenth of the maximum output in the best modern designs, such as Mordey, Ferranti, Kapp, Westinghouse, and Thomson-Houston transformers. With the "Hedgehog" type it does not exceed 0·07 at no load, and never exceeds about 0·8 even at full load. This ratio does not give the efficiency of the transformer, but, since it is dependent on the conductivity of the magnetic circuit, it forms a rough guide to the suitability of transformers for given work, and also is a useful figure of merit for determining the probable line and plant efficiency at light loads, especially when the secondary circuits are open.*

---

* It is now well recognised that a recording wattmeter is the most satisfactory instrument for measuring the true watts given out from a central station. The integration of the resulting diagram will give the total energy supplied in any given time, and the meter-readings will give a fair idea of the quantity of energy received by the various consumers during the same period (assuming the line to be non-inductive) ; hence it is possible to estimate the plant efficiency with considerable accuracy.

Table U.*

Efficiencies of Transformers Calculated from Observations made by a Dynamometer Wattmeter.

Size and description of Transformer.	Fractions of full secondary load.												
	0	0·025	0·05	0·1	0·2	0·3	0·4	0·5	0·6	0·7	0·8	0·9	1·0
	Percentage efficiencies.												
6,500-watt Westinghouse	0	61·8	75·9	85·7	91·9	94·0	95·1	96·0	96·3	96·6	96·8	96·9	96·9
4,500-watt Thomson-Houston	0	49·1	76·6	78·8	87·8	91·1	92·9	93·8	94·2	94·6	95·0	95·0	94·7
15,000-watt 1892-pattern Ferranti...	0	62·0	65·4	86·5	92·6	95·0	95·6	96·1	96·5	96·8	96·8	96·6	96·6
11,000-watt " "	0	65·5	79·0	88·1	93·4	95·0	95·7	96·0	96·2	96·3	96·1	95·8	95·5
6,000-watt Brush (Morley)	0	52·1	67·6	80·1	88·5	91·1	92·7	93·7	94·2	94·7	94·9	95·1	95·4
4,000-watt Kapp..	0	39·5	56·5	72·3	83·8	88·0	90·4	91·9	92·6	93·3	93·8	94·0	94·2
3,750-watt 1885-pattern Ferranti...	0	28·8	44·5	61·4	75·6	81·9	85·5	87·7	89·0	89·8	90·2	90·5	90·8
6,000-watt Swinburne "Hedgehog"	0	48·4	65·2	79·0	88·2	91·8	93·7	94·8	95·5	95·9	96·1	96·1	96·1
3,000-watt " "	0	39·3	56·2	71·7	82·9	87·6	89·8	91·2	92·2	92·8	93·2	93·4	93·5

* From the *Journal* of the Institution of Electrical Engineers, Vol. XXI., Part 101, p. 651.

The types of transformers in general use are not very numerous; indeed, there is not room for much variation, although, of course, the details differ widely. Broadly speaking, transformers may be divided into two classes, which differ fundamentally, viz., the closed magnetic circuit and the open magnetic circuit. The first includes the majority of designs, and has been demonstrated to be generally superior to the second, which is practically only represented by the " Hedgehog " transformer. The specific claim for the open-circuit converter is an assumed higher average efficiency. But Dr. J. A. Fleming, after an exhaustive series of trials,* finds that the closed type is the better, even

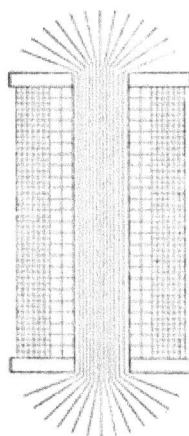

FIG. 90.—" Hedgehog " Transformer. The vertical lines in the centre with spiked ends represent the soft iron wire core.

on this count. With respect to light-load efficiency there is no doubt that the closed circuit is preferable. For the power-factor of the open-circuit type is so low that the wattless current is always large, and the " magnetising current," or primary current with secondary circuit open, appears to be practically prohibitive.

The " Hedgehog " transformer (Fig. 90) is, however, of commercial interest.

* See *Journal* of the Institution of Electrical Engineers, Vol. XXI., Parts 101 and 102.

It will be seen that the iron core is surrounded by the copper winding, and the soft wires of which it is built are turned over at the ends into hemispheres. It is a very cheap design, and finds favour from considerations of first cost.

It is fair to state that Mr. Swinburne explains the indifferent results obtained with the particular " Hedgehog " transformer

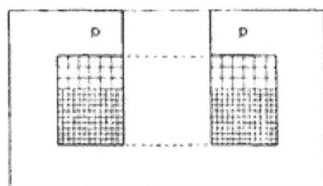

FIG. 91.—Westinghouse Transformer.

tested by Dr. Fleming by the fact that its winding compressed the soft iron wire core so tightly that the films of insulating oxide were partially broken, and the whole core acted as a more or less solid mass. This no doubt militated largely against the efficiency, and it may be that the open-type transformer in

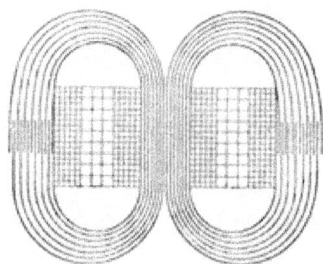

FIG. 92.—Ferranti Transformer.

Mr. Swinburne's hands will ultimately attain to as high a power. factor and all-round efficiency as the closed type ; but this is not the case at present.

The better-known closed-circuit transformers are illustrated diagrammatically in Figs. 91 to 100. The primary windings are suggested by fine cross lines, and the secondary by coarser ones.   Various methods of disposing of the windings—some separate and some interplaced—are also indicated, but they are not peculiar to any of the types, being practised as found convenient.   With the exception of the Ferranti

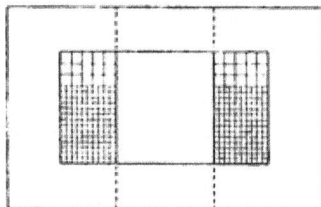

Fig. 93.—Mordey Transformer.

transformer, Fig. 92, which is built up of thin iron strips, all the types illustrated are made from thin stampings of soft iron, the gauges most in use being from No. 24 to No. 30 B.W.G.   Generally the design aims at reducing the external magnetic leakage to a minimum, thereby causing the whole of

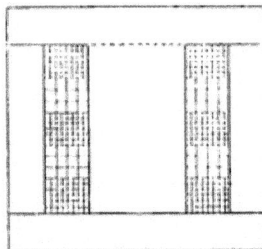

Fig. 94. – Kapp Transformer.

the lines of force induced by the primary current to pass through the secondary circuit, and *vice versâ*.   The methods of putting the windings in position will be easily understood.   The primary and secondary coils are first wound on "formers," and then thoroughly insulated both from each other and from the iron core. In Fig. 91 the pieces marked *p p* are bent back, and the coils are passed over the core; when as many stampings as are required are threaded, the plates are distanced with mica and firmly secured in position by the external casing. In Figs. 94,

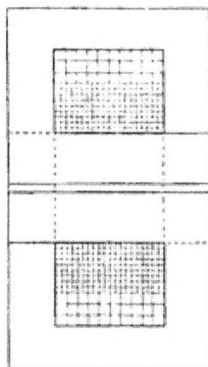

Fig. 95.—Gülcher (Esson) Transformer.

95, 96, 97 and 98, the coils are simply laid around the centre core pieces before the ends are placed in position. In Fig. 93 the coils are first placed in the centre of the larger plates, and the small punchings are put between consecutive plates, and thus,

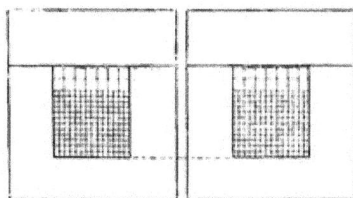

Fig. 96.—Snell and Kapp Transformer.

passing through the coils, form the core. In the Ferranti type the coils are slipped over the iron strips, which are then bent into shape and secured by the outer casing. In Fig. 99 the coils are dropped over the core, and then the plug pieces, *k k*, are put in position. In Fig. 100, which is the latest Thomson-Houston design, the coils and the core are placed together in the centre of the rectangular stampings.

The insulation spaces permit a free circulation of air— or of oil if the whole be immersed in a resinous oil, as is frequently done in modern designs for very high pressures.

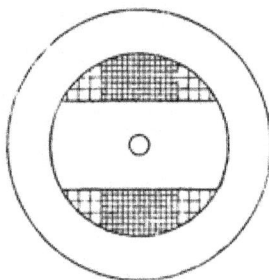

FIG. 97.—Rankin Kennedy Transformer.

By means of oil insulation converters have been built with a transformation ratio as high as 1 : 140, viz., from 50 to 20,000 volts ; and even this is not necessarily the limit if

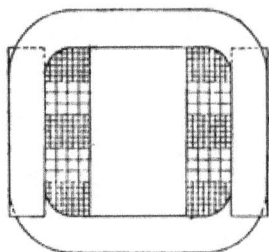

FIG. 98.—Brown Transformer.

special precautions are taken as regards the disposition of the high pressure coils, and especially of the leading-in and out wires, which are usually encased in glass tubes widely separated in order to prevent arcing. It must be recollected that the normal maximum pressure, which in this case should be about $20,000 \times \sqrt{2}$, or, say, 28,200 volts, determines the limit of disruptive discharge, and that if there be a resonant effect in the circuit, it is quite possible for this value to be considerably exceeded. No doubt the cumulation of the

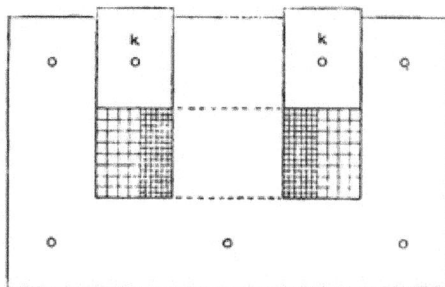

Fig. 99.—Thomson-Houston Transformer.

several harmonic pressures in a circuit explains the otherwise apparently inexplicable puncturing of insulation. Oil insulation is especially valuable in such cases, for the puncture closes as soon as the spark has passed, and the insulation is as strong as before. Extra high pressure transformers are at present

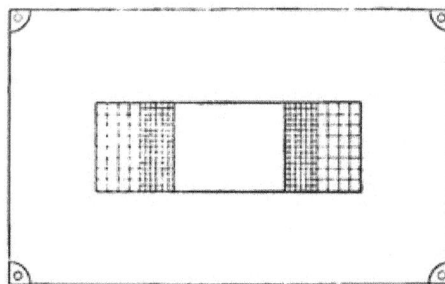

Fig. 100.—Latest Form of Thomson-Houston Transformer.

only in use for long-distance transmission, and are used as step-up converters at the power station, and as step-down converters at the distributing end of the line. By their use the main current is made as small as desired, and the extra high pressure is reduced to a safer limit before serving the distributing system (see § 38, p. 208).

The performance of a transformer is affected by size, relative quantities of iron and copper, magnetic density, copper loss, stray magnetic field, and resistance of the joints at right angles to the path of the lines of force. Speaking generally, high efficiency at low loads may be attained if the total copper loss be kept small and the magnetic induction be made relatively low—i.e., if there be a large quantity of both copper and iron; but this design is necessarily expensive owing to the large mass of metal. If a high efficiency be required only at nearly full load, then the copper loss may be larger and the cross-section

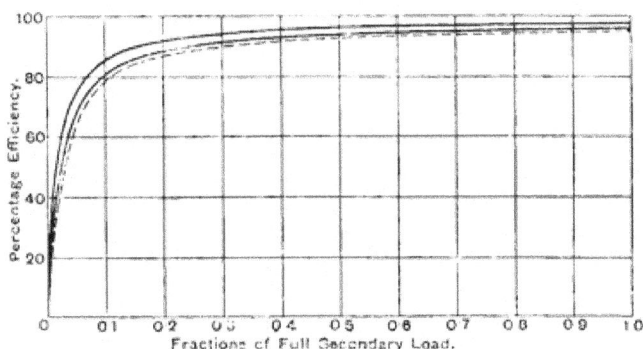

Fig. 101.—Efficiency Curves of Various Transformers.

Top Curve, 6,500-watt Westinghouse Transformer.
Middle Curve, 6,000-watt Mordey Transformer.
Bottom Curve, 4,500-watt Thomson-Houston Transformer.

Note.—If the Transformers were of the same power, the curves would probably be nearly identical.

of the iron less, the design being lighter and cheaper. If high average efficiency with cheap design be aimed at, a compromise must be made between these two extreme proportions. These remarks are applicable to any of the designs mentioned, and the results actually obtained in practice may be gathered from Table U, p. 189. The efficiency curves of some of the standard transformers are shown in Fig. 101.

§ 35. TRANSFORMERS IN PARALLEL; PRIMARY AND SECONDARY DROP OF PRESSURE; MAGNETIC LEAKAGE.

If a transformer be designed for parallel working—and this is the more frequent practice—then the aim is to make the mutual induction of the primary and secondary circuits a maximum, and to restrict the drop of volts to as small limits as possible—good practice being within 2 per cent. of the open-circuit secondary pressure. The " total secondary drop " at full load is affected not only by the secondary resistance, but also by the primary resistance and the stray magnetic field. The leakage effect increases with the output, for the disturbing effect due to the secondary ampere-turns is, of course, proportional to the secondary current, i_2. The total drop is also affected by the increase of value of the primary drop, $i_1 R_1$; for, if the pressure at the primary terminals be kept constant, the primary counter pressure decreases as the primary current increases. And as the secondary pressure is proportional to the primary counter pressure (not to the primary terminal pressure) the secondary drop is increased in proportion as this diminishes. The number of volts in the secondary drop due to the primary drop is given by the expression

$$\frac{T_2}{T_1} \times \text{primary drop, or } \frac{T_2}{T_1} \; i_1 R_1. \quad \ldots \quad (42)$$

In the transformer equations the suffixes o and f are severally used here to distinguish current or pressure at no load and full load ; T_1, T_2 = the respective number of turns in the primary and secondary coils; and R_1, R_2 = the primary and secondary resistances.

The relative importance of these three causes of drop in the pressure varies with different designs. In the Westinghouse type the magnetic leakage accounts for about one-half of the drop ; whilst in the Kapp transformer, which has interplaced windings, the leakage is very small (*see* Fig. 102). Dr. J. A. Fleming gives the following rule for estimating the number of volts

P

lost by magnetic leakage with a closed-circuit transformer.
Add together the product of secondary resistance and maximum
secondary current and $\frac{T_2}{T_1}$ times the product of primary resistance and current, after deducting the "magnetising current"
from the primary current at full load; subtract this sum
from the total observed drop, and the remainder is the number
of secondary volts lost by magnetic leakage.   Thus :—

Secondary volts lost by magnetic leakage

$$= ({}_oV_2 - {}_tV_2) - \{R_2\,{}_t i_2 + \frac{T_2}{T_1} R_1({}_t i_1 - {}_o i_1)\}. \quad . \quad (43)$$

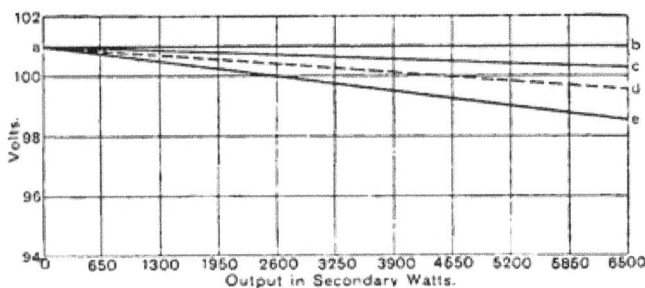

FIG. 102.—6,500-watt Westinghouse Transformer.

a b, Line of Maximum Pressure.   b c, Drop due to Secondary Resistance.   c d, Drop due to Primary Resistance.   d e, Drop due to Leakage.

The order of the "drop" curves is illustrated in Fig. 102,
which is taken from observations made with a 6,500-watt
Westinghouse transformer.

These curves indicate the lowering of the secondary pressure
for each of the three causes.   The top line, a b, shows the
open circuit secondary pressure of 101 volts; line a c marks
the loss due to secondary resistance; a d that caused by
primary resistance; and a e that by magnetic leakage.   The
maximum drop is about 2·4 volts, or a mean variation of 1·2
per cent.

If the primary and secondary currents be nearly in opposition to each other, *i.e.*, separated by a phase difference of 180deg. approximately, the resultant ampere-turns at any load are simply $i_1 T_1 - i_2 T_2$, where T_1 and T_2 are severally the number of primary and secondary turns. And with the secondary circuit open the excitation is simply due to $i_1 T_1$. This particular value of the primary current is called the magnetising current. It has been determined by Profs. Ryan and Fleming and others that the resultant excitation is constant for all loads; hence the primary current at no load is a criterion of the merit of a given transformer if the above-mentioned phase difference be preserved. The following figures from a 3,000-watt Gülcher transformer will give some idea of the value of the effective or magnetising ampere-turns. The primary turns are 1,000 in number, and $i_1 T_1 = 1,570$ at full load; the secondary turns are 50, and $i_2 T_2 = 1,500$; therefore the magnetising ampere-turns $i_1 T_1 - i_2 T_2 = 70$ (approx.), and the magnetising current $= 0·07$ ampere.

This law, however, is not true for open circuit transformers, or for those closed circuit types which have defective joints, for then there is a varying lag between the two currents, and the value of the expression $i_1 T_1 - i_2 T_2$ continually diminishes as the load increases. It will be remembered that the "Hedgehog" transformer was specially mentioned as having a low power-factor, the value being a minimum at no load and increasing towards the full output, but never exceeding about 80 per cent., whereas the best closed-circuit types have a power-factor of 100 per cent. at about one-tenth of their output.

It is theoretically possible to run transformers in series upon a constant current circuit, and to supply the secondaries with constant currents; but there are so few cases in which such a system of distribution would be economical that few transformers have been built for this purpose. It is sometimes desirable, however, to feed a secondary circuit with constant current while the primary is supplied at constant pressure, as occurs with arc lamps in series. Both the Westinghouse and Thomson-Houston Companies have devised such transformers; that of the last-named Company being shown in

p 2

Fig. 102A. It will be seen that the magnetic flux is carried partly through the side cores and partly across the air-space in the centre marked A. The proportions are so arranged that a decrease in the number of arc lamps, lowering thereby the secondary resistance, increases the reaction of the secondary circuit on the primary, and causes a greater leakage of magnetism across the gap A. This cuts down the induction in the secondary and increases the self-induction of the primary, thus proportionately decreasing its current.

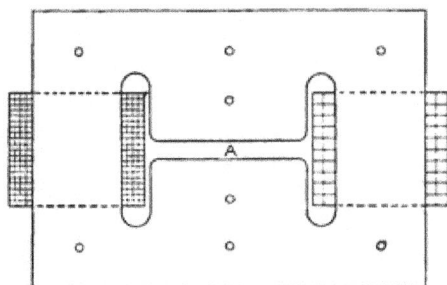

FIG. 102A.—Thomson-Houston Parallel Transformer with Constant Secondary Current.

§ 36. CURVES OF CLOSED MAGNETIC CIRCUIT TRANSFORMERS.

In examining the performance of a given closed-circuit transformer, the essential points to be determined are :—

(a) Pressure at primary terminals.

(b) Magnetising current.

(c) Primary and secondary resistances.

(d) Terminal secondary pressure at various loads.

(e) Secondary currents at loads corresponding to readings taken under (d).

The product of (a) and (b) gives the apparent watts consumed at no load. Since the power factor or ratio of true to apparent watts is then about 0·8, if this product be multiplied by 0·8 the true watts are known, and the iron core loss for *all loads* can be found by subtracting the primary coil watts with open secondary coil.

Thus, $\quad 0.8\, V_1\, _0i_1 - _0i_1{}^2\, R_1 =$ watts lost in iron. . . (44)

Then from the ratio of primary to secondary pressure at no load, and from the same ratio with various secondary currents, the corresponding primary currents can be calculated. And hence, knowing the resistances, the primary and secondary i^2R losses can be estimated. Then, drawing a total i^2R loss curve and a total loss curve (*i.e.*, adding hysteretic and eddy current losses) parallel to it, there are sufficient data to determine the efficiency at all stages of the load.

To illustrate this the following calculations are given with reference to a Ferranti 1892-pattern transformer of 15 H.P., having the following data :—

$\quad V_1 = 2,400$ volts (kept constant).
$\quad V_2 = 100$ volts nominally.
$\quad \sim\ = 82.5$ periods per second.
$\quad R_1 = 3.77$ ohms.
$\quad R_2 = 0.0092$ ohm.
$\quad T_1:T_2::24:1.$
$\quad _0i_1 = 0.103$ amperes.
$\quad _fi_1 = 4.837$ amperes.
$\quad _fi_2 = 113.85$ amperes.

On open circuit :— $_0V_1 = 2,400$; $_0V_2 = 100$; $_0i_1 = 0.103$;
The apparent watts $= _0V_1 \times _0i_1 = 2,400 \times 0.103 = 247.2$;
And loss in core $= 0.8 \times 247.2 - 0.103^2 \times 3.77 = 197.5$ watts.

At full load :— $_fV_1 = 2,400$; $_fV_2 = 97.8$; $_fi_1 = 4.837$;
$_fi_2 = 113.85.$

Loss in primary, $\quad i_1^2 R_1 = \quad 4.837^2 \times 3.77 \quad = 88.5$
,, secondary, $i_2^2 R_2 = 113.85^2 \times 0.0092 = 119.0$
,, core $\qquad\qquad\qquad\qquad\qquad = 197.5$

Total loss at full load $\qquad = 405.0$

Efficiency at full load $= \dfrac{11,134}{11,134 + 405} = 96$ per cent.

These calculations give the open circuit and full load points of the several curves. The intermediate points can be determined in a similar manner, and the whole series be plotted in succession.*

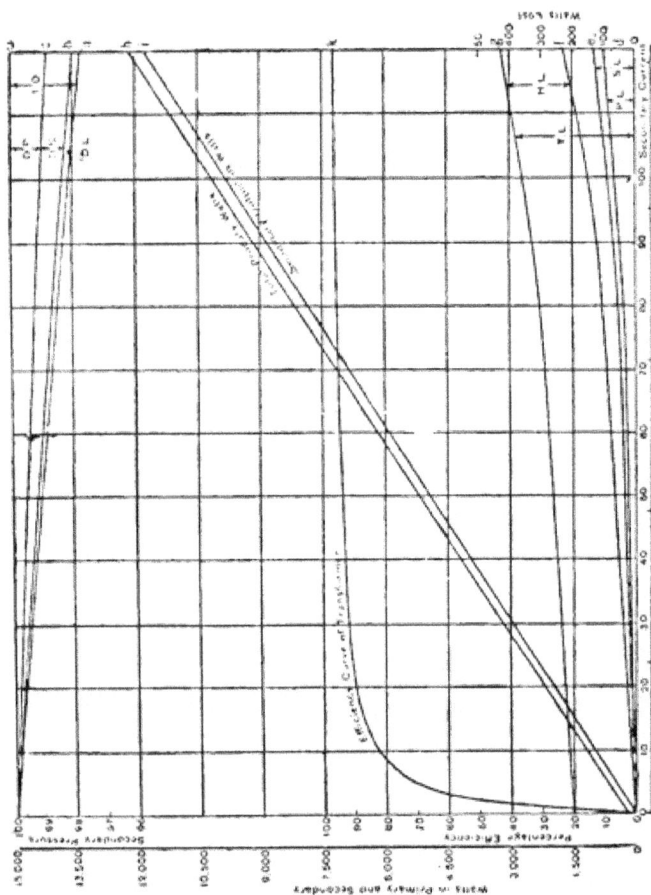

Fig. 103.—Curves of Ferranti Transformer, 1892 Type. Primary pressure, 2,400 volts. D.P., Drop due to Primary Resistance. D.S., Drop due to Secondary Resistance. D.L., Drop due to Leakage. T.D., Total Drop. P.L., Primary Loss. S.L., Secondary Loss. H.L., Hysteretic and Eddy Current Loss. C/, Total Copper Loss. T.L., Total Loss.

---

* Full particulars of this transformer will be found in Dr. Fleming's Paper on page 636 of Part 101 of *Journal* of Institution of Electrical Engineers, to which reference should be made for further information.

These results are illustrated in Fig. 103, in which

Curve a shows the total secondary drop,

,, b ,, secondary drop due to the primary and secondary circuits,

,, c ,, secondary drop due to the secondary circuit,

,, d ,, primary copper loss in watts,

,, e ,, secondary ,, ,,

,, f ,, total ,, ,,

,, g ,, loss (including hysteresis and eddy currents),

,, h ,, primary output in watts,

,, i ,, secondary ,,

,, k ,, efficiency.

All the curves are referred to the currents as abscissæ, the corresponding primary and secondary currents being plotted to suitable scales. The ordinates are made to different scales in order to condense the whole.

Some idea of the order and relation of the phases of pressure, current and resulting magnetism in a transformer may be gathered from the following :—

In a transformer there are—

(a) A wave of impressed E.M.F., $_1V_{imp}$, in the primary circuit.

(b) A primary current, i_1, lagging behind the primary E.M.F. by an angle ϕ, which may have any value between 0deg. and 90deg.

(c) A wave of magnetism lagging behind the primary current by considerably less than 90deg.

(d) A counter E.M.F., $_1V_m$ due to the inductive action of the magnetism and lagging 90deg. behind the magnetic wave

(e) An impressed secondary E.M.F., V_2, due to and measured by the rate of change of the magnetism, and hence lagging 90deg. behind it.

(*f*) A secondary current, i_2, in phase with the secondary E.M.F. if there be no self-induction in the circuit; lagging by an angle of any value between 0deg. and 90deg. if there be inductive action in the circuit; or leading if there be capacity in the circuit.

The relative positions of the several curves are roughly indicated in Fig. 104.

On an open circuit the counter E.M.F. of the primary is nearly in opposition to the primary impressed E.M.F. The

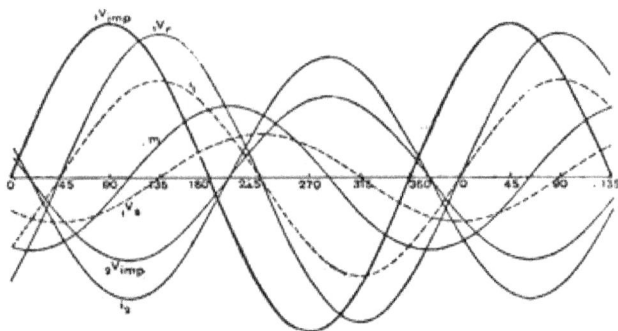

FIG. 104.—Curves of Transformer.

$_1V_{imp}$, Primary Impressed Pressure. $_1V_r$, Primary Resultant Pressure. $_1V_s$, Primary Self-induction Pressure. i_1, Primary Current. $_2V_{imp}$, Secondary Impressed Pressure. i_2, Secondary Current. m, Curve of Magnetic Induction.

Angle of lag between primary pressure and current assumed to be 45°; and secondary pressure and current taken in phase.

primary and secondary currents are very nearly in opposition at all loads, but most nearly so at heavy loads.

It is useful to remember that when the secondary coils are short-circuited the impedance of the primary circuit becomes sensibly equal to about twice its ohmic resistance, and there is a danger of burning out the winding.

§ 37 DESIGN OF TRANSFORMERS.

From the preceding practical and theoretical remarks it will be evident that the design of a transformer is from the commercial point of view a compromise between conflicting conditions of efficiency and cost. It is generally recognised that efficiency can always be secured if price be of no moment; but sound commercial engineering requires the annual cost of power wasted in a transformer to be equated to the cost of maintenance plus interest on capital outlay. This demands a special application of the economic law. There are, however, certain well-established limits for the permissible losses in copper and iron cores for different sizes of transformers at each of the more usual frequencies, and if these are observed the results arrived at will not be far from the best attainable. Some of these practical data are to be found in Dr. J. A. Fleming's Paper on "Experimental Researches on Alternate-Current Transformers," and as Dr. Fleming tested transformers by some of the best makers in England and America, the figures are, beyond dispute, characteristic of modern practice.

From the mean of a number of efficiency tests it appears that the copper losses for small transformers do not exceed 1·75 per cent. of the full load secondary watts, and for larger ones the loss decreases proportionately to about 1·25 per cent. The losses in the iron are also found to vary between 1·0 and 0·25 per cent. of the full load output. So that the total losses may be said to average from 3·0 to 1·5 per cent.

Bearing these figures in mind it is comparatively easy to find the permissible resistance for primary and secondary circuits, and also the number of cubic inches of iron for a given output (*see* Table L, p. 173). In the first instance, the following data will be known :—

V_1 = Primary Pressure (assumed constant).
$_0V_2$ = Secondary ,,   no load.
i_2 = Secondary current, full load, and
$T_1 : T_2 :: V_1 : _0V_2$.

The probable load factor should be calculated first of all in order to determine the *equivalent* current. Then R_1 and R_2,

the resistances of the primary and secondary circuits, may
be selected. Next the current densities may be chosen, the
usual values lying between 250 and 1,500 amperes per square
inch. The hysteretic and eddy current loss should next be
estimated, and then the copper losses and the drop can be
approximately determined, the values being chosen between
the limits given here. This will enable curves (a), (b), (c), (d),
and (e) (see Fig. 103, p. 202) to be plotted in the following
order :—

First plot curve (a) from the assumed value of the total drop,
and curve (c) from the calculated value of R_1 and the known
value of $_oi_1$, by equation (42), p. 197. Then plot curve (b)
from the assumed values of R_2 and $_oi_2$, and the remaining
space is the leakage drop. Then curves (d) and (e), the
copper loss in primary and secondary, can be calculated ;
and the iron core loss at all loads can be determined from
the selected total loss. Then curves (d) + (e) + core loss give
curve (g). The secondary output curve (i) is easily calculated, and then the primary (h) is found by adding the total
watts from curve (g) and plotting the result to a suitable
scale. The efficiency curve (k) is simply curve (i) divided by
curve (h).

The primary current for any load can be read from the curves
now plotted. The ratio of the turns in primary and secon-
dary circuits depends simply on the ratio $\dfrac{V_1}{_oV_2}$, say $\dfrac{2,000}{100} = \dfrac{20}{1}$.

The exact number of turns in each can be found thus :—

First, let the number of primary turns T_1 be determined,
then the effective value of the primary pressure

$$V_1 = \frac{2 \pi \sim N T_1}{\sqrt{2} \times 10^8} = \frac{4 \cdot 45 \sim N T_1}{10^8}, \quad . \quad . \quad (45)$$

which is the fundamental equation for transformers, and corresponds with equation (6), p. 18, for dynamos.

The only difficulty is to fix the ratio of N to T_1 when only the product of the two is known. For example let $\sim = 100$ and $V_1 = 2,000$;

then
$$N\,T_1 = \frac{2,000 \times 10^8}{4 \cdot 45 \times 100} = 4 \cdot 5 \times 10^5.$$

Unless there be data available it will require a series of trial calculations to fix the best ratio between these quantities.

The usual plan adopted in practice is to take a stock size core, and this fixes N, for $N = B''A$, where A is the area of cross-section and B'' the induction. The value of B'' is from 60,000 to 65,000. Hence T_1 is known.

If there be no fixed core to guide the design in respect of N, an approximation may be made from the permissible hysteretic loss (*see* Table L, p. 173) and the corresponding volume of iron for the given frequency. This is a tedious process, as it leaves the relation of the length of the magnetic path to its cross section an open question, and requires several trial calculations before satisfactory proportions are determined.

A few calculations, however, for different shapes of carcases will demonstrate the principles. This part of the design necessarily involves the consideration of the winding space, which is not a difficult problem when the gauges of wires are approximately known.

As soon as the number of primary turns T_1 is settled, that of the secondary is found by multiplying T_1 by the ratio of transformation ; or the secondary turns

$$T_2 = \frac{V_2}{V_1}\,T_1.$$

This description of transformer design is incomplete, but the Author feels that this class of plant is not likely to be taken in hand by any except specialists, owing to the many difficulties which arise in connection with the selection of various designs for specific work, such as parallel or series lighting, transmission or distribution of power; or, what is still more intricate, a combination of power and lighting work. It has been assumed in the calculations, as is customary, that the secondary turns are few in number, and therefore have no appreciable self-induction, and also that the secondary external circuit is a non-inductive one, such as a bank of incandescent lamps. Thus the drop of pressure follows simple laws and is easily pre-determined. But if an inductive load, as electric motors or arc lamps, is put in the circuit, then there is a counter-pressure of self-induction; and, the secondary current lagging behind the secondary pressure, the drop is increased in proportion. Again, if there be resonance in the circuit, such as that caused by a condenser in parallel with the primary mains, or if the power be transmitted through long concentric mains armoured with continuous metallic sheathing, the current may lead the pressure, in which case the "drop" will become negative, and the supply pressure will be raised. These two sources of trouble, being diametrically opposite in their effects, counteract each other, and hence it has been suggested that condensers should be used at central stations supplying power to inductive circuits. But the treatment of such problems demands a close acquaintance with the underlying theory, and it would serve no useful end to discuss at length this part of the subject. The engineer desirous of studying self-inductive and resonant effects is referred to Dr. Fleming's "Alternate-Current Transformer,"* in which the subject is treated both practically and mathematically. There are, however, some special problems which occur in alternate-current transmission, and these are referred to in Chapter VII. (*see especially* §§ 42 and 43).

§ 38. STEP-UP AND STEP-DOWN TRANSFORMERS.

In cases where the pressure in the transmission mains is from 4,000 volts upwards it is sometimes found convenient to employ step-down transformers to lower the line pressure at the sub-stations. And in polyphase current systems step-up transformers are sometimes used, because this class of alternator can be most economically designed for pressures of from 50 to 100 volts. It is, however, generally cheaper

---

* Published in "The Electrician" Series, in 2 vols., "Theory" and "Practice."

to increase the insulation resistance of the windings than
to use a step-up transformer, at any rate up to pressures
about 5,000 volts. The best authorities appear to differ as
regards the pressure at which step-up transformers become
advisable or necessary. For example, in the Rome-Tivoli power
plant the single-phase alternators are banked in parallel at
an omnibus pressure of about 5,000 volts, and at the Ponta Pia
sub-stations step-down transformers lower the pressure to 2,000
volts, at which the primary circuits of the house transformers
are fed. On the other hand at Lauffen-Heilbronn the polyphase-current alternators give 50 volts and 4,000 amperes
each, this pressure being raised by three-phase transformers to
5,000 volts. It is reduced at Heilbronn to 1,500 volts to feed
the distributing network.

Intermediate transformers are used primarily to economise
copper in the transmission mains by means of extra high
pressure, and they secure this important end with freedom
from risk to life and with immunity from breakdowns. For
it is evident that the distributing network can be easily
protected from the inroad of the extra high pressure by use
of the safety devices already mentioned ; and that an accident
to the step-down transformer only affects the primary circuits
of the supply transformers, which in turn are protected by
similar devices. The consumer is thus doubly secured from
an invasion of the high pressure. From the mechanical aspect
there can be no doubt that a transformer properly built and
buried in a sealed oil-tank is a far more lasting device than the
most carefully constructed high or low pressure alternator.

The use of intermediate transformers becomes of greater importance as the pressure of the transmission service is raised. The
experiments with three-phase currents between Frankfort and
Lauffen with pressures of from 20,000 to 30,000 volts establish
extra high-pressure working on a practical basis, and the difficulties arising from capacity and self-induction are not found to
present serious obstacles.

At Cassel an interesting variation in plant design comprises
single-phase alternators, step-up transformers, and alternate-
current motors at two sub-stations, each motor running two

continuous-current dynamos, which each feed a low pressure three-wire distributing network. And a similar scheme involving two-phase generators has been carried out by Messrs. Schuckert and Co. at Buda-Pesth.

For power purposes, since the load-factor will generally be at least 50 per cent., it will be possible to work the large extra high pressure transformers at more than 50 per cent. of their full capacity during the greater part of the time, and probably by carefully selecting the units the load-factor of each may be kept as high as 85 per cent. If this be the case the large transformers will generally be less costly to build per unit than ordinary transformers, which require a high average efficiency over a large range of output, and therefore must contain a liberal allowance of copper.

This is a weighty point in considering the use of step-up transformers, and must largely determine the advisability of their use for any plant.

The relative merits of low-pressure alternators and step-up transformers *versus* high-pressure alternators, are not easily estimated. Time is necessary to permit the accumulation of experience before this question can be satisfactorily answered.

CHAPTER VII.

ALTERNATE-CURRENT TRANSMISSION AND DISTRIBUTION.—SINGLE-PHASE.

§ 39. ADVANTAGES AND DISADVANTAGES OF ALTERNATE CURRENTS.

THE special points in favour of the alternate-current systems may be summarised as follows :—

(*a*) High-pressure machines are permissible, since the currents are not commutated, and, therefore, sparking difficulties are entirely obviated.

(*b*) Power is readily divided into units of any size and at any pressure by suitable transformers—either step-up or step-down.

(*c*) The regulation of pressure over a large area of supply is facilitated by parallel high-pressure primary circuits.

(*d*) The power in any of the branch circuits can be regulated by choking coils without serious loss—thus the pressure of a feeder, or the light of a single lamp, can be graduated without using a wasteful resistance.

(*e*) Alternate currents are generally more easily handled than continuous currents, and the sparks at breaking circuit are less destructive.

(*f*) Synchronous motors of large sizes have a high all-round efficiency, and necessarily run at constant speed.

(*g*) Asynchronous motors can be made without collecting rings and brushes, and are of very simple construction.

(*h*) Separately-driven alternators will run perfectly in parallel if the driving power and excitation be properly apportioned.

(*i*) Two alternators of unequal pressure may be coupled in parallel, and will give an intermediate pressure if the self-induction of the armatures be of suitable amount—too much or too little self-induction lowers the plant efficiency, renders the condition unstable, and prevents the load from being proportionately divided between the several machines.

The chief drawbacks to alternate currents are :—

(1) Transformers, especially small ones, are not efficient with low load factors, and therefore transformer sub-stations are required for economy.

(2) Synchronous motors are not self-starting, and will stop if temporarily overloaded.

(3) Asynchronous motors will not start against the designed running torque, unless special precautions are taken in the designing, and are only efficient at high loads.

(4) The lag between current and pressure in an inductive circuit, such as one loaded with motors, is productive of wasteful current in the copper circuits. This is specially the case when the total useful power is small.

(5) Synchronous alternators are more costly to build than dynamos of equal output.

(6) Alternate-current transformers fed with the primaries in series at constant current require the secondary circuits to be also fed with constant current.

(7) Free-running synchronous alternate-current motors will not work in series.

(8) Alternate currents are not suitable for electrolytic purposes.

(9) There is some difficulty in regulating the pressure of supply in circuits containing induction and capacity.

(10) Accumulators cannot be used either for storing power or steadying pressure.

These points will be severally developed in § 40 to § 51.

In this chapter only single-phase currents are considered, and hence the term alternate current will be limited here to this meaning.

§ 40. SERIES WORKING NOT POSSIBLE; SINGLE ALTERNATOR AND MOTOR.

In Chapter IV. it has been shown that the simplest case of transmission of power with continuous current is that in which one dynamo and one motor are run in series with each other. The same is the case with alternate currents; but series working is strictly limited to one alternator and one motor if synchronous machines are used. Properly speaking, series working is impossible with synchronous machines. A reversed alternator runs perfectly as a motor when driven off a similar generator if running in synchronism with it. The term synchronism implies that the periodic time of the alternations of the two machines is the same. Theoretically, it is sufficient that the frequency of one machine either coincides with or is equal to a small multiple of that of the other.

It has been already shown that the frequency is equal to the number of pairs of poles in the field-magnets multiplied by the number of revolutions of the armature (or field-magnets) per second. Therefore, if the machines be similar, the speeds must be the same when synchronism obtains; and if the numbers of poles in the two machines differ, the revolutions will vary inversely as the relative numbers of poles. From these considerations it is seen that a synchronous alternate-current motor

Q

is *not self-starting* ; and that the moving part must be started
and revolved until it falls into step with the generator. This
is most readily accomplished by running the generator slowly
until the motor has got into step with it, when the speed of the
dynamo may be gradually increased and the motor speed will
follow it with considerable persistence. This will be compre-
hended by reference to the following diagram.

In Fig. 105 let V be the curve of pressure of the alternator
running as a generator ; *v* the curve of counter pressure of a
similar machine running as a motor; and assume that, by means
of suitable adjustment of both field-magnet excitations, V is
greater than *v* by an amount sufficient to drive the current *i*
through the combined circuit. Then the work given out by

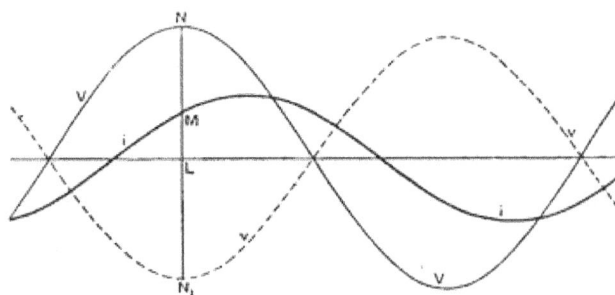

FIG. 105.

the generator at any instant is equal to the product of the corresponding ordinates of V and *i*, as $N L \times M L$; and the work
absorbed by the motor at the same instant is $- L N_1 \times M L$.
The sign of each curve is taken as positive when above the
time line and negative when below it. The total work done by
either machine in one complete cycle can be obtained graphically
by dividing the time line into any number of equal parts, when,
if the time of one-half period be equal to T, then the several
values of $V i \dfrac{T}{n}$ and $v i \dfrac{T}{n}$ can be readily found and plotted in
new curves whose areas will be proportional to the respective
outputs. Care must be taken to use the right signs, and to
recollect that the algebraic product of two negative quantities
is reckoned as positive ; hence the values of V *i* in the second

half of the generator curves are positive, although each of the factors is negative. The negative sign before the product $V i$ implies that the machine is *receiving power* from the circuit, whilst its absence shows that the machine is *giving power* into the circuit. The displacement of the current curve i, with reference to the resultant pressure $V - v$, is called the lag. The amount of the lag is usually measured in degrees, and is determined by the self-induction of the circuit.

§ 41. PARALLEL WORKING WITH ALTERNATE CURRENTS.

Alternate-current working is seen at its best when the alternators as well as the transformers are coupled in parallel. The term "parallel working," indeed, is usually held to imply this particular arrangement of the generators, and does not refer to the transformers, although these are generally coupled in parallel as well. The ordinary practice is to drive each alternator separately. In Continental practice two alternators are frequently coupled, one at each end of a turbine shaft, the whole forming one unit. Machines of unequal size work perfectly in parallel, dividing the load in proportion to their capacities, if the excitations be properly adjusted and the driving power be of suitable amount. And alternators of even widely different pressures may be coupled in parallel without risk of damaging either, the resultant omnibus pressure being intermediate between the two. But the load will not necessarily be divided according to the capacities of the two; and the machine with the higher terminal pressure will do work on that with the lower pressure during a portion of each period. Therefore, the power absorbed will be greater than is necessary, and may be excessive. In fact, the difficulty with parallel working is to proportion the load according to the several capacities of the generators. It is necessary to measure, by means of a *wattmeter*, the power given out by each machine, and to adjust the excitation and the steam admission until the load is properly distributed. An ammeter is of no use, since it gives no indication of the lag between the current and pressure, and, of course, the omnibus pressure is necessarily the same for all the machines. As already mentioned in § 32, p. 179, the engines (or turbines) should

Q 2

not have separate governors, but the throttle valve of each should be first properly adjusted and then fixed. The main supply of steam (or water) should be controlled by a single governor, affecting all the prime motors. Further adjustment of output must be made by the field excitation, the safest condition of working being given, not by the minimum current (*see* Fig. 88, p. 177), but by pushing the excitation beyond the lowest point of the volt-ampere curve, so that a slight rise of speed will increase the *armature* current, and the converse will bring it to its normal value.

Alternate currents lend themselves readily to all of the systems of transmission and distribution discussed in §24 (p. 122) with reference to parallel continuous-current working. That is, they are adapted for—

(*a*) Simple parallel and reverse parallel systems, either direct or with transformers.

(*b*) Two-wire feeder systems, direct or with transformers; and

(*c*) Multiple-wire feeder systems, direct or with regulators or transformers.

It is in the transformer systems that they compare most favourably with direct currents, for dynamotors cannot seriously compete with transformers, either in prime cost, efficiency, or durability. It is important to examine carefully the use of transformers, since they are likely to play a leading part in all large power plants in the future.

It is sufficiently clear that high pressure is essential in the transmission mains from economical considerations; and it is also evident that the distributing pressure must not exceed about 500 volts. It has already been shown how these two conflicting conditions are met in continuous-current working by means of regulators, dynamotors, and other devices; all, however, having moving parts, and, therefore, requiring attention more or less continuously, being always liable to breakdowns from mechanical defects, and subject, also, to wear and tear from mechanical friction. It would be unfair to the continuous-current systems to lay too much stress on these points, but they undoubtedly must weigh heavily in the judgment of engineers when comparing these two important classes of electric machinery.

In alternate-current working the connecting link between the feeders and the distributors is the transformer, which has been fully described in Chapter V. Reference is there made to the structural details and to the fundamental features of design. It remains to investigate the functions of the transformer, and to see how admirably it meets the many and varied requirements of power and lighting work. It has been shown that the construction is of the simplest character, involving no moving parts, and comprising merely a laminated iron core and two separate windings of copper wire, the cable being protected by an iron case commonly filled with resinous oil. From the engineering point of view, it is as near perfection as possible, a machine of the highest efficiency without moving parts, and requiring no attention. If properly designed for the work expected of it, there is no reason why it should not last for an indefinite period without repairs of any kind.

In parallel working the transformers may be banked at sub-stations, and any number may be coupled up to the distributors according as the load varies, the necessary connections being made by an attendant on the spot; or the switching may be controlled by magnetic devices worked automatically or from the central station. The load factor of each transformer must be kept as high as possible, though the load factor of the station may vary largely. To attain this end it is necessary to select the transformer unit so that the minimum load is about 75 per cent. of the maximum output of one transformer. As soon as the full load of one transformer is reached, a second is switched in. With this arrangement no transformer will run with a load factor of less than 50 per cent., at which the efficiency is about 95 per cent. If the maximum load obtains for only short periods it may be advisable to have some larger transformers in addition to the small ones, but this is a matter of detail which must be settled on the merits of each case.

Another method, which was largely practised in the early days of alternate-current working (and is still used in commencing a new plant), is to place a transformer in the premises of each consumer. It has several disadvantages. In the first place, the prime cost of small transformers is necessarily more per unit of output than that of large ones ; and, secondly, the size of the transformer must be chosen with reference to the maximum, and not to the average, load ; hence the all-day load-factor will of necessity be small in most cases, often in lighting plants not averaging more than 10 per cent. When this system is carried out on a large scale for lighting, the result is disastrous as regards coal consumption, which may reach as high as from 20lb. to 25lb. of coal per kilowatt-hour at the terminals of the lamps (one Board of Trade unit delivered) ; whereas the best continuous-current three-wire systems consume from 6lb. to 8lb., and sub-station alternate-current plants not more than from 10lb. to 15lb. This disparity is due to the efficiency of transformers being comparatively small at low loads, and also because the exciting currents must be supplied whether the secondary circuits are closed or open. Hence the primary circuits carry large idle currents, which are wasted in heating them and the transformers. The power thus lost has been variously estimated at from 10 to 15 per cent. of the total station power. This drawback, however, is not likely to be so serious in the case of power work, for with large motors the load-factor will generally be high, and the transformers can always be switched out of circuit when the motors are standing, lamps (if used) being supplied from a separate transformer. And in a case requiring a number of small motors, such as that of a manufacturing town or a large factory, a transformer sub-station can be easily arranged to deal with both light and power work.

§ 42. INCREASE OF RESISTANCE IN CONDUCTORS CARRYING ALTERNATE CURRENTS.

With periodic currents it is found that the distribution of the current is not uniform over the cross-section of a solid conductor, but is limited to the outer layers of the metal. This *skin effect* is the more marked as the frequency is raised. A solid or stranded conductor, therefore, offers a higher resistance to an alternate than to a continuous current, and the apparent resistance exceeds the measured ohmic resistance by an appreciable amount.

This virtual increase of the resistance of conductors was first discovered by Prof. Hughes during his experimental researches, and was brought to the notice of practical men by Lord Kelvin in his Presidential Address to the Institution of Electrical Engineers in 1889. It is now customary to speak of the combined ohmic and skin effects in an alternate-current circuit as the *virtual* resistance of the circuit. Mr. Mordey has had special opportunities for investigating this important feature, and he has formulated some useful tables which give the practical limits of the effect. Lord Kelvin's figures are used in Table V, and for convenience a current density of 450 amperes per square inch of cross-section has been selected. This gives a loss of 1 per cent. per mile with 2,000 volts, and 1·15 per cent. per 100 yards with 100 volts, these being useful data for primary and secondary mains respectively. The Table also contains the limiting powers for the various sizes of conductors with this current density at 2,000 and 100 volts respectively. The skin effects are calculated for 80, 100, and 133 ∿ per second, which are usual in this country. It is seen that the frequency varies inversely as the square of the diameter of the conductor for the same percentage increase of virtual resistance over ohmic resistance.

Table V.—*Increase of Resistance of Conductors in Alternate-Current Working.*

Diameter in inches.	Area in sq. in.	Increase over ordinary resistance.	Current at 450 amperes p r sq. in.	Watts at 2,000 volts.	Watts at 100 volts.	∿ per second.
0·3957	0·122	Less than ¹⁄₁₀₀°	55	110,000	5,500	
0·5905	0·274	2½	133	266,000	13,300	
0·7874	0·487	8	2·0	440,000	22,000	80
0·9842	0·760	17	...	...	...	
1·575	1·95	63	...	...	...	
3·957	12·17	3·0 times	...	...	...	
0·3543	0·093	Less than ¹⁄₁₀₀°	45	90,000	4,500	
0·5280	0·218	2½	93·5	197,00	9,850	100
0·7086	0·394	8 %	178	356,000	17,800	
0·8826	0·611	17½	...	...	...	
0·3013	0·071	Less than ¹⁄₁₀₀%	32	64,000	3,200	
0·4570	0·164	2½	74	148,000	7,400	133
0·6102	0·292	8	131·4	263,000	13,140	
0·7622	0·456	17	...	...	...	

These figures show that the increase of resistance is practically insignificant with conductors of less than 0·35in. in diameter with frequencies up to 133 ∿ per second; that with cables of 0·5in. in diameter it is only 2·5 per cent. of the calculated ohmic resistance, while with diameters as large as 0·75in. it is as much as 17·5 per cent. This latter perhaps gives the practical limit of size for stranded conductors for alternate currents. Larger conductors may be built up from insulated strands or tubes, or strips may be used. Probably a fair increase of resistance for ordinary practice will be 10 per cent.; when, if the calculated ohmic resistance absorbs say 10 E.H.P., the virtual resistance will absorb 11 E.H.P.

Examining the output columns, it is seen that the skin effect is not of much importance with the primary circuits; for, with a pressure of 2,000 volts and 80 ∿ per second, 440,000 watts can be transmitted with an increase of resistance of only 8 per cent. over the ohmic resistance, and with 100 ∿ 356,000 watts, the former corresponding to 220, and the latter to 175 amperes. These powers (590 and 465 H.P. respectively) are probably sufficient for single mains or feeders, and for exceptionally long-distance transmission the power can be increased by simply raising the pressure. And if higher pressure be prohibited, the number of feeding points can oftentimes be increased. Thus there is nothing in the virtual resistance effect to hinder the transmission of enormous powers with solid or stranded naked conductors, so far as the feeders are concerned.

In the low-pressure distributing circuits the virtual resistance effect, however, may be of importance, because the watts that can be transmitted with a given percentage increase of resistance vary in inverse proportion to the pressure. Thus, in a 100-volt circuit the load (under the same conditions as referred to in considering the primary mains, i.e., 10 per cent. increase) is only 13,000 for 133 ∿, and 18,000 for 100 ∿. If the percentage of increase of resistance be raised, then it may affect the pressure of supply. But it is evident from Table V that the drop from this cause will not be serious in ordinary cases.

§ 43. VARIATION OF PRESSURE IN CONDUCTORS HAVING SELF-INDUCTION AND CAPACITY.

The next point to consider is the effect of variations of load on the drop. A bank of incandescent lamps is practically a non-inductive load, and the drop can be estimated without difficulty from the known ohmic resistance and the small increase of resistance in the mains due to the frequency. But if the load be partly or entirely composed of arc lamps or motors, or other devices containing self-induction, then there will be a further drop in the pressure due to a counter electromotive force, which will vary in magnitude with the current. The difficulty is met to some extent in lighting stations by running separate circuits for arc lamps. Motors are not yet used to any appreciable extent in alternate-current circuits, and when used are run chiefly by day, when the lighting work is comparatively unimportant. And in the few cases where power work is sufficiently large to warrant the expense, separate circuits are set apart for the motors. The author believes that most of the Continental stations which have tried the dual load have ultimately found it advisable to use separate circuits (at any rate for the larger motors). For, although a varying pressure may be unimportant, perhaps, as regards motors, it is a serious matter with arc lamps, and still more so with incandescent lamps. The effect of capacity in a circuit is precisely opposite to that of self-induction, the condenser electromotive force being phasially 180deg. distant from the electromotive force of self-induction. And since both of these effects are functions of the circuit, they may under certain conditions be made to counterbalance each other to some extent. The capacity effect is found in concentric conductors and in single or concentric mains armoured by continuous metallic sheathing, and it is worthy of consideration whether the latter are not of special use for circuits supplying alternate currents to inductive loads. At any rate they should tend to cause a smaller drop of pressure than ordinary copper conductors in an inductive circuit.

Some idea of the order of these effects in the mains may be gathered from the following.

First, consider a concentric conductor carrying an alternate current, and assume that the alternator and the motors have neither self-induction nor capacity. Let the terminal pressure of the alternator be V. Now the capacity

$$C = \frac{K\,l}{2\,\log_e \dfrac{R_1}{R_2}} \cdot \frac{1}{9 \times 10^5}\ \text{microfarads}, \quad . \quad . \quad (48)$$

l being the length of the two conductors, R_1 the radius of the inside of the outer tube, R_2 the radius of the inner conductor,

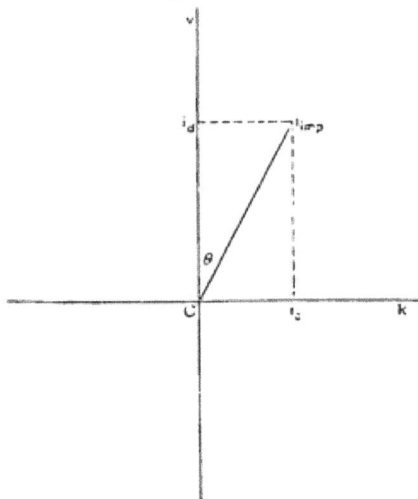

Fig. 106.—Showing the Effect of Capacity in an Alternate-current Circuit. O V is the impressed E.M.F.; O i_d, dynamic current; O i_{imp}, impressed current; and O i_c, condenser current; θ, angle by which resultant current leads impressed pressure.

and K the coefficient of specific inductive capacity of the dielectric in C.G.S. units. And the condenser current in practical units is

$$i_c = 2\pi \backsim V\,C\,10^{-6}. \quad . \quad . \quad (49)$$

i_c and V being effective values.

The condenser current is, as already stated, a quarter period in advance of the pressure. In Fig. 106, let O V represent V.

Then $O i_c$ represents i_c, $O i_d$ is the dynamic current, and, of course, is in phase with V. Through i_d, parallel to $O i_c$, draw $i_d i_{imp}$. At i_c in O K erect the perpendicular $i_c i_{imp}$, meeting $i_d i_{imp}$ at i_{imp} and join $O i_{imp}$. Then $O i_{imp}$ is the current that must be supplied by the alternator, and it is in advance of the impressed terminal pressure by an angle θ.

Fig. 107 is constructed in a similar way, and shows the lag ϕ due to self-induction.

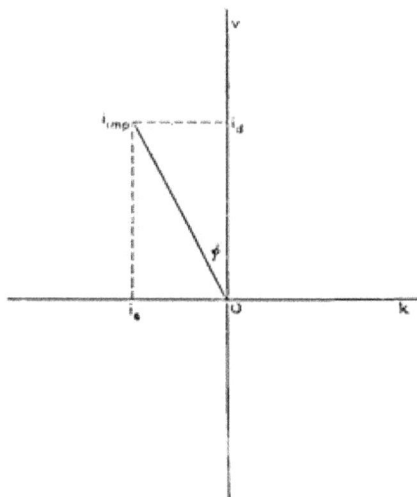

Fig. 107.—Showing the Effect of Self-induction in an Alternate-current Circuit. O V is the impressed E.M.F.; O i_d, dynamic current; O i_{imp}, impressed current; and O i_c, current due to E.M.F. of self-induction; ϕ, angle by which impressed current lags behind impressed pressure.

The result of capacity, then, is to cause the impressed current to lead the line pressure, while self-induction causes it to lag. This phase-difference introduces several important results. If it be positive, the terminal pressure of the generator will be raised, because the armature reactions will then tend to strengthen the field. This is seen to be the case by considering that the maximum pressure obtains when the armature coils

are midway between the field poles ; and therefore, if the maximum value of the current obtain before this position is reached,
the coils will help to magnetise the field magnets ; and, on the
contrary, if by reason of self-induction the current lags in phase,
it will help to decrease the magnetism.

The effects of self-induction and capacity with concentric
cables may be shown graphically, when the constants of the

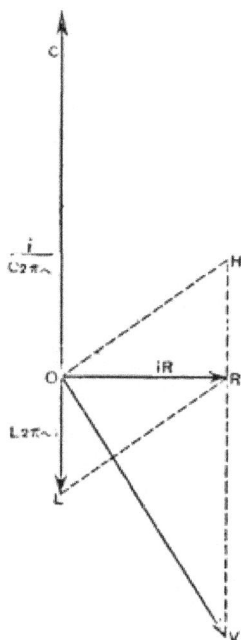

Fig. 108.—Showing the Effect of Capacity and Self-induction in a Circuit.

circuit are known, by combining the diagrams in the preceding
figures. And by suitable approximations the effects for a complete power plant may be estimated in the same manner. The
procedure is indicated in Fig. 108.

O C is the maximum value of the difference of pressure
between the condenser plates $= \dfrac{i}{C \, 2\pi \, \curvearrowleft}$, O L is the maximum

value of the counter pressure $= L i 2\pi \backsim$. O R is the resultant pressure which sends the current i through the inductive circuit of impedance $\sqrt{R^2 + L^2 4\pi^2 \backsim^2}$. The pressure at the terminals of the condenser is H V by the construction (for R H = O L, and R V = O C − O L). Then O V represents in phase and magnitude the impressed pressure required to maintain the current through the condenser.

The condenser effect is a maximum when the angle H O V is a right angle; or H V, the condenser pressure, has a maximum ratio to O V, the impressed pressure, when this is the case.*

In practice, with the insulations commonly used and the sizes of conductors most in vogue, the capacity of concentric mains between the inner and outer conductors appears to be about one-third of a microfarad per mile run. But the effect is far more serious between the outer conductor and the earth, being nearly ten times greater. It is, therefore, important, with high frequencies and high pressures, to take into consideration the capacity effects as well as the self-induction.

The impedance of a circuit carrying an alternate current is dependent upon the resistance, capacity, and self-induction of the line, alternators and load, and may be also affected by the mutual induction of a neighbouring circuit.

Self-induction and capacity in alternators and motors have already been referred to. In a long line the self-induction of the conductors may have an important effect upon the impedance. It gives rise to a counter E.M.F. which varies with the diameter of the conductors, the distance between their centres, and the intensity and the frequency of the current. The value in volts per 1,000ft. of double circuit per ampere is shown graphically in Fig. 109, which is adopted from the calculations of Mr. Chas. F. Scott.

The constants in the abscissæ have been calculated for a frequency of one $\backsim$ per second. The value for any frequency is thus found by simple multiplication. For example: Given a 100-ampere circuit at a frequency of 60 $\backsim$, supplying power

* *Vide* "The Alternate-Current Transformer," by Dr. J. A. Fleming. Vol. II., p. 386.

Distance between Centres.   Inches.

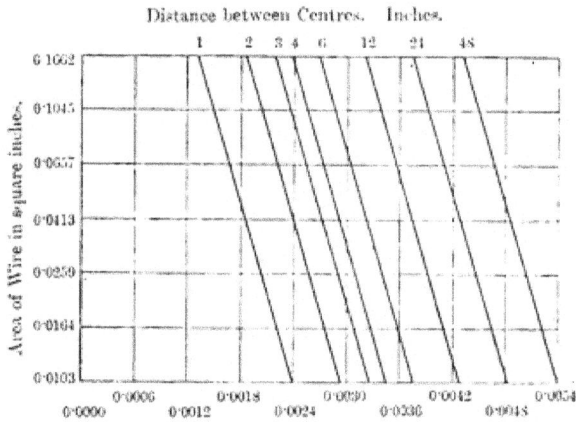

Fig. 109.—Inductive E.M.F. in Volts per Ampere per 1,000ft. of double circuit at 1 ∿ per second.

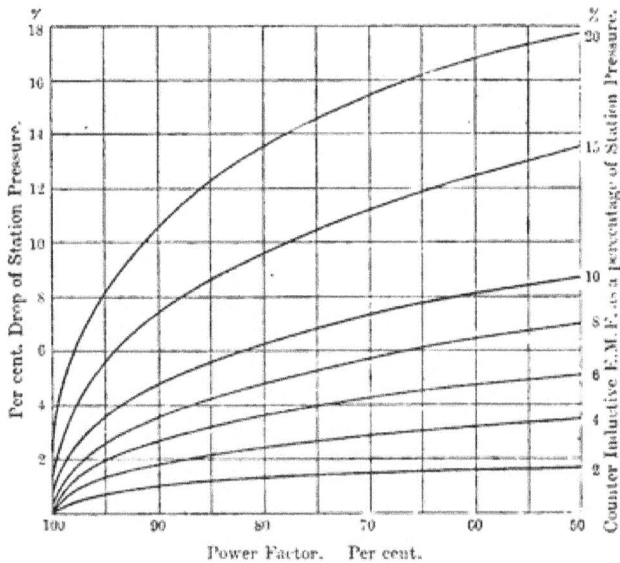

Power Factor.   Per cent.

Fig. 110.—Inductive Drop (as a percentage of Station Pressure) for different E.M.F.s and different Power Factors of Load.

to a distance of 500ft. with conductors of 0·1662 sq. in. in section, and 6in. between centres. The constant is 0·0027, and the inductive E.M.F. $= 0.0027 \times \dfrac{100}{1} \times \dfrac{500}{1,000} \times 60 = 8.1$ volts.

It is seen that the counter E.M.F. is least with small conductors lying as close together as possible. This condition is

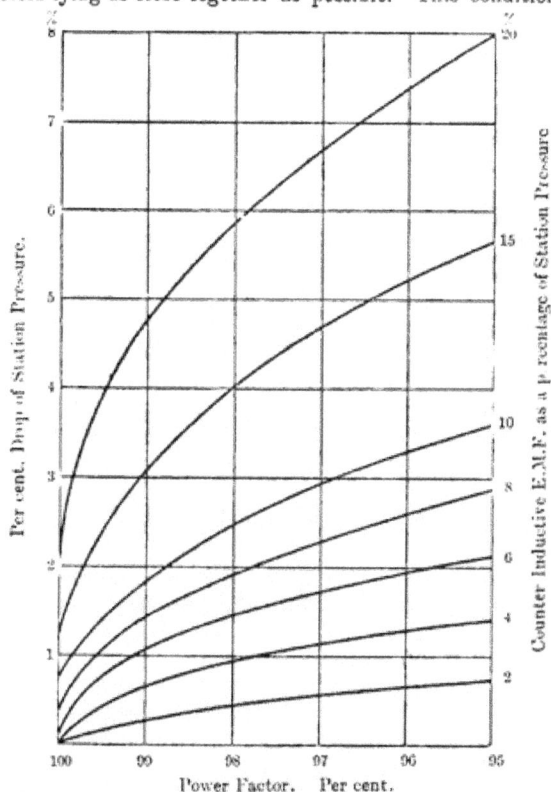

Fig. 110A.—Enlarged view of the commencement of Curves in Fig. 110.

best secured by using a concentric cable, when the effect will be reduced to a minimum.

From these curves it is possible to estimate the counter E.M.F. in most conductors with a fair degree of accuracy. To

predetermine the inductive drop of pressure between the generator and receiver terminals, it is necessary to know the power
factor of the circuit.   Then the inductive drop for any load
can be found by reference to Figs. 110 and 110A (also due to
Mr. Chas. F. Scott), after the counter E.M.F. is determined
from Fig. 109.

   In these two figures the counter E.M.F. and the drop due to
the self-induction of the line, alternators and load, are given as
percentages of the supply pressure.   The total drop on the
line is found by adding to this figure the fall of pressure due
to the virtual resistance of the conductors, as given in
Table V. p. 219.

   If two or more circuits are lying in close proximity to each
other, there will be mutual induction between them, i.e.,
they will act with reference to each other as the primary and
secondary circuits of a transformer.   This effect is noticed in
overhead lines when two or more circuits are coupled to
separate alternators, and is evidenced by a surging of pressure
sometimes sufficient to cause an easily discernible fluctuation
in the light of incandescent lamps at the end of long circuits
parallel to each other.   This is also the case with three-wire
circuits (either two- or three-phase).   The E.M.F. induced by
two circuits upon one another is not affected by the distance
between the conductors in each ; but simply depends upon the
relative positions of the two circuits, the intensity and direction
of the two currents, and the distance for which they are parallel
to each other.

   Some of the usual arrangements of a pair of two-wire
circuits, with the mutual induction effects measured in volts
per 1,000ft. of double circuit per ampere at frequencies of
133 and 60 ∿ per second, are given in Fig. 111.

   These data have been determined by Mr. Chas. F. Scott.   The
wires of the one circuit are marked by black dots, and those
of the other by rings.   The volts given refer to the E.M.F.
induced by each circuit in the other, and will be positive or
negative according as the two currents are in opposite or

the same directions. The diagram is directly applicable to a single-phase system or to a two-phase system with four wires, and can be readily modified to suit a three-wire system.

Mr. A. E. Kennelly has given much attention to the impedance of mutually inductive circuits, and has developed an elegant geometrical treatment, which is applicable to most practical problems, even when they involve resistance, self- and mutual-induction, and capacity. The subject is not suitable for discussion here. The student is referred to the technical Press for further information.*

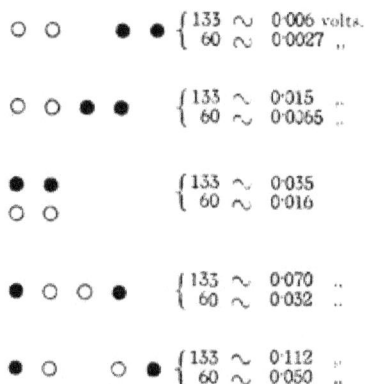

$$\circ \quad \circ \qquad \bullet \quad \bullet \quad \begin{cases} 133 \;\sim\; 0.006 \text{ volts.} \\ 60 \;\sim\; 0.0027 \text{ ,,} \end{cases}$$

$$\circ \quad \circ \quad \bullet \quad \bullet \qquad \begin{cases} 133 \;\sim\; 0.015 \text{ ,,} \\ 60 \;\sim\; 0.0065 \text{ ,,} \end{cases}$$

$$\begin{matrix} \bullet \quad \bullet \\ \circ \quad \circ \end{matrix} \qquad \begin{cases} 133 \;\sim\; 0.035 \\ 60 \;\sim\; 0.016 \end{cases}$$

$$\bullet \quad \circ \quad \circ \quad \bullet \qquad \begin{cases} 133 \;\sim\; 0.070 \text{ ,,} \\ 60 \;\sim\; 0.032 \text{ ,,} \end{cases}$$

$$\bullet \quad \circ \qquad \circ \quad \bullet \quad \begin{cases} 133 \;\sim\; 0.112 \text{ ,,} \\ 60 \;\sim\; 0.050 \text{ ,,} \end{cases}$$

Fig. 111.—Showing the number of volts per ampere per 1,000ft. of double circuit due to mutual-induction between two circuits in various arrangements. The wires of the two circuits are severally distinguished by rings and black dots.

§ 44. EFFECT OF CAPACITY UPON THE CHANGE RATIO OF TRANSFORMERS.

When transformers are used for testing alternators for insulation resistance, it is sometimes noticed that the pressure at the terminals of the transformers is far in excess of that due to the ratio of conversion. This is due to the capacity of the alternator. The effect is of some importance in practice. The theoretical principles on which it depends are difficult to

---

* Especially *The Electrician*, October 27, 1893 ; and the *Transactions* of American Institute of Electrical Engineers for April, 1893.

R

express without the use of coefficients and complicated equations. But some idea of the order and magnitude of the

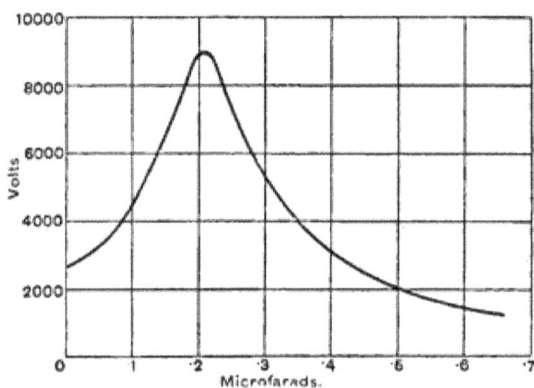

FIG. 112.—Curves showing the Variation in Pressure in the Secondary Circuit of a Step-up Transformer, by varying the Capacity in the Secondary Circuit. The experiments made with a No. 2 Siemens Transformer, and Current obtained from a W1 Siemens Machine, frequency 100~. The Exciting Current kept constant, and Alternator kept running at 750 revolutions during all the experiments.

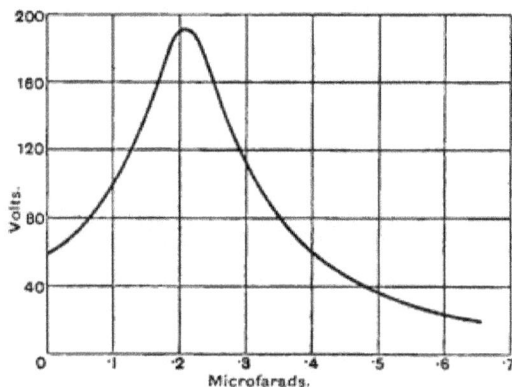

FIG. 113.—Curves showing the corresponding Variation in Pressure in the Primary (Low Pressure) Circuit in the same Transformer, by varying the Capacity in the Secondary Circuit.

increase of ratio of conversion will be gathered from the curves in Figs. 112, 113, and 114, taken from "The Alternate-

Current Transformer in Theory and Practice," by Dr. J. A. Fleming, Vol. II., pp. 396–398.

The point to be specially remarked in connection with power work is that long concentric conductors have considerable capacity (about 0·33 microfarad per mile run), and hence may readily cause effects of the order shown in the experiments. They will be most marked with step-up transformers of large size. It is also important to notice (see Fig. 114) that there is a critical value of the capacity for the given self-induction of the circuit which will produce a maximum rise of pressure, and that the ratio of conversion is steadily increased as the capacity increases (to in this case 0·65 of a microfarad).

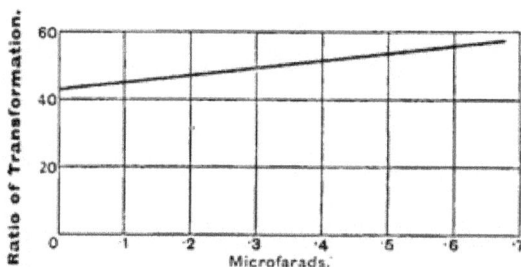

Fig. 114.—Curves showing the Dependence of the Ratio of Transformation of Pressure between Primary and Secondary Circuit on Capacity of Secondary. Deduced from the experiments illustrated in Figs. 112 and 113.

It is, therefore, theoretically possible to compensate for the drop of pressure at the far end of a feeder by putting a suitable capacity in the circuit.

§ 45. COMPENSATING FOR DROP OF PRESSURE IN CONDUCTORS.

There are several methods in use at present for compensating the variations of pressure in the feeders or in the transmission mains. The most common consists in simply adjusting the excitation by hand or automatically, the generators being separately excited; another consists in compounding the alternators in a manner similar to that adopted with dynamos, the alternators being either self-excited or else partly self-excited and partly separately excited.

R 2

Hand regulation by a resistance in series with the field-coil circuit is easily understood, and there is no need to make any special reference to automatic devices whose function is simply to vary the resistance of the exciting circuit.

The Brush Company, Messrs. Ganz and Co., Messrs. Easton, Anderson and Goolden, and others, make automatic regulators of this description, and the author has spent much time in designing similar devices, but in his opinion they cannot be regarded as complete successes, and certainly are not adapted for use in power plants where there is no skilled supervision. Compound winding, however, is a practical device. By

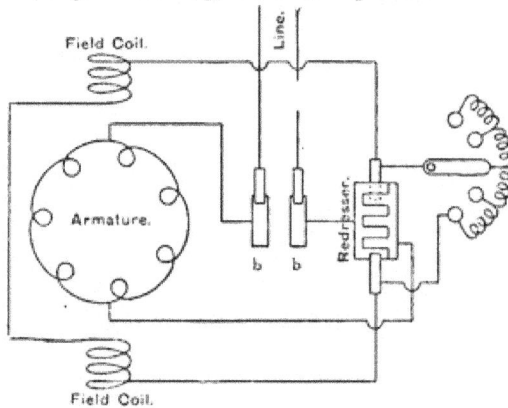

FIG. 115.—Thomson-Houston Method of Compounding Alternators.
b b, Collector rings.

means of it alternators can be made to compensate for any predetermined drop, and therefore to meet most of the regulation difficulties in the feeders. The arrangement adopted by the Thomson-Houston Company is to partly excite the field from a dynamo with a steady current and to provide a variable excitation by redressing a portion of the main current and passing it through a pair of coils opposite to each other, or through coils on each of the field poles. It is clear that, since the armature current is redressed, the excitation due to it will be proportionate to the load.

The arrangement is shown in Fig. 115. The two main collecting rings are shown at b b. The "redresser" consists of

two circular gun metal castings, having each half as many
projections as there are pole-pieces in the field; both being
insulated. One of these is permanetly connected to one of
the collecting rings and the other to the free end of the
armature coils as shown. Two brushes, resting on projections
belonging to opposite rings of the redresser, form the terminals
of the exciting circuit. To reduce sparking at the instant of
short-circuiting the field coils when the brushes rest on the
same rings, a suitable resistance is placed in shunt to the field,
and this serves in addition to control the proportion of current
in the exciting coils, and therefore to vary the compounding
within the limits of the design. This is a very complete and
convenient device. It is characteristic of the methods generally

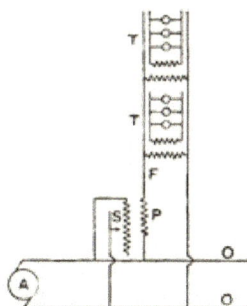

Fig. 116.— Kapp's Method of Regulating the Drop of Pressure at the end
of a Feeder.

adopted by Messrs. Ganz and others. Alternators thus compounded will work in parallel as readily as separately-excited
machines.*

Another very simple plan is to place the primary of a
suitable transformer in series with the feeders, and to excite
the compound coils from its secondary current, redressed.
Since this current is proportionate to the armature current,
it is clear that the compounding can be arranged to any slope
of characteristic.

A device patented by Mr. Kapp is interesting in this
connection. It is shown in Fig. 116.

---

* See letter by Mr. E. M. Mix, *The Electrician*, March 23, 1894.

One coil, P, of a small transformer is placed in series with the feeder, and the other coil, S, is coupled across the station mains. By inserting more or less of the coil S in the circuit by a shifting contact, a few auxiliary volts are put into the feeder by the inductive action of the shunt-coil. This regulator may be designed to add a sufficient number of volts to compensate for any desired drop. It may be applied to each feeder independently, although all the circuits are working off one alternator. This is a distinct advantage, and makes it a most convenient device for power plants. The Westinghouse Company use a similar arrangement, which they call a "booster."

§ 46. MEASUREMENT OF PRESSURE AT THE END OF FEEDERS.

The pressure at the termination of the alternate-current feeders may be read by means of two transformers, without the aid of pressure wires run back to the station. The general

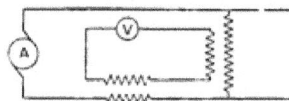

FIG. 117.—Diagram of Station Voltmeter Connections for Reading Pressure at the far end of Feeder.

arrangement is as follows :—The voltmeter is in series with the secondary circuits of two transformers, one of which has its primary in series with the feeder, and the other its primary as a shunt across the feeder. It is illustrated in Fig. 117.

The pressures of the two transformers are opposed to each other. The secondary pressure of the transformer in shunt is proportional to the station omnibus pressure, V ; and that of the transformer in series to the primary current strength i_m, and therefore to $i_m R$, if the resistance be constant. The voltmeter will consequently read proportionately to $V - i_m R$, which is the expression for the pressure at the far end of the feeder. This principle, with important modifications, is employed by Messrs. Ganz and Co. Its application is shown

in diagram in Fig. 118, the adjustable resistance in the
exciter circuit being also shown. The feeders are shown
at L_1 L_2, the shunt transformer at T_1, and the series
transformer at T_2. The secondary circuit of T_1 is closed
through the solenoid of the regulator S, the resistances R_3, R_1
and R_2. The secondary of T_2 is closed through the resistance
R_2. The transformers are joined up so that their electromotive
forces are in opposition as regards producing a current in the

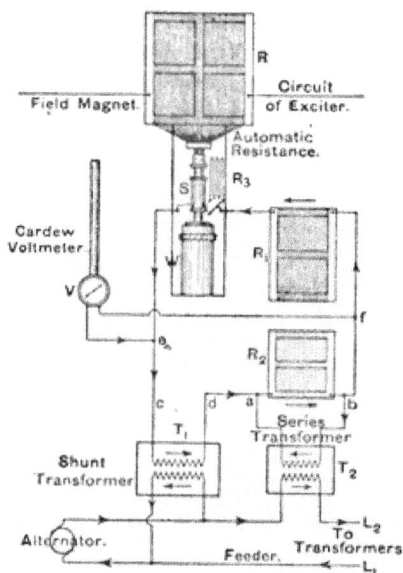

FIG. 118.—Ganz and Co.'s Method of Compensating for Drop in Feeders.

solenoid S. The terminal pressure of T_1 is evidently proportionate to the station pressure. The difference of pressure
between the points a and b, the terminals of R_2, will be proportionate to the feeder current; and so the pressure between
the points e and f will be equal to $V - i_m R$, the pressure at the
termination of the feeders. The intensity of the current in the
solenoid circuit will determine the excitation, and hence the
pressure at the sub-station can be kept constant for variation

of drop. It will be seen that a Cardew voltmeter is used to
read the primary pressure. This is made possible by a suitable
ratio of conversion in the transformers. It will be evident that
these methods of reading the sub-station pressure are only
applicable when the virtual resistance of the feeders is
practically constant, or varies according to some easily determined law, in which case the voltmeter can be specially
calibrated to include these effects.

§ 47. COMPENSATORS, OR REGULATORS.

The Thomson-Houston Company use, in particular cases, a
compensator, whose functions are similar to those of the "regulator" referred to on page 135, with two-wire continuous-
current feeders and multiple-wire distributors. This may be of
use for power work in cases where the regulation of the supply

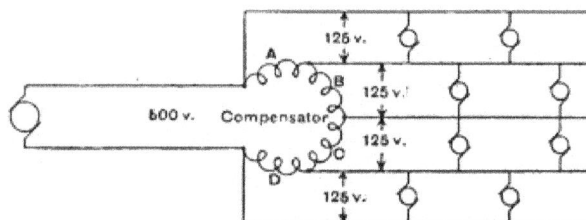

FIG. 119.—Prof. Elihu Thomson's "Compensator" System applied to
Power Circuit.

pressure presents difficulties. The plan of connection is shown in
Fig. 119, the "compensator" being simply a choking coil, with
its circuit divided into as many parts as there are sub-circuits.
Each of the distributing circuits is assumed to be at 125 volts.

When the load is evenly divided between the circuits the
compensator will carry practically no current, owing to
the impedance of its coils, which are wound round soft
iron wire. But if the load be removed from one of the four
circuits, the compensator coil, say coil A, in parallel to it will
act as the primary of a transformer, and induce in the remaining coils of the compensator, B C D, a secondary current which
will act in the outer circuit in the same direction as the

main current.   The coils B, C, D will supply $\frac{i}{4}$ amperes, i being the current for two motors (assuming the circuits to be equally loaded), and the alternator will supply $\frac{3}{4} i$ amperes. If the two circuits A and B are opened, then the coils C and D will supply $\frac{i}{2}$ amperes, and the alternator an equal number.

The regulating capacity of the compensator is, therefore, seen to be in this case equal to four times the current it is designed to carry.   It is evident that small differences of load between the various circuits are readily compensated by this ingenious device, which may be defined as a combination of a transformer and a direct system of distribution.

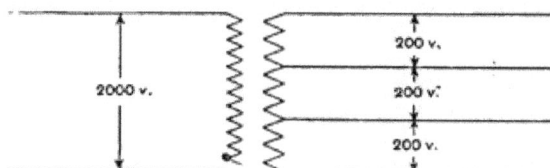

Fig. 120.

As far as the author is aware, this method of multiple-wire distribution with alternate currents has not been applied to power work, although it has been used to some extent for lighting in America.   The evident objection to it is that it brings the high pressure of the primary circuit into the consumers' premises, and, therefore, constitutes a source of danger.   A better arrangement for power work, when it is desired to use a multiple-wire distributing system, is to divide the secondary circuit of a large transformer at the sub-station into as many circuits as may be required, and at suitable pressures for the work (they need not be of the same value)—*see* Fig. 120.   This separates the high-pressure mains from the distributing circuit, and the total secondary pressure can be made of any desired value.   A transformer thus loaded will work as well as if the secondary circuit were not divided, and it will maintain the same ratio of pressure between the distributors, neglecting the drop due to the resistance, if the primary circuit, as a whole, adjusts the power according to the output. Difficulty, however, may be found if more than one transformer be used, unless both the primaries are coupled in parallel, and also both the secondaries; or, better still, if the primaries are coupled in parallel and the secondaries in series. This latter arrangement requires, for four-wire distributors, three transformers, the middle wires being coupled to the junctions of the secondaries.

It must be understood that these proposals offer no advantages as regards the transformers: they simply permit the use of a multiple-wire distribution, which is a point of some importance in power work.

§ 48. COUPLING SEPARATELY-DRIVEN ALTERNATORS IN PARALLEL.

Before an alternator is coupled in parallel it should be ascertained to be as nearly as possible in step with those already in circuit. The practical methods of judging this condition vary with the types of machines and their relative sizes and speeds, but in most cases there is no difficulty. If the alternators are of similar make, all that is necessary is to raise the pressure to the omnibus voltage, and to notice that the frequencies are approximately the same before closing the parallel switch. At the instant of making circuit there may be a comparatively large rush of current, and then the bank of alternators will drop into step, and the load may be adjusted between them, if necessary, by their excitation. The author has seen Mordey and Gülcher alternators put parallel without further precautions; but, of course, the lamps pulsated until the machines were in step.

It is customary, however, in most lighting stations, to use a phase indicator. This usually consists of two transformers, one of the primaries being coupled to the alternator running the load, and the other to the incoming machine. The secondary circuits are coupled in series with two incandescent lamps of suitable pressure. The arrangement of details is as shown in Fig. 121.

The procedure is usually as follows :—The incoming alternator, A_2 say, is run up to proper speed, and its field excited so that it gives the terminal pressure of the omnibus bars. Sometimes it is coupled to the station lamps or an artificial resistance, but this is not essential in most cases. The synchroniser switch W is then closed, and the lamps l_1 and l_2, being fed with the joint currents from the transformers T_1 and T_2, will be black or luminous according as the alternators A_1 and A_2 are in similar or opposite phase. The instant the two are in step the lamps will become steadily black or luminous (according to the coupling up of the secondaries), and then the main switch m_2 may be closed without disturbing the lamps on the outer

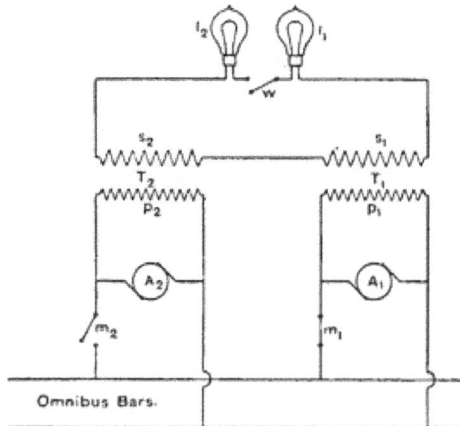

FIG. 121.—Diagram of Phase Indicator, or Synchroniser.

circuit. During the short interval of time required for the machines to get into step, the lamps pulsate with luminous "beats" as the lagging machine overtakes the leading one. The synchroniser can be coupled successively to any pair of the alternators.

It is found convenient with the Ganz and with most of the Continental alternators to run the incoming machine on an artificial load to approximately the same current as it is required to give on the main circuit before using the synchroniser and closing the main switch. This precaution seems to be only necessary with armatures of large self-induction, in which the terminal pressure varies considerably with changes of current. In England it is not customary to use an artificial load.

The Thomson-Houston Company insert a variable impedance coil in the armature circuit of the incoming alternator in order to check any in-rush or out-rush of current. The method of using these impedance coils with the synchroniser is shown in Fig. 122.

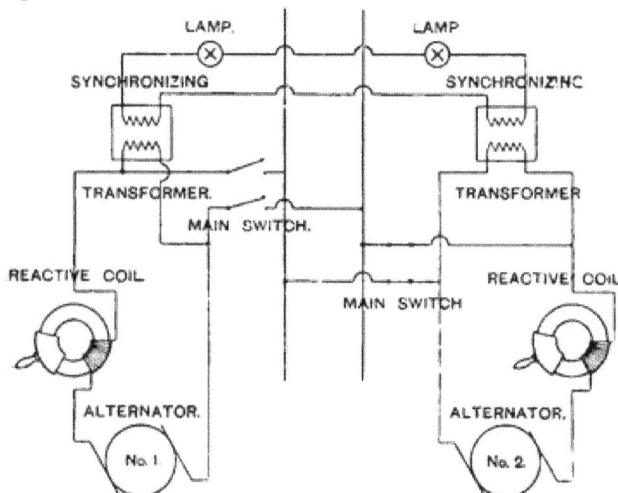

Fig. 122.—Arrangement of Impedance Coils used in putting Thomson Alternators in Parallel.

The impedance coil is shown separately in Fig. 123. It consists of a laminated iron ring, with a few turns of well-insulated wire wound round about one-eighth of the circumference. A copper sheath is arranged so as to be capable of rotation about the centre. The sheath is of such width as just to surround the coil. When the winding is covered, the device corresponds to a transformer with the secondary circuit short-circuited, and the impedance is then very slight ; as the sheath is gradually removed the impedance increases ; when the stationary coil is entirely uncovered the device has its maximum effect. In some forms the impedance coil is short-circuited when the sheath completely encloses it. These impedance coils are useful for balancing loads between machines and for varying the pressure of feeders; they generally act the same as rheostats with continuous currents, but have the advantage of consuming less power.

Parallel working is without doubt assisted by using a moderate rate of frequency, say not more than about 100 $\backsim$; but there is no gain in reducing it below about 50 to 60 $\backsim$. It is also necessary to synchronise the engines as well as the alternators if the running is to be perfect. This points to the use

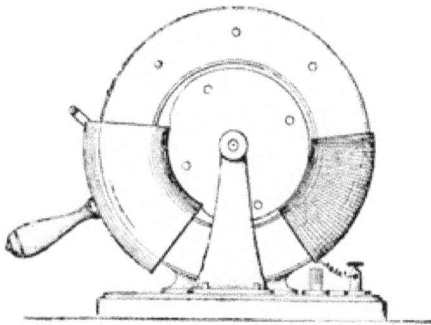

Fig. 123.—Impedance Coil, or Dimmer.

of high-speed engines with flywheels and moving parts of small inertia, so as to readily lead or lag as the conditions of working may require. For the same reason the engine should not be governed directly, as the better the governor the greater the difficulty of varying the rate of speed. The station governor should be designed to control the steam admission to the bank of engines, and the load should be distributed proportionately by varying the excitation. A "trip" governor should be used on all engines of this class, to check racing in the event of a belt breaking or slipping off, a fuse blowing, or other similar accident

§ 49. ALTERNATE-CURRENT MEASURING INSTRUMENTS.

It has already been stated that the energy of an alternate-current circuit is proportional to the $\sqrt{\text{mean square}}$ value of the instantaneous values of the current, and it has also been stated that this $\sqrt{\text{mean square}}$ is equivalent to that of a continuous current that will produce the same heating effects. Therefore, alternate-current measuring instruments are calibrated by direct currents, and the readings on the scales are accordingly proportional to the $\sqrt{\text{mean square}}$ of either continuous or alternate currents.

A few hints of a practical nature will assist engineers in selecting suitable instruments.

Some of the most practical forms of instruments in general use are tabulated in Table W, the names of typical designs being given.

Table **W.**—*Examples of Practical Instruments suitable for Power Stations.*

—	Voltmeters.	Ammeters.	Wattmeters.
Electro-dynamic	...	Siemens' Dynamometers.....	Kelvin, Swinburne
Electro-magnetic	Elihu Thomson, Evershed, Dolivo, Ayrton & Perry	Evershed, Elihu Thomson, Kelvin, Ayrton & Perry, Dolivo	...
Electro - thermal	Cardew, Holden	Holden, Ayrton and Perry ...	...
Electro-static ...	Kelvin's Multicellular, Swinburne, Ayrton & Mather ...	...	...

It is desirable, if possible, to have direct-reading pressure instruments, which are directly applicable to the circuits, without non-inductive resistances in series with them, both because the details are simpler and also because the energy wasted in the resistance is often considerable. For this latter reason electro-thermal, or hot-wire, instruments, although very convenient for reading low pressures, say up to 150 volts, are objectionable if kept in circuit. For high-pressure work the most practical instruments are of the electro-static type, for they absorb scarcely any power.

If hot-wire instruments are used to read high pressures, it is advisable to have a voltmeter transformer with a suitable ratio between the primary and secondary. Thus, with a 2,000-volt primary and a 100-volt secondary, a convenient ratio of conversion will be $\frac{1}{20}$. This transformer method is more expensive in first cost than direct-reading with a multicellular voltmeter, and it also wastes more power, probably absorbing about 30 watts in the hot-wire and 30 in the transformer.

Current-measuring instruments are generally kept in circuit, and hence it is important that they should waste little power. For reading large currents electro-magnetic instruments of low resistance are perhaps as good as any of the other types. The electro-dynamic method has the disadvantage of requiring mercury contacts. The hot-wire principle is used with success for measuring current, but the instruments are necessarily delicate, and the author does not advise them for power work as a rule. The electro-chemical principle is not generally applicable for ordinary commercial purposes.

Ohmmeters are useful when erecting station plant, but they they are not necessary for every-day work.

Wattmeters, or power meters, are of the utmost importance in all stations where economy of prime power is an object. One should be placed in the circuit of each alternator, to indicate the power given out by the machine. This enables the attendant to divide the load between the several alternators according to their capacity, and thus to work with the smallest excita tion. One of the best forms of this instrument for station use is Lord Kelvin's engine-room wattmeter. The appearance of the interior of the instrument is shown in Fig. 124. It has a main circuit of a double rectangle of copper rod. The pressure coils are made of fine wire wound in the shape of a pair of spectacles, and are in series with a non-inductive external resistance.

The general arrangement of the movable shunt coils, the suspension springs, and the gravity adjustment for calibrating are shown in Fig. 125. Each of these shunt coils has about 1,000 turns of insulated wire of approximately 1,000 ohms

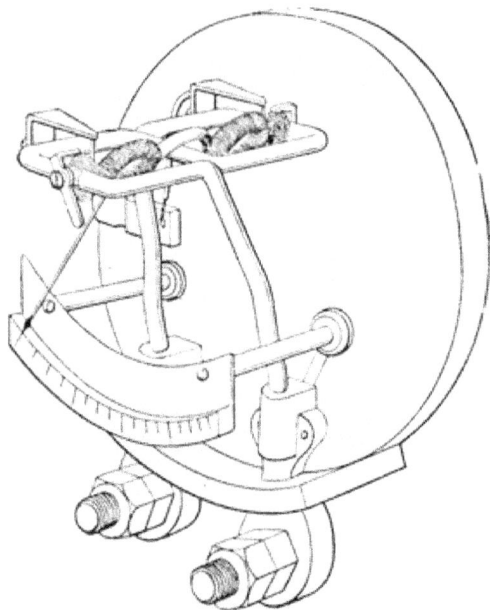

FIG. 124.—Lord Kelvin's Engine-room Wattmeter. General view of instrument, with case removed.

resistance. The scale has nearly uniform divisions, and is graduated to read directly in watts or kilowatts, as required.

Siemens' dynamometer-wattmeter is also a practical instrument, but as usually designed absorbs more power than Lord Kelvin's.

If a wattmeter be made recording it answers the purpose of an ergmeter, for the load curve is easily integrated by a planimeter, and thus a continuous record of the station output may be kept without much expense or trouble.

It is not necessary to put the full primary pressure on the terminals of the shunt-coil of the wattmeter; a suitable transformer may be used to reduce the pressure to a convenient fraction, and hence a small movable coil, carrying a mere trace of current, may be used, and will add considerably to the accuracy of the instrument.

Ergmeters, or energy meters, are not yet in common use, and are not likely to be much used for power purposes.

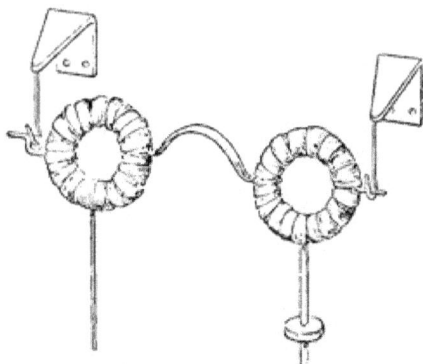

FIG. 125.—Lord Kelvin's Wattmeter. Fine-wire Shunt Coils, showing details of suspension. Controlling springs removed.

§ 50. PRECAUTIONS TO BE OBSERVED IN ALTERNATE-CURRENT WORKING.

Owing to induction and capacity effects, it is usually necessary in alternate-current circuits to observe certain precautions in starting and stopping alternators, in opening and closing circuits, and in adding or withdrawing feeders from omnibus bars. It must be recollected, however, that these effects are largely the result of the self-induction of the alternators, and that generally there will be no difficulty when the armatures have comparatively small reactions, as is the case with the Mordey alternator, for example.

s

The chief danger to be apprehended is a sudden variation of pressure, which may be sufficiently large to break down the insulation of the circuit at one or more points or to seriously affect the load. This is due both to the change of conversion ratio in the transformers and to the variations of impressed pressure in the alternators.

The precautions to be observed are :—

Never open a high-pressure circuit containing self-induction or capacity, without first slowing the alternator or weakening its excitation, or reducing the load by an impedance coil.

Never switch an alternator into parallel with another machine without first adjusting its excitation so that the *coupled* alternators will give the proper omnibus pressure.

Never switch an alternator out of parallel without adjusting its excitation (and that of the remaining machines if necessary) so as to leave the omnibus pressure unaffected.

An alternator of large self-induction requires to be excited to a higher pressure than that of the omnibus bars before coupling in parallel, as the terminal pressure will fall considerably when the circuit is closed. In taking such a machine out of parallel the excitation may be reduced until the current is nearly zero; or the speed may be lessened until the same end is attained. The switch under these conditions may be opened without a dangerous spark.

If, however, the conversion ratio of the transformers be affected by the capacity of the circuit (as occurs with step-up transformers and concentric mains), then the incoming alternator with large self-induction must be under-excited, in order to bring it into such a condition that it can be coupled to the active machines without affecting the pressure at the far end of the feeders. And, conversely, in taking it out of parallel the excitation must be raised a little, in order to avoid serious sparking on opening the circuit. These effects have been noticed in practice, and are explained by an alteration of the capacity and self-induction of the circuit affecting the conversion ratio of the transformer.

§ 51. EARTHING THE PRIMARY AND SECONDARY CIRCUITS OF TRANSFORMERS; SAFETY DEVICES.

In transformer work there is a chance of the primary pressure invading the secondary circuit. If this occurs there may be danger to life. And this is the more serious, since, in most cases, the leakage is entirely unsuspected by the consumer. To obviate this danger various arrangements have been suggested, and some have met with the approval of the Fire Insurance Offices and the Board of Trade. The safest plan, perhaps, is to ground one of the primary cables close up to the station (as is done by the London Electric Supply Corporation), and to place between the secondary circuit and earth a device which will ground the secondary circuit the instant the high-pressure current enters the consumer's circuit. This will cause a large primary current to flow, and hence the primary fuses will blow and cut off the premises. The danger to life is entirely obviated, unless the transformer or other device has a large electro-static capacity, but this is never the case with ordinary transformers.

There are three protective devices in general use: Major Cardew's "Mouse Trap," Mr. Kent's metallic sheath, and the Thomson-Houston film cut-out.

The Cardew safety device is shown in Fig. 126. It consists of two brass plates placed near together, but insulated from each other. Between them is a strip of aluminium foil attached by ebonite pins at one of its ends to the bottom plate. One plate is connected to earth, and therefore to one of the primary mains, and the other to the secondary circuit. As soon as a leak occurs between the primary and secondary circuits of the transformer there is an electro-static pressure between the plates, and when this equals a definite number of volts the free end of the foil is attracted and makes contact with the top plate. This grounds the secondary circuit, and the primary current blows the primary fuses.

Mr. Kent's arrangement consists of an earthed metal ring, properly slit to avoid eddies, placed between the primary and

s 2

secondary windings.   Any fault of insulation in either circuit
can be made to ground the coils and blow the main fuses.

The film cut-out of Prof. Thomson is merely a piece of prepared
paper between two metallic contacts.   This paper is pierced
when the pressure between the contacts exceeds a predetermined amount.   Thus the paper insulation will easily stand
the secondary pressure of, say, 100 volts, but will break down
instantly it is subjected to the high pressure of the primary.
In practice one of these film contacts is connected to each of
the secondary mains.

FIG. 126.—Major Cardew's Earthing Device, for protecting Transformer
Circuits.

Mr. Ferranti has devised a very ingenious method for protecting transformer circuits from the risk of partial earth
(always a source of danger, since a second ground on the same
circuit will cause a dead short-circuit on the secondary, and
may cause a fire).   This method requires the middle of the
secondary circuit to be permanently earthed.   It is perhaps
more useful at present as a testing device for partial grounds,
as the Fire Offices' regulations do not generally allow any part
of the consumers' circuit to be earthed.   But for power
plants, where these restrictions do not obtain, the device is
most useful.   The arrangement is shown in Fig. 127.

It will be seen that there are two small transformers across the secondary mains, with a ground wire from their point of junction. Their pressures are arranged in opposition, and normally no current flows through them. If, however, another earth, full or partial, is made in one of the secondary mains, the corresponding transformer will cease to produce current, and, the balance of pressure being upset, a current will flow in

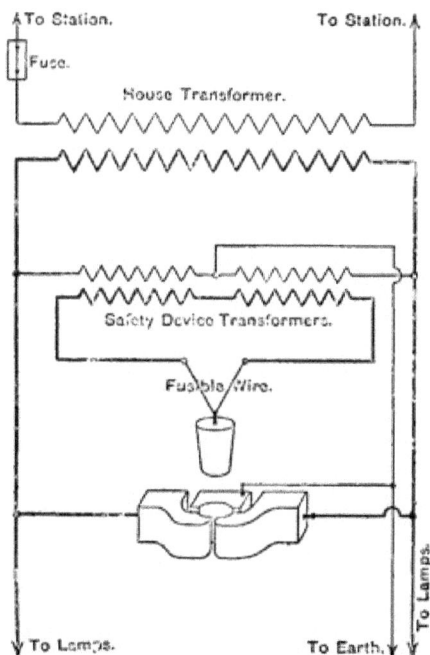

FIG. 127.—Ferranti " Earthing Device," for Transformer Protection.

their secondary circuit, and will melt the fuse, thereby allowing a conical plug to drop into the split cup shown in the diagram. This short-circuits the secondary circuit and blows the primary fuses.

It is urged strongly by most of the leading electrical engineers that the secondary circuits should always be earthed as well as the primary. If this were generally practised it would be impossible for a plant with defective insulation to remain in connection with the high-pressure circuit, for the main fuses would blow out as often as the main switches were closed. With the existing compulsory regulations, it is possible for a bad earth to exist on an installation until a second earth reveals its existence, possibly by a fire.

CHAPTER VIII.

POLYPHASE ALTERNATE-CURRENT WORKING.

§ 52. DEFINITIONS; TWO-PHASE CURRENTS.

THE term polyphase is applied to circuits in which two or more alternate currents, of the same wave length and direction, succeed each other at regular intervals. In practice, the number of impulses is usually limited to two or three, and the systems are then severally distinguished as two-phase and three-phase. An alternate current (single or multiple), when used to excite field magnets, produces a rotating magnetic field; that is, the axis of the field rotates with reference to space, though the windings are stationary. This is seen to be the case by considering an ordinary two-pole series-wound motor which is supplied with a single-phase alternate current. The poles will change from north to south polarity at every half period, and the axis of the magnetic field may be regarded as rotating, making one complete revolution in each period. Such a motor is, therefore, said to have a rotary magnetic field, and on the Continent is called a Drehstrom motor. A continuous-current motor, however, is not adapted for working with alternate currents of the usual frequencies, since the self-induction of the field-magnet circuit is too large.*

If two separate currents, with phase intervals of 90deg., be supplied to a four-pole motor, in which the armature consists of an iron core with windings short-circuited, either through brushes or else by rings of metal at each end, a rotary magnetic field in the stationary part will result. And this will induce currents in the closed coils of the armature whose direction will be such as to cause a magnetic field

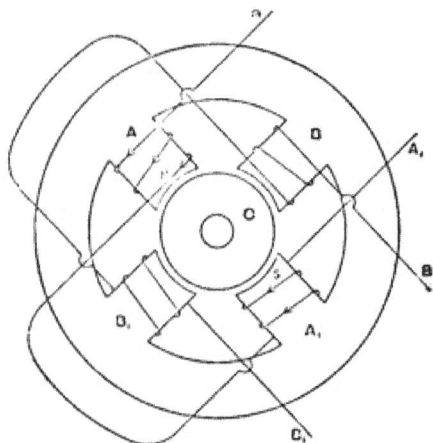

Fig. 128.—Four-pole Two-phase Motor.

tending to stop the rotation of the main field. This will be understood by examining Fig. 128, which represents a four-pole two-phase motor, having neither commutator nor brushes.

There are two alternating exciting circuits, A A$_1$ and B B$_1$, in quadrature.

<hr/>

* Attempts have been made, but with little success, to overcome this difficulty by laminating the magnets, lowering the frequency, and modifying the windings, so as to reduce the self-induction as much as possible, and to limit the hysteretic and eddy-current losses. *Vide* "The Distribution of Power by Alternate-Current Motors," by Aibion T. Snell ; *Proc.* Institution of Electrical Engineers, Vol. XXII., part 106.

Suppose that A and A_1 be magnetised so as to cause lines of force to pass from A to A_1 through the armature C. Then B and B_1 will be neutral, for there will be no current in them at this moment.

Next, the current in A A_1 will gradually die away; and that in B B_1 will steadily rise in an equal ratio but opposite direction until the magnetic flux is from B to B_1, and A and A_1 are neutral.

This cycle will be repeated, but with this alteration—the current in A A_1 will rise to a negative maximum (assuming it to have been positive before), and the flux will be in the direction of A_1 to A, and in the next wave from B_1 to B. Carefully noting these changes, it is seen that the magnetic field has rotated once for a complete $\sim$ in each of the circuits; and therefore the speed of the field rotation is $\dfrac{N}{p}$, where N is the number of $\sim$ per second and p is the number of pairs of poles in one circuit. In the case referred to, since $p = 1$, the revolutions of the magnetic field will be the same as the number of $\sim$.

In a two-phase dynamo the closed-circuit armature is replaced by a suitable electro-magnet, excited by a continuous current. Two-phase periodic currents are induced in the two circuits A A_1 and B B_1, the phase difference between them being 90deg.

Mathematically the currents in the two circuits may be expressed thus :—

That in A $A_1 = K \sin \alpha$,

„ B $B_1 = K \cos \alpha = K \sin \left(\alpha - \dfrac{\pi}{2} \right)$,

which shows the phase difference between the two currents. Now K is a constant for the particular machine, and α is a measure of the angular motion of the field. It will be seen that the current in A A_1 is a maximum when α is equal to $\dfrac{\pi}{2}$ and $\dfrac{3}{2} \pi$, and is zero when α is equal to 0 and π. The converse of this is the case with the current in B B_1.

Two-phase current circuits are usually worked with four wires—two for each circuit—but three wires, one having a section 1·41 times as great as that of each of the others, may be used. The latter arrangement is possible because the large wire will act as a common return for the two currents, the sum of whose instantaneous values is 1·41 times that of either. This will be seen on reference to Fig. 129, in which the full line curves relate to the impulses acting in the directions A A₁ and B B₁; and the dotted line curves to those in the directions A₁ A and B₁ B.

Curves A and B show the current-fluctuations in the two circuits, and D that in the common return. The magnetic field

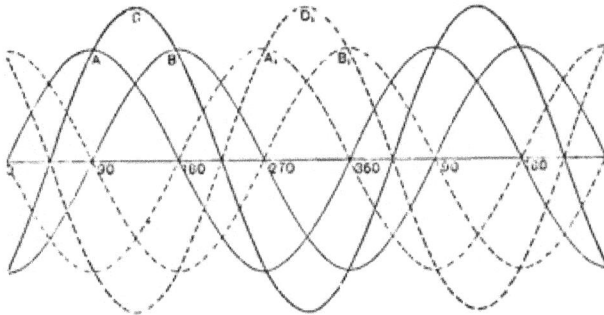

FIG. 129.—Diagram of Two Periodic Currents with a Common Return. Phase Difference of 90 deg.

will be caused by A and B alone, and, therefore, will be proportional to their sum at any instant. When A is a maximum B is zero, and the induction at this moment is proportional to the current in A alone, and is equal to, say, $i \sin \frac{\pi}{2} = i$. When A and B are equal, as at 135deg. of field rotation, the induction is proportional to $2 i \sin 135\deg. = i\,1\cdot4$. Thus the current varies in intensity between 1 and 1·4.

To illustrate the fluctuation of the field excitation the sign of the negative curves in Fig. 129 may be changed, and they may be plotted above the time line as if they were positive in value (see Fig. 130).

This diagram is justified by the consideration that each circuit in the motor contains at least two coils wound in opposite directions, and thus produces poles of different signs.

In Fig. 130 the excitation fluctuation is shown by the top curves, which may be regarded as having reference to one of the poles of the rotary field, say the north pole.

Since the theoretical mean variation of the exciting current is as much as 15 per cent., it may be supposed that the magnetic induction also varies by nearly as large a percentage, but this is not the case. These curves are deduced upon the supposition that the exciting circuits have no self-induction, whereas, on the contrary, they have much. And, further,

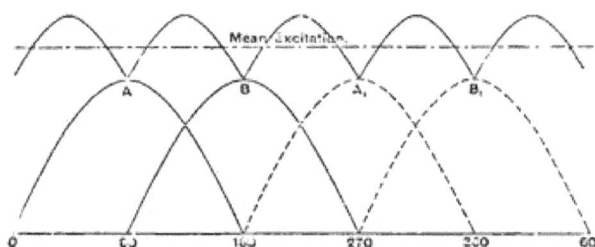

Fig. 130.—Diagram of Excitation Variation with Two-phase Current. Phase Difference of 90deg.

the resultant field cannot vary in intensity between such wide limits as those indicated in the diagram. Indeed, it is safe to assume that the field will be nearly steady as long as the effective values of the two currents are equal and their sum is constant. Armature reactions introduce serious complications, for their effect depends partly upon whether the current leads or lags the impressed electromotive force, and partly on the amount of the phase difference between the pressure and current. As already pointed out, when considering single-phase alternate-current machines, armature reactions tend to magnetise the field when the current leads and to demagnetise the field when it lags. The effect will be a maximum at starting, will decrease as the armature speed increases; and will be *nil* if the speeds of armature and field coincide, *i.e.*, if the motor works synchronously.

§ 53. THREE-PHASE CURRENTS.

Three-phase current working presents several advantages,
but entails complications in the design of the machine, and
also in the regulation of the pressure, because the three circuits are connected.   This is demonstrated in the following
abstract from a paper by the author entitled "The Distribution of Power by Alternate-Current Motors," which was
read before the Institution of Electrical Engineers in the spring
of 1893.

Fig. 131 represents diagrammatically the phases of three
equal periodic currents separated by intervals of 120deg. ; the

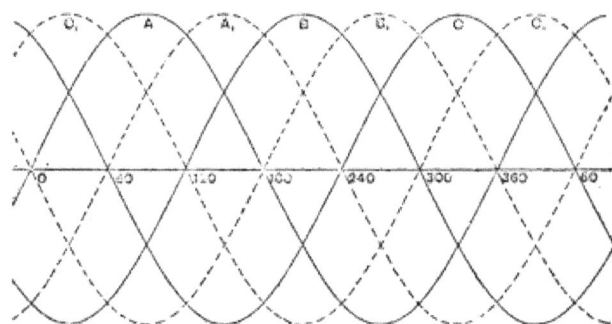

Fig. 131.—Three Equal Periodic Currents separated by Phase Intervals
of 120deg.

full and dotted lines referring to the direction of the impulses,
as already explained for Fig. 129.   If A, B, C represent the three
currents, then their instantaneous values are severally given
by :—

$$A = K \sin a,$$
$$B = K \sin (a - \frac{2}{3}\pi),$$
$$C = K \sin (a - \frac{4}{3}\pi).$$

Now, $K \{ \sin a + \sin (a - \frac{2}{3}\pi) + \sin (a - \frac{4}{3}\pi) \} = 0$, as is easily
proved ; and, therefore,

$$A + B + C = 0.$$

It may be seen that the algebraical sum of the instantaneous values of the three currents is equal to zero under all conditions; for, even when one of the quantities is equal to nought, the remaining two are equal, and, being of opposite sign, cancel each other.

Three-phase current working is accomplished by parallel or series coupling of the circuits, the two methods being severally

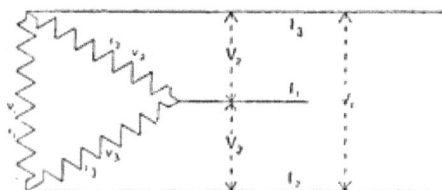

Fig. 132.—Parallel, or Closed Circuit, or Triangle Three-phase Coupling.

known as the *triangle* and *star* systems. In the following diagrams large capitals are used to distinguish the mains, and small letters to denote the windings. Suffixes are used to mark the phase order of the circuits.

The *parallel* or *closed circuit*, or *triangle* coupling, is shown in Fig. 132. Let I_1, I_2, I_3 be the effective values of the several

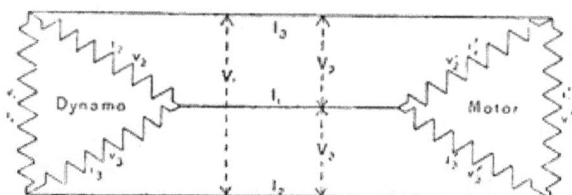

Fig 133.—Three-phase Dynamo coupled to a Motor.

currents flowing in the mains, and i_1, i_2, i_3 the corresponding effective currents in the coils. Also, let V_1, V_2, V_3 be the effective pressures at the terminals of i_1, i_2, i_3. And assume there is no self-induction or capacity. Then, if the load be equally distributed between the mains, we may assume

$$I_1 = I_2 = I_3, \ i_1 = i_2 = i_3, \text{ and } V_1 = V_2 = V_3.$$

Also, since i is in phase with v, and I lies 30deg. removed from i, it follows, if the mains be fed with a combined three-phase current (as in Fig. 133), that the current in one main will differ in phase by 30 deg. from the pressure between it and the two other mains. This follows from the

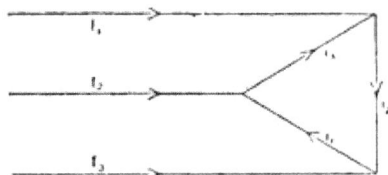

Fig. 134.—Diagram of a Three-phase Circuit.

geometrical relationship of the coils and the mains. And the effective value of the current in each of the mains is equal to 1·732 times the effective current in each of the coils, when the circuits are equally loaded. In Fig. 134 let the currents in the mains and in the coils be considered positive when flowing

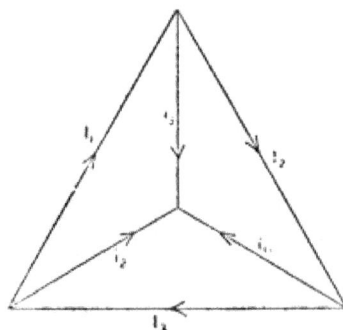

Fig. 135.—Diagram of a Three-phase Circuit.

in the direction indicated by the arrows. The phase and magnitude relations of these currents will then be represented by Fig. 135; and if this diagram be supposed to revolve uniformly around its centre, the length of the projections of the sides on any fixed straight line will represent the instantaneous

values of the corresponding currents; for it is plain that
these projections satisfy all the necessary conditions, viz. :—

$$\left.\begin{array}{l} I_1 = i_2 - i_3 \\ I_2 = i_3 - i_1 \\ I_3 = i_1 - i_2 \\ I_1 + I_2 + I_3 = 0 \\ i_1 + i_2 + i_3 = 0 \end{array}\right\}$$ All instantaneous values—*i.e.*, the lengths of the projections of the corresponding sides of Fig. 135.

Again, the sides of the diagram are proportional to the effective values of the corresponding currents; and, if the load
be equally distributed between the three mains, then for
effective values,

$$I_1 = I_2 = I_3, \quad i_1 = i_2 = i_3; \text{ and } I_1 = 2\,i, \sin 60\deg.\,;$$
$$\text{or, generally, } I = 1.732\,i.$$

This proof is due to Dr. W. E. Sumpner, Professor of
Electrical Engineering at the Battersea Polytechnic.

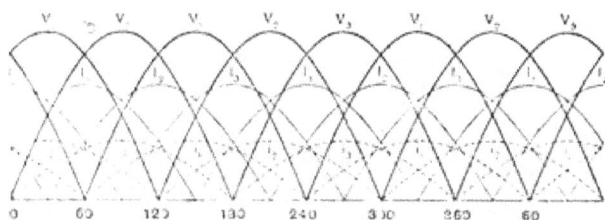

Fig. 136.—Diagram of Relative Position of Curves of Currents in Mains
and Coils of a Closed-type Combined Three-phase Circuit, with no Self-
Induction or Capacity.

By a similar diagram the relation between the effective
pressures on the coils and mains in the *open* or *star* arrangement can be shown to be $V = 1.732\,v.$ (*See* Fig. 137.)

The relative positions of the phases of currents and pressures
in the mains and coils of a combined three-phase circuit of the
closed type with no self-induction are shown in Fig. 136, V and *e*
having the same value and coinciding in phase. It will be seen
that the maxima of the currents in the mains, I, are always

midway between those of the coils, i, and that the phase difference between them is 30deg.

The *series, open circuit*, or *star* coupling is shown in Fig. 137

As before, let there be no self-induction or capacity, and let $I_1 = I_2 = I_3$; $i_1 = i_2 = i_3$; and $V_1 = V_2 = V_3$. Now, since the mains, I, are in series with the coils, i, and there is by hypothesis no self-induction or capacity, $I_1 = i_1$, $I_2 = i_2$, and $I_3 = i_3$. But the pressures between the mains are not the same as those at the coil terminals, and v will be in advance of V by 30deg., and will be numerically equal to $\dfrac{V}{2 \sin 60}$, or $v = \dfrac{V}{1\cdot732}$, therefore,

$$V = 1\cdot732\ v.$$

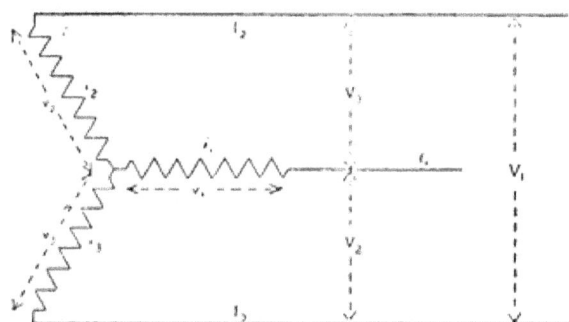

Fig. 137.—Series, or Open Circuit, or Star Three-phase Coupling.

The phases of current and pressure in a combined three-phase circuit of the open type are shown in Fig. 138, I and i having the same value and coinciding in phase.

The conclusions thus arrived at are true only on the assumptions (a) that the coils themselves have no self-induction, and (b) that both the coils and the mains are equally loaded. The first condition is never found in practice, and the second is only likely to obtain with small motors, and then only in an

approximate degree. The effect of self-induction is to cause
the current in the coils to lag behind the pressure at the
terminals.

To measure exactly the power in a three-phase system is
difficult ; but it can always be done by measuring the work
performed in each of the separate circuits, and adding the
quantities.

In Fig. 136, the power $= 3\,i\,v = 3 \times V\, \dfrac{I}{1\cdot732} = V \times I \times 1\cdot732$, if
the three circuits be equally loaded and there be no self-
induction or capacity.

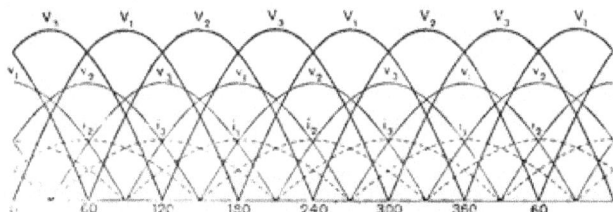

Fig. 138.—Diagram of Relative Position of Curves of Pressures and
Currents in the Coils and Mains of the *Open-type* Combined Three-phase
Circuit, with no Self-Induction or Capacity.

In Fig. 137, also, the power $= 3\,i\,v = 3 \times I\, \dfrac{V}{1\cdot732} = V \times I \times 1\cdot722$,
on the same assumptions.

So it appears, if there be no self-induction and an equal load
in each circuit, that the number of amperes in one of the mains
multiplied into the pressure between two mains into $1\cdot732$
gives the power in watts. If there be self-induction, the above
quantity must be multiplied by the cosine of the angle of lag
between the current and the pressure. The energy absorbed
by a motor, therefore, will be expressed by

$$V \times I \times 1\cdot732 \cos\phi,$$

where ϕ is the angle of lag.

T

Dr. W. E. Sumpner suggests that when the three circuits are equally loaded the power can be measured by one wattmeter (*see* Fig. 139). Put the current coil in one of the mains, say I_2, and take two readings, one with the pressure coil coupled between I_1 and I_2, and one with it coupled between I_2 and I_3. The two readings will be found to be of the same value, and the power will be equal to their sum.

Fig. 139.—Method of Measuring Power in Three-phase Circuit, with Mains equally Loaded.

When the load is unequally distributed, two wattmeters are required (*see* Fig. 140). Place the current coils in two of the mains, say I_1 and I_3, and couple one pressure coil between I_1 and I_2, and the other between I_2 and I_3. The power is then the sum of the two wattmeter readings.

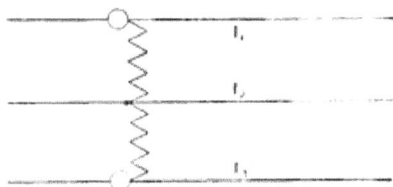

Fig. 140.—Method of Measuring Power in Three-phase Circuit, with Mains unequally Loaded.

This method of measuring power is applicable whatever the law of variation of the current, and however unequally loaded the mains may be.

It has been shown on page 258 that with combined three-phase currents, even when neglecting the self-induction of the coils, there is a constant phase difference between the current and the line pressure. This, although not in itself a direct loss, causes difficulty in measuring, regulating, and controlling the currents. Hence it has been found expedient to work the dynamo circuits unconnected and to use six or more separate coils. The currents from these are combined by a suitable transformer so as to convert the secondary currents into a combned high-pressure rotary current, with phase differences of 120deg. This high-pressure current is reduced at the motor end of the line by another transformer and subdivided as required.

It is not a simple matter to represent the intensity of the magnetic field by a diagram, for the magnetism is not merely the result of the exciting current in the field coils, but is also largely affected by the magnitude of the induced currents in the closed coils, which vary from instant to instant with the fluctations of load. There are two periodic currents acting in the closed coils—one of high frequency, equal to the number of the pairs of field poles multiplied into the number of revolutions per second ; and one of lower frequency, which depends simply on the difference of speed rotation between the revolving field and the rotating coils.

The low-frequency current produces the torque, and is highest at starting, when the *slip* of the armature is greatest. It is expressed numerically by $\dfrac{N - n}{60}$, where N = the number of revolutions of the magnetic field per minute and n = that of the rotating coils. For example, if $N = 2,400$ and $n = 1,920$, then the frequency will be 8 per second. In determining the excitation for polyphase motors and dynamos it is, therefore, necessary to make two calculations, one for full and one for light load, just as with direct-current machines.

It is clear that it is not possible to represent the changes of the rotary field magnetism by a general diagram. Fig. 141, however, shows diagrammatically the relative position of the current and pressure curves in a combined three-phase circuit, with the resultant excitation, assuming the armature reactions to have no effect—*i.e.*, the field and coils are supposed to rotate at nearly the same speed. An arbitrary angle of lag of

T 2

30deg. between the pressure and the current in the exciting
coils has been assumed, and the curves have been drawn to
suit the sine law. The pressure is shown in V V, and the
current in $i\,i$. The resultant excitation is indicated in curves
R R, the limits of which are $2\,i$ and $1.732\,i$, a mean difference
of about 8 per cent.

This is true only when self-induction and armature reactions
are neglected ; if these be taken into account, it is sufficiently
clear that the field is approximately constant, and that is all
that is necessary in practice.

The torque will vary with the ampere-turns and the number
of lines of force in the circuit. The torque $= \dfrac{C\,i_{a}\,N}{K_{1}}$, where i_{a}
is the current in the short-circuited coils, C is the number of

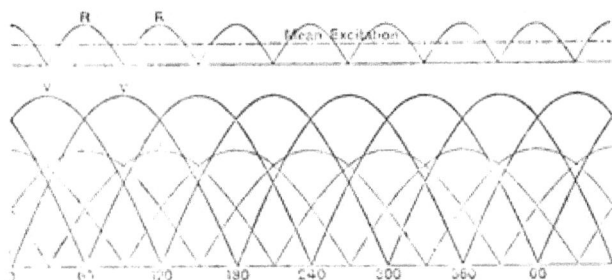

Fig. 141.—Diagram of the Excitation Fluctuations in a Combined
Closed-type Three-phase Circuit.

turns, K_{1} is a constant, and N is the magnetic flux in the
resultant field.

The magnetism is determined by the excitation, and the
number of turns of wire is, of course, fixed for a given design;
hence, to increase the torque, it is necessary that the moving
part rotate more slowly—that is, the *slip* must be increased.
This raises the frequency, and consequently the magnitude,
of the current in the closed coils, thereby weakening the
resultant field and lowering the counter electromotive force of
the exciting coils, and allowing more current to flow in them.
But, since the resultant magnetism decreases as the current in

the closed coils increases, there is a point at which the torque is a maximum. This corresponds to a definite line current, which is not necessarily the starting current, and hence these motors do not necessarily exert their greatest effort at starting.

To obtain maximum torque at starting, it is necessary to insert resistances in series with the closed-circuit coils. The function of the resistances is to determine the critical current, thus limiting the armature reactions ; and also to regulate the lag in phase between the currents in the closed and exciting coils.

In order to avoid large rushes of current when closing the line circuit, it is advisable to insert variable resistances in

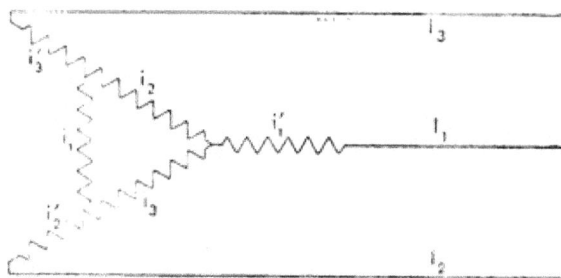

Fig. 142.—Combined *Series* and *Parallel* Three-phase Couplings.

series with the closed windings. This is specially necessary with large machines (*see* § 59, p. 293).

In order to obtain a more constant exciting current than that given by the arrangement shown in Figs. 132 and 137, Mr. Dobrowolski has devised a very pretty combination of the open and closed type windings, which reduces the mean variation of the excitation to 3·5 per cent. The winding is shown diagrammatically in Fig. 142, and is known as the double-linked winding. The coils indicated by the symbols i'_1, i'_2, i'_3, are severally wound in two parallels, 15deg. removed from the closed coils i_1, i_2, i_3. The complications involved are considerable, and the gain perhaps not commensurate with them ; yet for large machines the device may prove useful.

There are some important differences to be noticed between the *series* and *parallel* connections. It has been shown that with the *series* coupling $r = \dfrac{V}{1\cdot732}$ (*see* Fig. 137), and that the currents in the coils and mains are the same ; and in the *parallel* device (*see* Fig. 132), that each coil carries a current equal to $\dfrac{I}{1\cdot732}$, and that the pressure at the coil terminals corresponds to the pressure between the mains. These differences are suggested by the terms *parallel* and *series*, which are in this respect more apt than those of *triangle* and *star*.

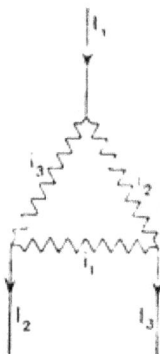

Fig. 143.
Diagram showing Ampere-turns with Parallel Coupling.

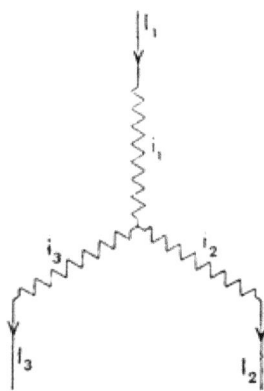

Fig. 144.
Diagram showing Ampere-turns with Series Coupling.

Now, the magnetising effect exerted by a given number of turns of wire, and a definite effective current in the mains, will be different with the two kinds of windings.

Consider Figs. 143 and 144, which severally represent the *parallel* and *series* couplings, and the direction of current at the instant when it is entering by one main and returning equally by the two others. Let n be the number of turns of wire in one coil. Then the total ampere-turns in the *parallel* device is $n\,I$, and in the *series* arrangement $\dfrac{3}{2}\,n\,I$. Therefore, the magnetising effect, and also the self-induction, of the *series*

device is greater than that of the *parallel* for a given current
and a fixed number of turns. The total power absorbed by the
two circuits may be made the same by a proper adjustment of
pressure.

§ 54. RELATIVE MERITS OF TWO-PHASE AND THREE-PHASE CURRENTS.

Experts differ as to the relative merits of two- and three-
phase currents. As regards the alternators and motors there
is, perhaps, not much to choose between them. Machines
designed for similar outputs, whether synchronous or non-
synchronous, are found to have about the same efficiencies,
starting torque, weight, and apparent watt consumption. The
three-phase machines, however, are the better, according to

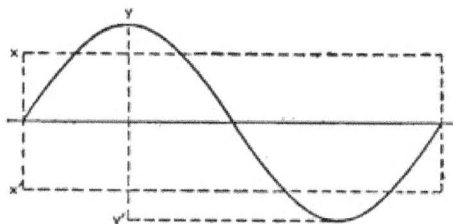

Fig. 145.—Single-phase System.

$x x' =$ Effective Pressure of 3,550 volts (*see* Table X, p. 268).
$y y' =$ Maximum Pressure of 5,000 volts.

experiments made by Mr. Kolbein, the chief engineer of the
Oerlikon Works. As regards the winding details, perhaps the
two-phase type is the simpler. The main difference between
the two systems lies in the relative weights of copper required
in the conductors for given conditions, and the ease with which
the currents may be handled, regulated and controlled.

The first experiments with three-phase working brought to
light difficulties in the regulation of pressure between the
three mains at the points of supply, when the currents in the
three circuits were not approximately the same. These, though
since to some extent overcome, are greater than those with two-
phase currents. The difficulties lie chiefly with the design of

Table X.—*Comparative Table of Ratios of Weights of Copper required with different systems to deliver 50,000 watts at Receiving Station under the conditions of Constant Maximum Pressure and Constant Effective Pressure at Receiving Station, with an total line loss of 5,000 watts or constant drop of 500 volts.*

System.	Conditions. M.P. = Maximum Pressure. E.P. = Effective Pressure.	Current in amperes.	Effective volts at works.	Effective volts at power station.	Line loss in watts.	Drop in volts.	Resistance of each main in ohms.	Ratio of weights of copper.	Weight of copper per when the economic line is considered.
Continuous current, 2 wires	M.P. 5,000 v., line loss 5,000 w.	10·0	5,000	5,500	5,000	2 × 250 = 500	25·0	1·0	1·0
	E.P. ditto	10·0	5,000	5,500	5,000	2 × 250 = 500	25·0	1·0	1·0
	M.P. 5,000 v., drop 500 v.	10·0	5,000	5,500	5,000	2 × 250 = 500	2·0	1·0	1·0
	E.P. ditto	10·0	5,000	5,500	5,000	2 × 250 = 5 0	2·0	1·0	1·0
Single-phase Alternate current, 2 wires	M.P. 5,000 v., line loss 5,000 w.	14·12	3,550	3,906	5,000	2 × 178 = 356	12·5	2·0	1411
	E.P. ditto	10·0	3,550	3,550	5,000	2 × 250 = 5 0	25·0	1·0	1·0
	M.P. 5,000 v., drop 500 v.	14·12	3,550	4,050	7,100	2 × 250 = 500	17·8	1·41	1411
	E.P. ditto	10·0	5,000	5,500	7,100	2 × 250 = 500	25·0	1·0	1·0
Two-phase Alternate current, 4 wires	M.P. 5,000 v., line loss 5,000 w.	Each circuit. 706	3,550	3,906	5,000	2 × 178 = 356	25·0	2·0	1411
	E.P. ditto	50	5,000	5,500	5,000	2 × 250 = 560	50·0	1·0	1·0
	M.P. 5,000 v., drop 500 v.	7·6	3,550	4,050	7,100	2 × 250 = 500	35·6	1·41	1411
	E.P. ditto	50	5,000	5,500	7,100	2 × 250 = 500	50·0	1·0	1·0
Two-phase Alternate current, 3 wires	M.P. 5,000 v., line loss 5,000 w.	Outer 10·0 / Return 14·12	2,500	2,794	Two outers 2,940 / Return 2,050	2 × 147 = 294	Each outer 14·7 / turn 10·3	2·91	1706
	E.P. ditto*	7·06 / 10·0	3,550	3,962	2,940 / 2,060	2 × 206 = 412	29·4 / 20·6	1·478	1205
	M.P. 5,000 v., drop 500 v.	6·30 / 7·05	5,00	5,590	2,940 / 2,060	2 × 295 = 560	50·0 / 41·25	0·737	0853
	E.P. ditto*	10·0 / 14·12	2,500	3,660	5,000 / 3,550	2 × 250 = 500	25·0 / 17·75	1·704	1706
		7·05 / 10·0	3,550	4,000	5,000 / 3,550	2 × 250 = 500	35·5 / 25·0	1·204	1205
		65·0 / 7·06	5,000	5,500	2,500 / 1,770	2 × 250 = 500	50·0 / 35·4	0·853	0853
Three-phase Alternate current, 3 wires	M.P. 5,000 v., line loss 5,000 w.	8·1	3,550	3,752	5,000	1 × 202	25·0	1·5	1212
	E.P. ditto	5·73	5,00	5,235	5,000	1 × 288	49·8	0·725	0866
	M.P. 5,000 v., drop 500 v.	8·1	3,655	4,050	12,200	1 × 500	61·8	0·610	1212
	E.P. ditto	5·73	5,000	5,500	8,700	1 × 500	86·5	0·435	0866

(263)

* Effective pressure measured (a) with respect to the working pressure in each circuit, (b) with respect to the component pressure. In the final column is given the necessary weights of copper, when the economic law is considered (to the exclusion of both line loss and line drop, the equivalent current being taken as in column 3.

alternators and transformers. Two-phase currents, if worked with *two separate circuits*, are as easily dealt with as single-phase currents, and serve readily most purposes to which single or poly-phase currents are usually applied. They are specially convenient for distribution. It may be safely inferred that two-phase currents are better adapted for a combined service of light and power work than three-phase currents, which are better adapted for running with equally loaded circuits, and therefore are not generally so suitable for lighting work.

For simple transmission of power between two distant points, however, the requirements are different, and the three-phase current system is generally cheaper than either a continuous-current or a one-phase or two-phase alternate-current system.

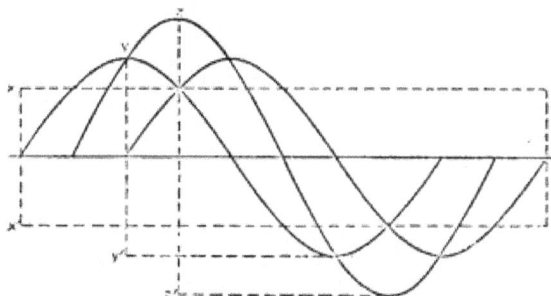

Fig. 146.—Two-phase System. Three Wires.

$x\,x'$ = Effective Working Pressure of 3,550 volts.
$y\,y'$ = Maximum Working Pressure of 5,000 volts.
$z\,z'$ = Maximum Component Pressure of 7,100 volts.

This interesting and important problem may be considered in three ways, each of practical value :—

(1) On the basis of maximum difference of pressure between conductors ;

(2) On the basis of effective pressure at the distributing station (or motor terminals, perhaps) ;

(3) On the basis of the economic law.

The ratios of the weights of copper under these conditions are stated in tabulated form in Table X, p. 268, in which, to give a practical character to the figures, a definite power has been selected for transmission to the receiving station. Constant line loss and constant fall of pressure have also been considered.

In the final column the economic law has been considered to the exclusion of both a definite line loss and line drop, the equivalent current being taken as in column 3. In practice, the weights will be modified by local considerations; but the main ratios will hold good, and the figures given form, therefore, a reliable guide between the relative merits of the systems as affected by the weight of copper.

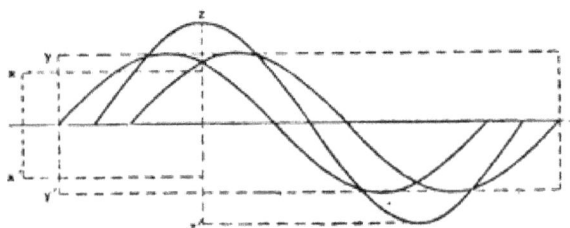

FIG. 147.—Two-phase System.   Three Wires.

$x\,x'$ = Effective Working Pressure of 2,500 volts.
$y\,y'$ = Effective Component Pressure of 3,550 volts.
$z\,z'$ = Maximum Component Pressure of 5,000 volts.

The significance of the difference between effective and maximum pressure is at once apparent if Figs. 145, 146, 147, and 148 are examined.

In these figures the *electrical centre of gravity* is shown graphically. In any electrical circuit of good or bad insulation there is a datum pressure with reference to the variations of pressure in the circuit. For example, in a continuous-current circuit the absolute pressure of the negative pole of the dynamo will have a constant ratio to that of the earth's surface at the particular site, and the difference of pressure between it and the positive pole will simply affect the absolute pressure of the positive pole. And, if the negative pole be grounded, the datum pressure will be that of the earth's surface.

Again, with an alternate-current circuit, with both poles insulated, the cyclical changes will alternately raise and lower the absolute pressure above and below zero pressure, which corresponds to that of the earth's surface. Thus, in a circuit carrying a current of 5,000 effective volts pressure, the maximum pressure between any two points of the circuit will be 5,000 $\sqrt{2} = 7,050$ volts, while the absolute pressure will be only 3,525, positive or negative. The stress through the insulation to earth will, therefore, be only 3,525 volts.

But if the circuit be grounded, say, at one of the main terminals, the absolute pressure between the other terminal

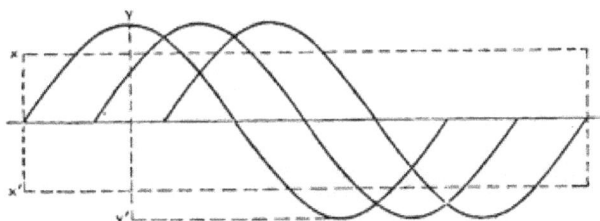

Fig. 143.—Three-phase System.

$x\,x' =$ Effective Pressure of 3,550 volts.
$y\,y' =$ Maximum Pressure of 5,000 volts.

and earth will be 7,050 volts, and therefore the insulation all over the high-pressure circuit will be subjected to this stress, although the effective working pressure will remain as before.

This possible change in the stress upon the insulation of a circuit carrying an alternate current has led to the general practice of earthing one of the primary mains, usually at both the generating and distributing circuits. The maximum stress is thus accepted, and suitable precautions adopted to safeguard against accidents; and, as already pointed out in §51, p. 247, additional security is gained because a fault in the other main, at any point, at once blows the fuses and disconnects the faulty section.

With two-phase three-wire circuits it is permissible to earth
the middle wire at the generating station, which then is at
zero pressure. The same precaution is observed with the
"star" coupled three-phase system (see Fig. 155, p. 277). With
the "triangle" three-phase system it is not permissible to
earth either of the mains, since the earth would then form
part of the circuit.

It appears, then, that, from considerations affecting prime
cost of line, the three-phase system is cheaper for long-distance
transmission than either single-phase or double-phase ones ;
and that continuous-current systems are the cheapest of all,
but are not admissible for extra high pressures.

The selection of a system for long-distance transmission
of power will mainly depend upon the maximum permissible
pressure. If it be possible to work at such pressures as are
feasible, in the engineer's judgment, for continuous-current
dynamos (coupled in series or otherwise), then there can be
no doubt, from considerations affecting the line, that this
will be the cheapest and best system. But if extra high pressure be deemed advisable, then the three-phase system will
generally prove the cheapest.*

In connection with this question it is important to notice
that the engineers responsible for the electric power plant
being laid down at Niagara have finally, after mature consideration, decided upon a four-wire two-phase system. The
reasons for their choice have been given by Prof. George
Forbes, in a Paper entitled "The Electrical Transmission
of Power from Niagara Falls," read before the Institution of
Electrical Engineers on November 9, 1893. The details of
the scheme have been strongly criticised by English experts,
and the choice of two-phase currents is open to doubt.
When it is considered that Buffalo, the nearest large city, is

* The generating plant may perhaps be a two-phase one, with suitable
step-up two-to-three-phase transformers, as proposed by Mr. Scott, of the
Westinghouse Company. And it may also prove economical to transform
the phase again at the receiving station. The ingenious proposals of
Mr. Scott, however, have not yet been verified in practice ; but there seems
to be no doubt of their feasibility and usefulness (see § 56, p. 278).

fifteen miles distant, it is clear that the weight of copper will be a very important item in the cost of the whole plant, and, in the author's judgment, a three-phase system would prove more economical in first cost and give as good pressure regulation as the two-phase. For it must be recollected that there is a line difference of pressure of 20,000 volts, and hence step-down transformers will have to be used at every point from which power is taken. The *main line* drop will, therefore, be as difficult to regulate with two-phase as with three-phase currents; and variation of pressure of submains of the distributing systems can be as readily controlled by using *independent* secondary circuits from three-phase transformers as from *independent* secondary circuits of single or two-phase transformers. Prof. Forbes's argument that two-phase currents are

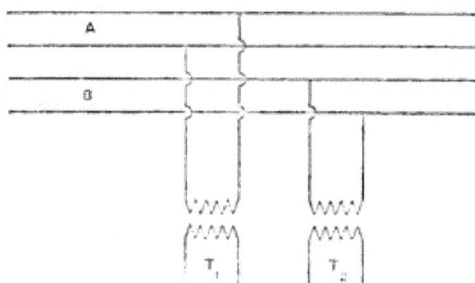

Fig. 149.—Four-wire Two-phase Circuit with two Ordinary Transformers, T₁, T₂ one being Coupled to each of the Circuits, A, B.

more easily *redressed* than three-phase currents is probably sound, but the fact that for large powers no form of alternate current has yet been made unidirectional in a practical manner cannot be ignored.* The use of a frequency of 25 ∽ per second is a departure from the best practice of to-day (*see* § 58, p. 281).

§ 55. POLYPHASE-CURRENT TRANSFORMERS; DIAGRAM OF CONNECTIONS.

The principle of polyphase-current transformers is the same as that of single-phase transformers. With two-phase currents

---

* *See* § 60, p. 297.

it is usual to couple single-phase transformers across each of
the circuits, as shown in Fig. 149, in which there are four conductors of equal section.   If three were used, the middle one
having 1·4 times the area of each of the outer ones, the transformers could still be coupled as above; but the pressure

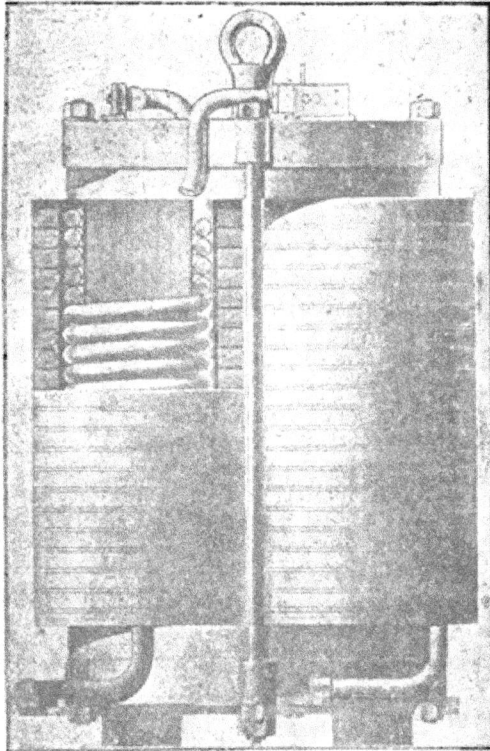

Fig. 150.—Dobrowolski Three-phase Transformer.

variation would be greater than with the separate arrangement, especially if they, or the conductors, or the machines,
possessed much capacity or self-induction.

With three-phase currents it is usual to employ combined
transformers, as it is important to balance the load on the

three circuits, but three separate single-phase transformers
may be used. Various forms are in use, such as that shown
in Fig. 150, which shows the Dobrowolski type. One of the
methods of coupling the connections is given in Fig. 151.

This figure also illustrates one method of winding the primary
and secondary circuits. Owing to the phase differences between
the currents in the coils of the inter-connected circuit and the
pressure between the mains, there is a loss of pressure of about
13 per cent. in the ratio of conversion as compared with a
single-phase transformer. Hence the number of turns in the

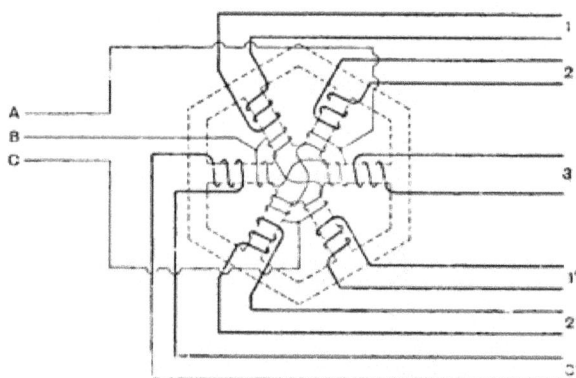

FIG. 151.—General Diagram showing one of the Methods of Winding the
Coils of a Three-phase Transformer. A, B, C: High-pressure mains and
coils. 1, 2, 3, 1', 2', 3': Low-pressure circuits, which may be combined or
separate. The phases are the same in opposite coils. The iron core
is indicated by the dotted lines.

primary circuit must be increased in this ratio. The efficiency
is, perhaps, not quite so high as that obtainable with the best
designed single-phase transformers, but it certainly exceeds 96
per cent. at full load, and the curve can be made to any slope
by suitably proportioning the weights of iron and copper as
already explained in § 37, p. 205, when considering ordinary
transformers.

A three-phase transformer weighs less than three separate
single-phase transformers of an equal aggregate power, the

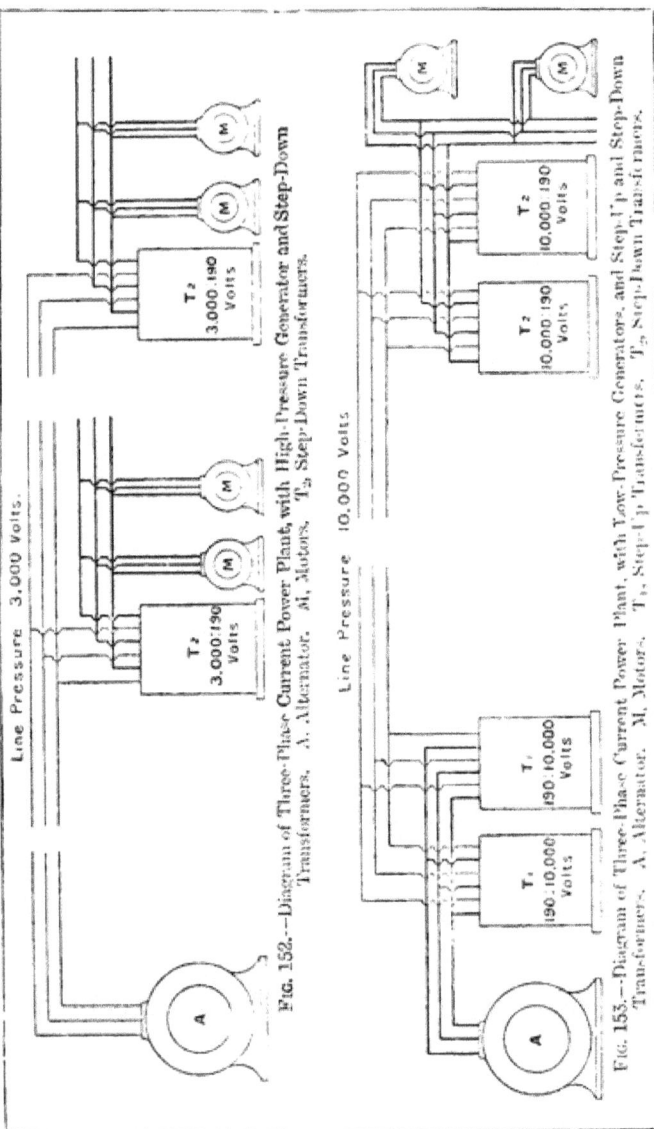

FIG. 152.—Diagram of Three-Phase Current Power Plant, with High-Pressure Generator and Step-Down Transformers. A, Alternator. M, Motors. T₂, Step-Down Transformers.

FIG. 153.—Diagram of Three-Phase Current Power Plant, with Low-Pressure Generators, and Step-Up and Step-Down Transformers. A, Alternator. M, Motors. T₁, Step-Up Transformers. T₂, Step-Down Transformers.

ratio of weight being about $3:4$. This increase of weight efficiency is partly due to the magnetic flux dividing in a similar manner to the current. The flux in one leg is in the opposite direction to that in the other two. But the flow in one of the two increases, while that in the other decreases. The cycle is thus continually changing, each leg in succession becoming the return for the magnetic flux in the other two.

FIG. 154.—Diagram of Connections of a Three-phase Circuit.
A, Alternator.   M, Motor.   T₁, Step-Up Transformer.   T₂, Step-Down Transformer.

The weight is also reduced to some extent, because the design permits of the framing and yokes being made lighter in proportion to the cores than is possible in the single-phase type.

In high-pressure transformers the high-pressure coils are generally inter-connected, and the ends are usually protected by glass tubes to a considerable height above the level of the coils. The low-pressure coils have separate terminals to permit of coupling up to separate circuits.

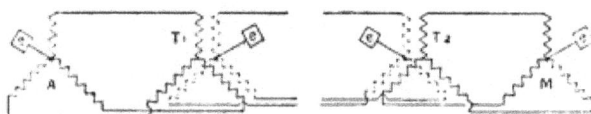

FIG. 155.—Diagram of Connections of a Three-phase Circuit.
A, Alternator.   M, Motor.   T₁, Step-Up Transformer.   T₂, Step-Down Transformer.   e, Earth Plates.

Table Y, p. 285, gives some interesting data of Oerlikon three-phase transformers, designed to work at line-pressures up to about 5,000 volts.

Three-phase transformers work in parallel just as effectively as single-phase ones. The general schemes for a transmission plant with polyphase currents, with and without step-up

U

transformers, is shown in Figs. 152 and 153. The details
are clearly illustrated. In the high-pressure system only one
transformer is shown, but a bank of transformers might be
arranged if necessary.

Two methods of arranging the connections of three-phase
circuits with step-up and step-down transformers are given in
Figs. 154 and 155.

§ 56. COMBINATION OF TWO- AND THREE-PHASE CIRCUITS.

It has already been shown that three-phase currents present
advantages for transmission as distinguished from distribution,
and that two-phase currents are specially adapted for serving
sub-divided circuits. A combination of the two systems is
thus suggested and necessitates a *phase conversion* at the

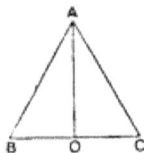

Fig. 156.—Two- to Three-phase Diagram.

distributing centre. Various inventors have worked at this
problem, and some have attempted to change the frequency as
well, for this is also of great importance in long-distance transmission of power. A solution of the latter has not yet been
published in a practical form, although it is in the air. But
the former has been apparently solved by Mr. Chas. F. Scott,[*]
the chief electrician to the Westinghouse Company. The
method is simple and practical, and is an ingenious application
of the fact that two alternate pressures of different phase in
series with each other do not give a combined pressure equal
to the sum of the components.

Thus, if in Fig. 156, A O and O B severally represent in
phase and magnitude two pressures in quadrature to each

---

[*] Prof. S. P. Thompson has independently developed the same idea.
See British Association Paper entitled " Some Advantages of Alternate
Currents " (*The Electrician*, August 24, 1894, p. 481).

other, and also in series, the resultant pressure will be represented in direction and magnitude by the hypothenuse A B of
the right-angled triangle A O B. If A O and O B be so proportioned that the angle A B O is 60deg., the relationship
is evidently unaffected. The triangle A O B, Fig. 156, is constructed to suit these conditions, and forms the half of the
equilateral triangle A B C, whose sides severally represent in
magnitude the three pressures acting in a three-phase circuit.
Now, from the construction, and from what has already been
demonstrated in § 52 and § 53, it is evident that the pressures
represented by the lines A O and C B may be combined to
give the two pressures represented by the lines A C and A B
or *vice versâ*.

The three-phase diagram A C B may also be drawn as in
Fig. 157, in which A O is the resultant of A C and A B. If the

Fig. 157.—Two- to Three-phase Diagram.

coils producing the pressures A O and B C be supposed to be
each wound on separate ordinary type single-phase transformers,
and if there be two separate coils wound outside each of them,
a o and *b c*, the arrangement may be diagrammatically
represented by Fig. 158.

Then a three-phase system may be supplied from the
terminals A, B and C, or a two-phase from those marked *a, o,*
and *b, c*. (In the diagram, the transformers may be imagined
to be feeding three-phase transmission mains at high pressure.)
The arrangement is simple and effective, involving no mechanical
details. It simply requires suitably-wound transformers at the

u 2

phase conversion station, where the pressure may be raised or
lowered as required.

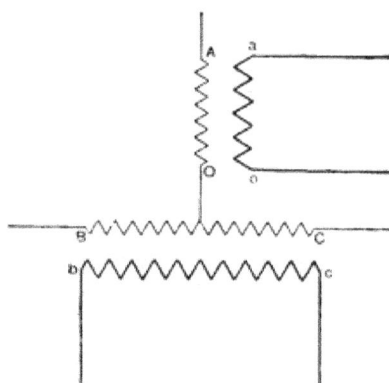

Fig. 158.—Diagram of Two Single-phase Transformers combined to give
Two- and Three-phase Currents.

The number of turns in the coils can be adjusted to give
any pressure between the several mains; but if the pressure

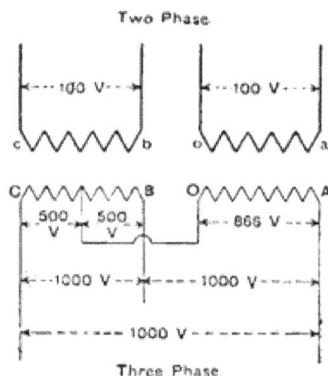

Fig. 159.—Diagram of Pressure Distribution in Two- to Three-Phase
Transformers.

between the three mains of the three-phase part be of the same
value, as is usual, then the number of turns in the two-phase

coils must be in the ratio of $2 : \sqrt{3}$; the coil $c\,b$ having, say, 200 turns, and the coil $a\,o$ 173. This ratio is determined by the phase difference of 30deg. between the line and coil pressure (*see* § 53, Fig. 137, p. 260).

The distribution of pressure will be seen from Fig. 159.

This system enables **a** three-phase line to feed three-phase motors direct at the line pressure, two-phase motors through transformers, and lighting circuits in two instead of three units. These various operations are indicated in Fig. 160. If the transformer which supplies direct from the terminals (C B, $c\,b$, Fig. 159) be loaded and the other be on open circuit, a single-phase current can be supplied at normal pressure, independently of the other circuit.

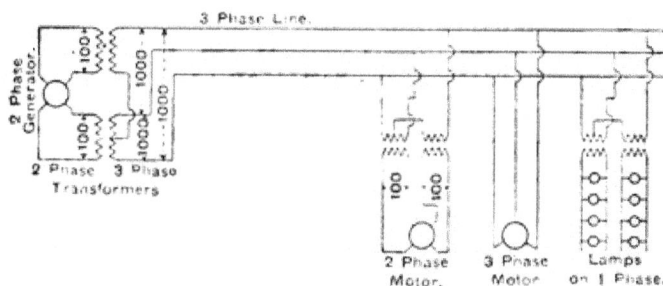

Fig. 160.—Diagram of Two-phase Alternators, Three-phase Line, Two- and Three-phase Motors, and Lamps in Single-phase Circuits.

But if the transformer (A O, $a\,o$) which supplies through the *three* mains be loaded and the other be on open-circuit, then a single-phase current can be supplied at a pressure equal to only 87 per cent. of the normal. Two mains on one side will be in parallel. The self-induction of the idle transformer will not affect the circuit, for the current entering at the middle of the winding divides equally in opposite directions through the two halves of the coil, and thus completely neutralises the self-induction. The ohmic resistance of the winding has to be overcome, but this loss is more than compensated by two of the mains being in parallel.

These conditions of running are not likely to be used under
ordinary circumstances. Yet one of them may prove of great
advantage in the case of damage to part of the system—for
instance, if one side of the alternator be damaged by lightning
or other causes. This feature may be made the means of
minimising accidents, especially if the alternators are of the
single-phase type coupled rigidly in quadrature.

The efficiency of two transformers arranged for converting
from two-phase to three-phase, or *vice versâ*, is said to be about
$\frac{1}{2}$th per cent. less than when the same transformers are used
with single-phase currents at corresponding loads.

§ 57. SYNCHRONOUS POLYPHASE MOTORS.

Polyphase alternators make excellent synchronous motors,
and are said to keep step better than single-phase machines,
the *link* between the rotating magnetic field and the revolving
part being more flexible. The line current is fed to the
revolving part of the motor, and the stationary part is constructed of solid cast iron, with or without windings.

If there be no windings on the stationary magnets the
torque is due to induced currents in the non-laminated
pole pieces. This design is not adapted for a high efficiency
at starting, and therefore it is usual in most cases to provide
secondary windings to assist the initial torque. This need not
involve any complicated device.

Synchronous polyphase motors are specially suitable for very
large units, since the bulk of the machine consists of cast iron.
They are therefore cheaper than induction polyphase motors of
the same capacity. And, what is still more important, they
cause no lagging current in the line except at starting, and
therefore have a power factor of 100 per cent. This type of
motor is likely to be used largely in the near future. Its
general appearance is indicated in Fig. 161.

A good example of large synchronous polyphase motors is
afforded by a plant erected early in 1894 at Taftville Cotton
Mills, Conn., U.S.A., by which power is transmitted for a distance of about $4\frac{1}{2}$ miles.

There are two 250-kilowatt alternators, delivering power to
the line at a pressure of 2,500 volts. They run at 600 revolutions per minute, and work perfectly in parallel. The motors
are of similar construction, and are separately excited by
3-kilowatt exciters driven by belting. They are coupled to
the main shafting by clutches, and are allowed to acquire the

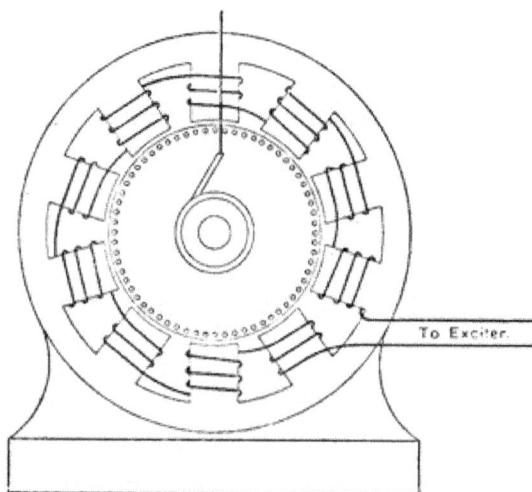

To Exciter.

Fig. 161.—Diagram of Synchronous Polyphase Alternator, with
Windings on the Field Magnets.

speed of synchronism before the load is applied. In every
respect the plant is a great success.

§58. PRESENT PRACTICE IN POLYPHASE
ALTERNATORS AND MOTORS.

Some of the most recent designs of polyphase alternators
and motors are shown in this section. They are typical of
present practice.

Polyphase generators for low pressure are frequently designed
to work at a pressure of 190 volts between the mains, and

are used with step-up transformers. High-pressure generators
and motors which will work at line pressures of from 5,000
to 7,000 volts are generally used for power transmission when
the line pressure need not exceed, say, 7,000 volts. The frequency adopted is from 50 to 65. The lower value is being

Fig. 152.—Oerlikon Low-Pressure Polyphase Alternator.

gradually adopted on the Continent, and will soon become the
standard. This frequency is too low for the most economical
conditions, and 65 or 70 $\sim$ would be better, especially for
the transformers, but the practical consideration of slow speed
with few poles in generators and motors determines the lower
value (*see* Table M, p. 174).

Fig. 163.—Oerlikon High-Pressure Polyphase Alternator, with Exciter attached to the Shaft.

Fig. 165.—Oerlikon Asynchronous Polyphase Motor.

Table Y.—*Oerlikon Three-phase Transformers, designed to work up to 5,000 volts line pressure.*

Type.	3,301	3,302	3,303	3,304	3,305	3,306	3,307	3,308	3,309	3,310	3,311	3,312	3,313
Kilowatts	2.5	5	8	10	15	20	35	50	65	80	100	140	200
Efficiency	90%	92%	93%	94%	94%	95%	95%	95%	96%	96%	96%	97%	97%
Weight in lbs.	330	440	620	770	1,140	1,340	2,000	2,650	4,200	5,000	5,750	6,600	8,8_0
Primary amperes in each circuit {1,732 v. between mains	0.8	1.7	2.7	3.3	5.0	6.6	11.6	16.6	22.0	27.0	33.0	46	66
{3,464 „ „	0.85	0.85	1.35	1.5	2.5	3.3	5.8	8.3	11.0	13.5	16.5	23	33
{5,196 „ „	0.6	0.6	0.9	1.1	1.6	2.2	3.9	5.5	7.3	9.0	11.0	12	22

Table Z.—*Low-Pressure Oerlikon Three-Phase Alternators, Star Couplings. 50 ~ : 110 Volts per Coil, 190 Volts between Mains.*

Type.	300	301	302	303	304	305	306
Kilowatts	19	32	66	100	135	205	350
Amperes per circuit	27	97	200	333	410	620	1,080
Horse-power absorbed	30	50	100	150	200	300	500
Efficiency	85%	87%	96%	91%	92%	93%	95%
Revolutions per minute	750	600	400	350	260	190	150
Weight in tons {With exciter	1.6	2.3	4.5	6.45	7.85	11.8	17.7
{Without exciter	1.37	2.06	4.1	6.1	7.5	11.2	17.0

(286)

Table A.A.—*High-Pressure Oerlikon Three-Phase Alternators, 50 ~.*

Type.	3,000	3,001	3,002	3,003	3,004	3,005	3,006
Kilowatts	19	32	66	100	135	200	342
Amperes per circuit {1,732 v. between mains	19.0	10.6	22.0	33.3	450	66.0	114.0
{3,464 „ „	6.3	5.3	11.0	16.6	22.5	33.0	57.0
{5,196 „ „	3.1	3.5	7.3	11.1	15.0	2.0	38.0
Horse-power absorbed	30	50	100	150	200	300	500
Efficiency	85%	87%	90%	91%	92%	92%	93%
Revolutions per minute	750	600	500	800	375	250	190
Weight, with exciter, in tons	1.0	2.56	3.5	5.0	7.0	9.0	13.6

In cases where power work alone has to be considered, and
the generators have to be coupled direct to slow-speed engines,
it is found convenient to decrease the frequency to even 25 or
20 ᵜ per second ; but these are exceptional conditions.

The Oerlikon Company build excellent polyphase machines.
Their low-pressure alternator is shown in Fig. 162, and their
high-pressure alternator in Fig. 163. The performances of list

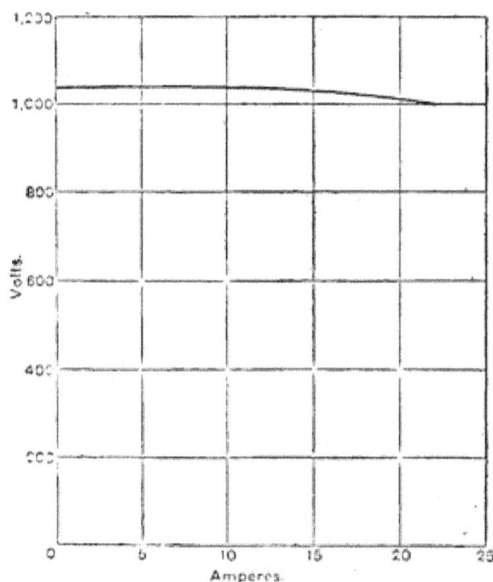

FIG. 164.—Characteristic Curve of a 100-H.P. Three-Phase Oerlikon
Alternator.

machines of these types are given in Table Z and Table A A,
and a characteristic curve of a 66-kilowatt alternator, 1,040
volts per coil, is given in Fig. 164. The drop practically
follows Ohm's law, and capacity and self-induction effects, if
present, apparently cancel each other.

Types of the Oerlikon Company's normal three-phase motors
are shown in Figs. 165 and 166 ; and one of the Allgemeine

Electricitäts Gesellschaft in Fig. 167. Some tests of the small motors are given in Table B B, p. 289. It is seen that the efficiency, although not equal to that of first-rate continuous-current motors of similar output, is sufficiently high to make the machines of great commercial value, especially when the

Fig. 166.—Oerlikon Asynchronous Polyphase Motor.

absence of commutator and brushes and all the troubles and risks incidental to them is borne in mind. The Oerlikon Company adopt as standards a pressure of 110 volts per coil, equal to 190 volts between the mains (series coupling, see Fig. 137, p. 260), and a frequency of 50 ∿ per second.

Table B B.—*Tests of small Oerlikon Three-phase Motors, 50 ~.*

	Size of motor.					
	¼ H.P.	½ H.P.	1½ H.P.	3 H.P.	6 H.P.	9 H.P.
Revs. per min., empty	1,450	1,450	1,450	1,475	975	970
Revs. per min., full load	1,320	1,335	1,350	1,380	910	900
Starting torque, in lb. feet	57	65	137	123	570	690
Torque at full load, in lb. feet	264	440	1,105	1,200	2,700	4,350
Pressure in volts	60·5	60	57	63	61	67
Amperes in each circuit, empty	1·3	4	7	7·5	20	25
Amperes in each circuit, full load	2·5	8	13·8	18	59	86
Power absorbed at no-load in watts	165	--	346	404	--	--
Power absorbed at full-load in watts	325	670	1,900	2,940	6,620	10,400
Output in brake H.P.	0·23	0·49	1·5	3	7·1	11·5
Output in watts	170	368	1,140	2,208	5,300	8,600
Efficiency	52%	55%	60%	75%	80%	83%
Weight of motor complete in lbs.	92	132	220	330	608	925

Some idea of the data of American three-phase motors may be gathered from Table C C, which is taken from the Columbian Exposition Supplement to the *Engineering Review*.

Table C C.—*Data of American Three-phase Motors.*

	Size of motor.				
	¼ H.P.	½ H.P.	1 H.P.	5 H.P.	50 H.P.
Apparent consumption in watts	230	518	985	4,300	40,200
Amperes per circuit loaded	1·4	4·0	8·0	36	280
Volts between mains	60	60	60	60	60
Frequency	50	50	50	50	50
Speed (unloaded)	2,380	1,490	1,490	1,490	745
Speed (loaded)	2,300	1,400	1,375	1,395	725
Efficiency	...	71%	75%	84%	91%
Number of poles	2	4	4	4	8
Weight in pounds	396	1,386	2,068	5,390	26,400
Amperes per circuit (unloaded)	...	...	4·5	15	150
Amperes per circuit at starting	...	...	...	50	400
Total apparent watts at starting	...	...	...	5,650	50,000

Messrs. Johnson and Phillips have built for use at the Sheba Gold Mines some two-phase alternators which embody an invention of Mr. Gisbert Kapp s. The construction (*see* Fig. 168) is similar in general appearance to that of Mr. C. E. L. Brown's machines, the field magnets being of similar design. The armature coils, however, which are, of course, arranged around the outer and stationary part of the machine, are not symmetrically placed with reference to each other, but the top half of

Fig. 167.—Allgemeine Electricität« Gesellschaft Asynchronous Polyphase Motor.

the coils is set with an angular lead of 90 degrees with respect to the lower half. The two form separate circuits with currents in quadature, and can be used as distinct machines if required.

One of the advantages claimed by the inventor for this arrangement is that an accident to one of the circuits will probably leave the other uninjured, and so prevent a complete

breakdown of the machine. In fact, an alternator thus constructed is equivalent to two single-phase machines of half its

FIG. 168.—Messrs. Johnson and Phillips' Two phase Alternator. 16 Field Poles and 14 Armature Coils in two sets of seven each, with a phase difference of 90 degrees between them.

output. The design tends to decrease the cost of construction
and to economise space.

§ 59. WINDING; STARTING TORQUE; POWER FACTOR.

One of the chief reasons for using polyphase currents is that they admit of motors which are both self-exciting and self-starting; and in many cases the costly and troublesome brushes and collecting rings may be done away with.

To obtain the best results it is necessary to reduce the exciting current to a minimum. This is accomplished by

FIG. 169.—Grooves for Windings of Polyphase Machines.

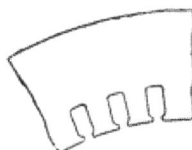

FIG. 170.—Grooves for Windings of Polyphase Machines.

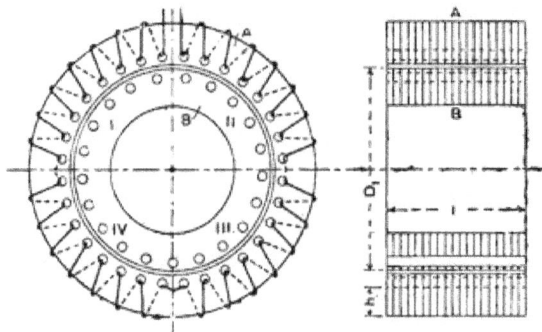

FIG. 171.—Diagram of Construction of Polyphase Motors, showing Holes for the Windings.

decreasing the air gap to a mere mechanical clearance by bedding in iron the windings on the stationary and rotating parts. The coils are sometimes laid in grooves as in Figs. 169 and 170, and sometimes wound in tunnels as in Fig. 171.

Grooves are, perhaps, the best, since they cause less magnetic leakage. When the motor is designed without brushes, the line current is supplied to the fixed part. This design is chiefly applicable to small motors. For large designs it is usual to arrange the exciting coils in the revolving part of the machine. This necessitates brushes and rings, it is true, and consequently simplicity of design is departed from. The chief point gained is that the losses due to hysteresis, eddy-currents, and self-induction are practically limited to the relatively small moving part of the machine; whilst the massive stationary portions are magnetised by currents of frequency equal only to the difference between the speed of the revolving coils and that of the rotary magnetic field.

When large polyphase motors are designed to work with the line current in the stationary part, it is usual to connect the ends of the revolving coils to collector rings on the shaft, and to close them through separate resistance coils. In this way the speed and torque of the motor can be regulated with comparatively small waste of power.

In the most recent designs the resistance coils are built into the rotating part and are thrown in and out of circuit by a lever actuating a device similar to a friction clutch. It is important to notice that a mere impedance coil will not answer the required purpose, since its self-induction causes a lag in the current, and therefore may increase rather than diminish the starting current. A non-inductive resistance coil must be used, the function of which is merely to regulate the current induced in the closed coils. It is found that there is one value of this current which corresponds to a minimum line current for each torque, and therefore gives the motor a maximum power-factor—a point of immense importance when a number of motors are running off a common power station.

By suitably proportioning the motor, the initial torque can be made as large as required, but, if high efficiency and good speed regulation are required it should not usually exceed about three times the running torque.

The relation between torque and current is well defined in Figs. 172 and 173, which are taken from a paper on Polyphase Transmission* by Dr. Louis Bell.

---

* *Electrical World* (New York), March 17, 1894.

x

In Fig. 172 is shown the relation between the static torque and current for a 10-H.P. (A) and for a 5-H.P. (B) three-phase motor.

Curve A_1 shows the effect of varying the resistance in the short circuited coils when the pressure is kept constant.

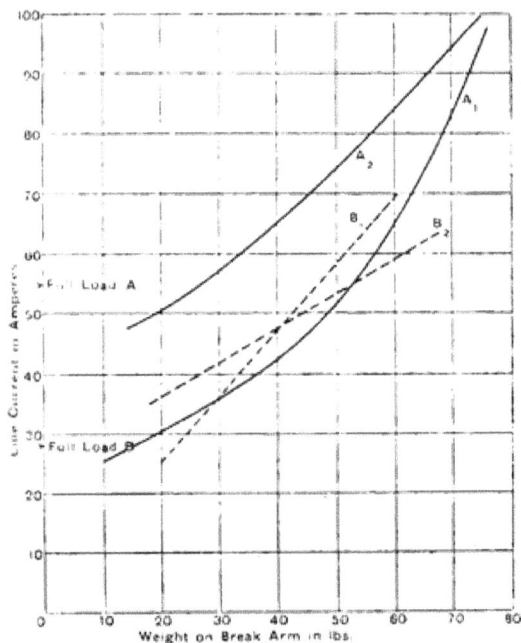

Fig. 172.—Curves showing Relation between Torque and Current in Polyphase Motors.

Curve A_2 shows the same with variable pressure for a constant resistance adapted to give a large torque.

The full load torque for the particular motor is 35 pounds on the given brake.

Curves B_1 and B_2 are similar curves for the smaller motor whose full-load torque is 17·5lb.

It will be noticed that each motor gives the running torque
at less than the full-load current, and that at maximum
current each gives roughly 50 per cent. more than the
designed full-load torque.   There is no rush of current at
starting.

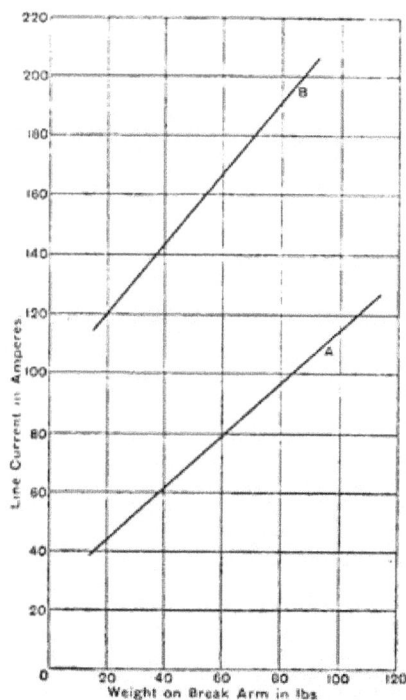

Fig. 173.—Curve A : Relation between Current and Torque with a
Properly Adjusted Resistance in Series with the Short-Circuited Coils.
Curve B : The same, with no Resistance.

Fig. 173 shows two curves taken from a 10-H.P. motor.
They connect amperes in the line with torque.   In B no
resistance is placed in the closed-coil circuits.   In A a carefully adjusted resistance is added, with the result that the
current for a given torque is reduced to about one-half of that

x 2

required in B. The importance of a starting resistance is obvious.

The initial torque of a well-designed continuous-current series-wound motor is about six times the running torque; and, therefore, a polyphase motor requires to be, relatively, of about twice the capacity for similar starting power; yet, since the polyphase design is the more compact, there is practically no

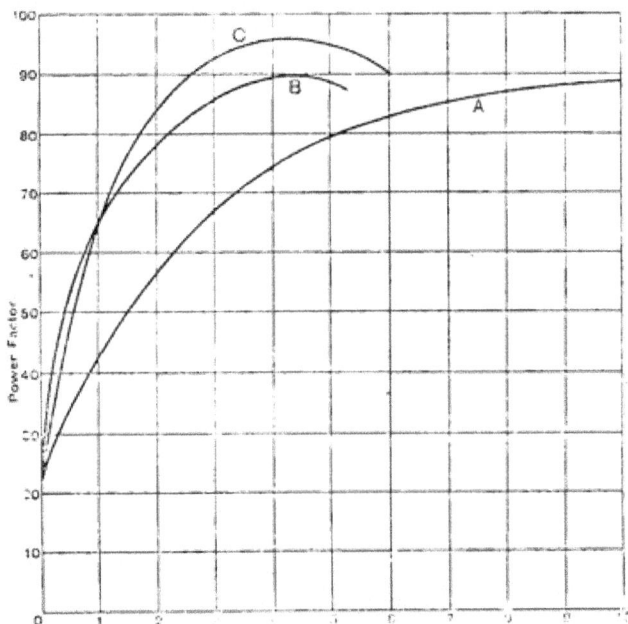

Fig. 174.—Curves showing Relation between Power Factor and Output in Polyphase Motors.

difference in the floor space occupied. Experience undoubtedly shows that the polyphase machine is the cheaper to build and the less costly to maintain in good running order. The power factor of polyphase motors varies with the load and within wide limits; but at from half to full load it appears to average from 75 per cent. to 91 per cent. There is thus no difficulty in running a number of motors upon the same circuit.

The variation of the power factor of a four-pole 15-H.P. motor at 50 cycles is shown in Fig. 174 by curve A; that of a 5 H.P. motor of similar design by B; and that of a 5-H.P. motor, specially designed to give a high power factor, by C.

It is sufficiently evident that the requirements of commercial working are met in these machines.

§ 60. COMBINATION OF POLYPHASE AND CONTINUOUS CURRENTS; RECTIFIED CURRENTS.

It is sometimes urged that alternate currents are not suitable for charging accumulators. In a sense this is true,

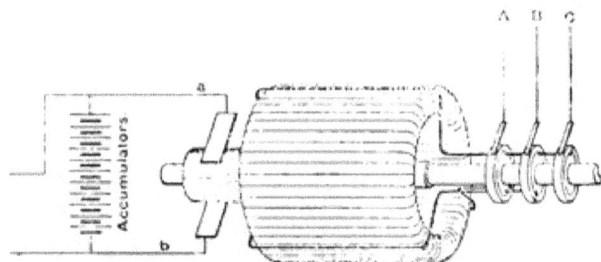

Fig. 175.—Diagram of Gramme Armature, designed to work as a Polyphase Continuous-Current Converter.

A, B, C, Polyphase Mains. *a*, *b*, Continuous-Current Mains, coupled for charging accumulators. The field magnet coils may be excited separately, or in shunt to the accumulators.

but there are two ways in which they can be utilised for this purpose. They possess some interest in a transmission of power plant. Most batteries are useful for dealing with lighting during the light load periods.

If an ordinary two-pole continuous-current dynamo have three connections made to the armature coils at intervals of 120deg.

each, and these be joined to rings in connection with the mains of a three-phase system, the armature will run as a motor (assuming the field to be separately excited), and a continuous current may be collected at the commutator, which can be utilised for charging accumulators.

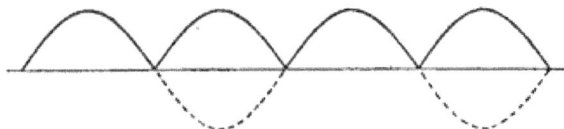

FIG. 176.—Rectified Alternate Current.

This machine may be described as a polyphase continuous-current converter. It is a practical device. Most dynamos can be readily adapted for this purpose, but if a high efficiency of conversion be required it will be necessary to make a special design. A general idea of the arrangement may be gathered from the diagrammatic sketch in Fig. 175.

FIG. 177.—Ferranti Rectifier, as used at Portsmouth.

Another method of using an alternate current for charging batteries is to "rectify" the negative waves. The resultant current may be diagrammatically represented as in Fig. 176, in which the dotted curves below the time line have been rectified. The curves show the instantaneous values of the current (or pressure), and the effective value is $\sqrt{\frac{1}{2}}$ of that of the maximum ordinate.

The reversal of sign is accomplished by a two-part commutator running synchronously with the alternator, and driven by a small synchronous alternate-current motor. (*See* the Thomson-Houston self-exciting alternator, Fig. 115, p. 232.)

Fig. 177 shows one of the Ferranti rectifiers used at Portsmouth for running arc lamps in series on a 10-ampere circuit having a frequency of 50 ∿ per second.

A rectified current appears to be admirably adapted for serving arc lamps, because the rhythmical waves of which it is composed assist the feeding of the carbons.

It is possible that rectifiers may prove to be useful details of a power plant.

CHAPTER IX.

ELECTRIC TRANSMISSION OF POWER IN MINING OPERATIONS.

§ 61. INTRODUCTION.

PERHAPS the most obvious field for the utilisation of electricity in power work is to be found in mining. At all events, the first practical plants were applied for mining purposes, and by far the greater number of large electric motors running in Great Britain to-day are used in this connection. The superiority of electricity as compared with compressed air or hydraulic power for driving dip pumps and other underground machinery was recognised as soon as the electric motor was discovered. But it was not until about 1885 that the new plant was well enough engineered to be cheap and reliable. The first application of sufficient magnitude to demonstrate indisputably the possibilities of the electric motor for mining work was made at Messrs. Locke and Co's, St. John's Colliery, Normanton, by the General Electric Power and Traction Company (Immisch and Co.), to the designs of the author. The motor was of about 60 brake horse power, and worked a set of ram pumps raising 120 gallons of water per minute through a vertical head of nearly 900 feet. It was an assured success from the start, and although modifications and improvements have since been introduced in many of the details, the general method still obtains for continuous-current mining work. The application of polyphase currents has opened up fresh possibilities, and largely extended the scope of electric power work, by obviating the need for commutators and brushes, and therefore removing the chief objection to the use of motors in collieries. The author confidently regards polyphase motors as essentially the machines of the future for mining work.

In the previous chapters this subject has been necessarily discussed in a general way, so as to include all varieties of plants. This chapter is confined to the consideration of mining work and its special requirements.

There is a wide difference between collieries and metalliferous mines in the character of the work ; and, indeed, the local conditions of different mines vary so largely as to require special treatment in almost every case. But the main difficulties peculiar to mining work, i.e., which do not usually obtain with surface plants, may be classified under three heads :

(a) "Falls" from roofs and sides, and "creeping" of the floors.

(b) Water, either continuous or intermittent.

(c) Explosive gases.

These may occur singly or conjointly, and evidently require properly selected plant and careful disposition of mains, junction boxes, switches, cut-outs and machines. What is successful at one pit, or part of a pit, is not necessarily so at other places ; and hence various methods are employed— some good and some indifferent. Only an engineer experienced in this class of work is able to specify with fair probability of success the best kind of material and most suitable method of erecting plant for each case. Much defective work has been done from want of practical acquaintance with collieries and mines, although the plant supplied has been usually good of its kind. That bugbear of competition, the lowest tender, has also much to answer for here as well as in other departments of electrical engineering;

and the practice of asking contractors to tender to their own specification instead of to that of an independent consultant has largely contributed to the breakdowns, which are far more frequent than need be. The absurdity of such a course is rendered the more apparent when it is considered that however up-to-date the mining engineer may be, he cannot possibly have such a close acquaintance with the quality and classes of electric cables, different *points* in dynamos and motors, and the various details of an electric plant, as the expert who gives his entire time to such work. And, moreover, the expert is not hampered by patents or special types of machines, etc., as a contractor is very likely to be, especially if he is also a manufacturer.

It is proposed to consider here some of the chief difficulties peculiar to mining work, and to examine the methods most in vogue to meet them, and also to suggest, as far as possible, the best plant and its most effective disposition for working under various conditions.

It is convenient to treat this part of the subject under the following divisions :—

(1.) Engine House and Equipment.

(2.) Conductors, Shaft, and Underground Cables, Junction Boxes, Switches and Fuses.

(3.) Motors and Driven Machines.

These will be considered in separate sections.

§ 62. ENGINE HOUSE AND EQUIPMENT.

The engine-house will generally be upon the "surface," although in some cases, where a steam-engine is already at work near the pit bottom, it may be necessary to erect a dynamo below ground. Electric plant frequently plays so small a part in the main machinery of a colliery that it is not permissible to provide a separate house for the dynamo. In such cases the winding-engine house or the fan-engine house will probably be selected. If the fan-engine has a margin of power, and is kept

running at a constant speed, it may be possible to couple the
dynamo to it through a fast and loose pulley. But this is not
a common occurrence, nor is the arrangement always advisable.
The fan forms a steady continuous load, and its engine should
be carefully proportioned, both as regards size and speed, to give
the most efficient running. And further, since the safe working of a colliery depends to a very large extent upon the ventilation, many mine managers very properly refuse to allow fan-
engines to be put to any other work than that of driving the fan.

It may be assumed, then, that the dynamo (or dynamos) will
be driven by an independent engine. In the early days of
electric work in mines it was frequently necessary to use any
odd engine that happened to be available at the time; and
hence some very curious combinations. The author has seen
a modern efficient dynamo coupled by belting to a large slow-
speed engine of antiquated type, rendered still more inefficient
by being supplied with wet steam at less than 40lb. pressure,
so that the high efficiency of the dynamo was practically
annulled.

Managers, however, are beginning to recognise the saving in
coal and steam made by using high-pressure boilers and quick-
running engines, and the electrical engineer has now little
difficulty in arranging for suitable steam plant. Having regard
to the fact that the fuel burnt at collieries is usually the most
unsaleable there, and frequently almost dust, mechanical
stokers are of special use. Vicars, Proctors, or any of the
well-known types, answer the purpose admirably, and will be
found to effect economy in labour. The type of boilers will
depend upon the class of water and upon local circumstances.
Generally, however, Lancashire boilers, with cross tubes, give
the best all-round results, and have the advantage of being
easily set and repaired by ordinary labour. The working
pressure may vary from 80lb. to 120lb. per square inch;
whereas multitubular boilers, although admirable for raising
steam at short notice, are comparatively difficult to clean out,
and require much more careful attention and skilled labour to
effect repairs. The same remark applies to water tube boilers
in connection with this class of work.

In most cases where the engineer has a free hand, and the pressure does not exceed, say, 500 volts, he will advise steam dynamos—*i.e.*, engine and dynamo directly coupled (*see* Fig. 178). This arrangement gives a positive drive, reduces the space occupied to the smallest dimensions, effects considerable saving in first cost, and gives increased facility for repairs ; while a stand-by set can be compactly placed ready for instant use at a moderate increase in the cost of the running plant. Granting the use of a high-speed engine, the question of open or closed type arises. There are various arguments in favour of each. The open type (*see* Fig. 178) has the advantage that all the rods and bearings are in full view, and

Fig. 178.—Steam Dynamo and Open-Type Engine.

can be inspected while the engine is running, and also that it is readily got at to make repairs. But it is likely to be affected by dust, and is to some extent liable to accidental damage from external causes.

The closed type (*see* Fig. 179), on the contrary, is completely protected from dust and chance external damage, and the crank shaft splashes into a lubricant in the crank chamber, thus ensuring lubrication so long as the oil is maintained at the proper level. Messrs. Willans and Robinson supply with their well-known closed-type engines a sight gauge, which shows continuously the height of the lubricant in the chamber. The speed of the engine will be selected with reference to its dimensions and that of the dynamo to which it is to be coupled. The number of revolutions will be large compared with those to which mining managers are accustomed, being from about 300 to 500 per minute. The question of single or double-acting engines is an open one, and both sides are admirably championed. Perhaps both types, if properly made, are equally good.

The type of dynamo will depend very much upon the class of work. The discussion of the various windings—compound, series, shunt and separate excitation—in § 14, page 59, gives the special features of each.

FIG. 179.—Holmes-Willans Steam Dynamo and Closed-Type Engine.

The pressure, as already suggested, will not be much higher than about 500 volts. A number of manufacturers build excellent machines, and it is easy to select a suitable dynamo or motor if a fair price be paid.

The conditions governing the choice of steam and electric plant are similar to those detailed in § 1, page 7. In plants where the work comprises pumping, hauling, coal-cutting and lighting, the generators should be capable of parallel working, feeding omnibus mains. But if there are only large motors, each coupled to, say, pumps or other independent machines, then it may be advisable to run separate pairs of mains to each motor, and to have a separate generator to each circuit. If the loads permit of one type of motor and one type of generator, a couple of spare machines will be sufficient to cover all. For driving pumps series winding is usually preferable for both dynamos and motors if a continuous-current system be adopted.

If polyphase currents be used, the same suggestions are applicable, with suitable reservations. The pressure at the generating station may be restricted to about 500 volts, and then transformers will not be necessary for stationary motors of from 10 B.H.P. and upwards. But in coal-cutters and rock-drillers this pressure is too high for the men to handle the machines with comfort, and it will be advisable to use step-down transformers to reduce it to about 50 volts. At this pressure a "shock" will be impossible, and the machines can be safely worked even in the dampest places.

In some cases of long-distance transmission it may be advisable to use a much higher pressure in the "line" between the generating station and the step-down transformers. The question then arises whether the line pressure should be reduced to, say, 500 volts at the pit bank, be supplied at this pressure to large stationary motors, and be further reduced for small or portable machines. These are questions, however, which must be decided for each case upon its merits, and a mine manager will do well to take the opinion of an independent expert before signing a contract for plant.

§ 63. SHAFT CABLES.

The classes of conductors suitable for the severe conditions of mining work are not very numerous, and differ chiefly in the kind of insulation, and the means employed to protect them from mechanical injury and from the effects of water.

Broadly, there are two kinds of insulation which have stood the test of time, viz., vulcanised rubber and bituminous compounds.* The former is costly and is thus better adapted for the lighter classes of cables, say up to strands of 19 15 S.W.G.,

---

* These are fully discussed in § 18, p. 84.

which, owing to their position, do not require to be covered
with a coating of lead or armoured with steel strip or galvanised
iron wire. The latter class is cheaper and hence is more suitable
for large trunk mains ; but, since it rapidly loses its insulating
properties if in contact with water, it is necessary to encase
the compound with lead, which should be armoured to protect
it from injury. The best form of bituminous cable is made
concentric (see Fig. 180). The centre conductor is doubly
guarded, so that if the metal sheathing be cut through, only
the outer conductor is earthed, and the cable as a whole is
workable until an opportunity occurs for repairing the injury.

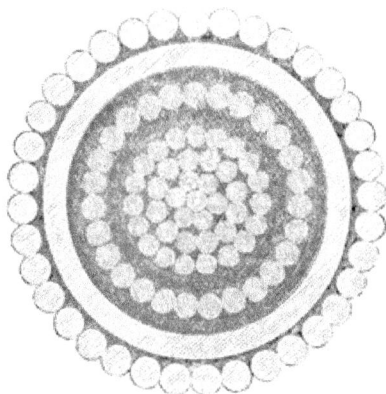

Fig. 180.—Armoured Concentric Cable, with Lead-Covered Bituminous
Insulation

The author finds it expedient, even in large plants, to limit
the size of each conductor to, say, a cable of 19, 14 S.W.G. strands;
and, if this be of insufficient cross-section for the current, to
duplicate the conductors and connect them to omnibus bars.
With this provision, an accident to one conductor does not
cause a complete stoppage, for the damaged cable can be
disconnected until the fault is localised and repaired. (It is
oftentimes a matter of impossibility to make repairs during
the drawing of coal, and hence the importance of providing
spare conductors.) Another advantage is that small mains
are relatively light, and are, therefore, easily handled and

fixed in position. And, further, the cost of repairs to the smaller sizes of cables is generally much less than to the larger ones, owing to the greater ease with which the joints can be made.

The methods of running cables down shafts and fixing them in position differ in almost every case. Pit shafts are usually of circular form, sometimes driven through the live rock, sometimes built round at intervals with courses of brick or stone laid on cast-iron cribs, sometimes faced with heavy timber walls, and occasionally lined with cast-iron sections through quicksands or strata heavily charged with water. If wooden "*conductors*" are used to guide the cages, heavy cross bearers of timber are built into the sides of the shaft at intervals to carry them. When iron rope guides are used the shafts are practically void of timbering.

These different conditions evidently demand different methods of supporting the cables.

In the first place, it is absolutely essential to guard against falling coal and stones, either from the cages or from the sides of the pit. The best position for placing the cables will, therefore, be largely determined by the shape of the cage, and whether its sides are covered up so as to restrict the falling of pieces of coal to the open ends. Again, stands of iron pipes already placed may limit space still further. In general it is not advisable to run electric cables near to pipes, because workmen repairing the latter may damage the cables. Another consideration of importance is whether the up-cast or down-cast pit is the better adapted for the purpose. If the " winding " be done from the down-cast, as is usual, it may be convenient to use the up-cast. But care must be taken that the up-cast air is not laden with any erosive vapour, such as is sometimes the case when the exhaust from an underground engine and the gases from a furnace burning coal containing a large percentage of sulphur, are turned into the up-cast. The author knows of one case in which a pair of lead-covered cables run in such a pit were corroded through in dozens of places in the course of a few months. In some soils there are found erosive agents that attack iron and rapidly rust it away, and

Y

in others a similar action takes place with lead. Generally it has been the author's practice to use the down-cast shaft; but there is no rule, and in all cases it is possible to ensure success by making suitable arrangements.

The plans adopted for supporting cables in shafts differ widely. In all cases, however, where it is desired to give the work a permanent character, the cables must be entirely encased in wood, run in iron pipes, or heavily armoured. These three methods are not always equally applicable, for local conditions may make one of them the cheapest or most desirable. If there is a set of, say, 4in. pipes already erected in a shaft, and it is desired to run a pair of, say, 19/16 S.W.G. cables down them, and the depth is not great, a pair of vulcanised rubber mains may be suspended in them from earthenware insulators. The weight on the supports will not usually be great, for the cable will generally bend from side to side, and thus may sometimes nearly carry its own weight if sufficient slack be allowed. In running cables down pipes it is advisable to insert a small flexible iron rope, and use this to haul in the insulated conductor, which should be payed out from the drum, perpendicularly into the centre of the pipes, over a wheel of sufficient radius. This method prevents damage to the covering of the cable. The interior of the pipes should be scoured to clean off sand from the cores, or the insulation will inevitably be abraded during the hauling-in.

If the pipes be of large diameter, and the depth too great for the mechanical strength of the cable, it must be supported at intervals. This may be done in several ways, but it will be necessary to break the continuity of the iron pipes in order to insert wedges, clamps, or similar devices.

It may be better, then, to adopt a wooden casing which will support the cable along its whole length. In using wood, however, care must be taken that the section is of sufficient strength, and that the "lengths" are spiked firmly to the pit sides. The grooves may be cut in the face, as shown in Fig. 181; or in the sides of the casing, as in Fig. 182, which illustrates a method used most successfully by the author in a number of mines.

The covering boards may be held in position by coach screws. This enables an examination of the cables to be made at any spot—which is oftentimes a convenience. The grooves should be cut an easy fit for the cable, which will then be perfectly

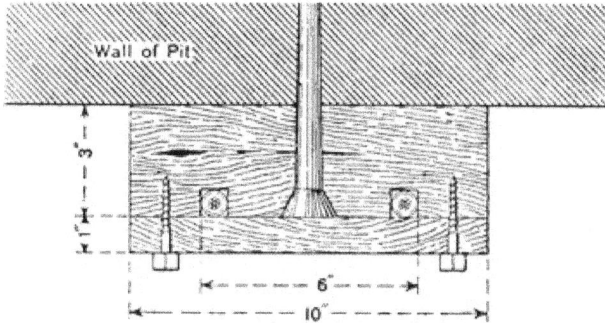

FIG. 181.—Wood Casing for Pit Work.

supported, and only a light attachment at the top will be necessary. It is always advisable, however, to loop the top ends securely round earthenware insulators firmly carried at or near to the surface.

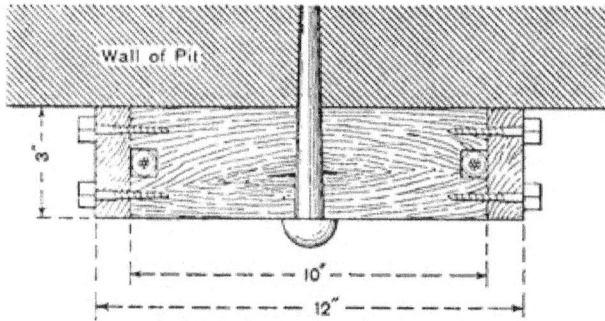

FIG. 182.—Wood Casing for Pit Work.

Concentric cables lead-sheathed and armoured are so stiff that they may be clamped directly to the pit sides, and then form strong compact work. As they are very heavy, there is need for strong brake gear on the surface to pay them out by,

Y 2

as they cannot be fastened in position until the whole length
is down the pit. It is advisable with very heavy cables to
lash ropes at intervals, and pay them out at the same rate,
thus relieving the top end of the cable of part of the weight.
This class of cable is steadily finding favour for pit work.

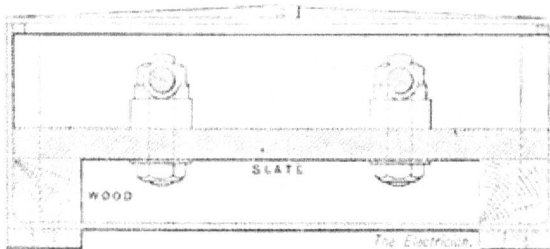

Sectional Elevation.   One-third full size.

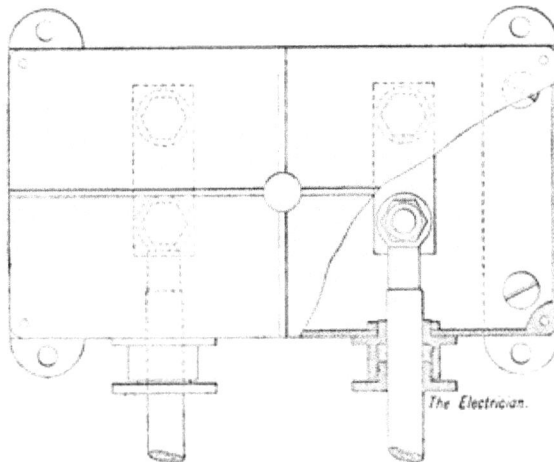

Plan (part section).   One-third full size.

Fig. 185.—J. Davis and Son's Junction Box, for use in Mines.

On no account should joints be permitted in new shaft
cables, although it may be necessary to make repairs in old
ones.   A cable with continuous insulation is much more likely
to stand the wet and exposure of a pit than one with joints
made with a different dielectric from the rest of the insulation:

the more so as the joints will probably be defective from the beginning. And it should be recollected that jointing is an extremely difficult operation in a shaft, owing to falling water, the cramped position, and the strong draught of air which together render soldering an impossibility in many cases, and often make the insulation uncertain owing to the presence of moisture. Of course it is possible, assuming time to be allowed, to provide special apparatus which, in the hands of skilled men, will enable excellent work to be done ; but breakdowns usually occur unexpectedly, and frequently have to be made good by the resident staff with such simple materials and facilities as may be at hand.

In most cases the author prefers to arrange junction boxes at the top and the bottom of shafts, so that the shaft cables can be severally tested independently of the surface and underground work. These boxes must be housed in dry places, secure from mechanical injury. For this purpose there should be sufficient free length of cable at top and bottom to allow the boxes to be placed well away from the drawing decks and pit bottom. An effective junction box manufactured by Messrs. J. Davis and Son is shown in Fig. 183.

§ 64. UNDERGROUND CABLES ; JUNCTION BOXES ; SWITCHES AND CUT-OUTS.

Each "district" is usually supplied from an independent pair of conductors coupled at the pit bottom to an omnibus bar or to a separate pair of shaft cables.

These conductors are, therefore, of small size, not often exceeding 19 strands of No. 16 S.W.G. They are carried on the "sides" or "roofs," being supported by wooden cleats or earthenware insulators, or buried in the "floor." The choice of the three methods will be determined by the peculiarities of the "level" in which they are run. The class of insulation will depend both upon the method of erection and the difficulties to be dealt with.

In dry roads, well supplied with timber props and cross bearers, a light vulcanised rubber cable, carried in wooden cleats

or on insulators, will prove to be cheap and efficient. An occa-
sional fall of roof or sides will not as a rule cut it if a little
slack be allowed between the supports; and "creeping" of the
floor will not affect it. Indeed, in the majority of cases where
long lengths of small conductors, about 7/16 S.W.G., have to
be carried to the "faces," this is the only practicable method,
from the simple consideration of first cost.

Wood casing is out of the question for "road-work," except
for the lighting circuit near the pit bottom, not only because
of its cost, but also because of the instability of most of the
"roads."

In main engine planes which have no "weight" upon them,
permanent work may be made of the erection; and it will
generally pay to bury the cables beside the metals and to
cover them loosely. But a rough wooden trench with a covering board should be laid to receive the cables, unless they are
armoured, for otherwise an unlucky blow from a pick may cause
a breakdown. If heavily armoured concentric cables, as shown
in Fig. 179, p. 308, are used, it is sufficient to bury them from
about 6in. to 12in. below the surface of the road, well to one
side, where they will not be disturbed during repairs to the
metals and roads. If conductors are carried along "travelling"
roads, it is specially necessary to secure them from damage
at the hands of the miners, who are sometimes inclined to get
"lightning" from them by means of their picks. If the road be
dry and in a fairly settled condition cables may be laid in fine coal
dust, covered by rough boards 1in. thick, and then protected by
a layer of coal dust of about 6in. deep. This forms a very safe
and cheap bed for hemp-braided vulcanised rubber cables, or
for lead-covered cables served or braided with hemp. It is
specially serviceable in travelling roads, or where the conductors are laid for a temporary purpose.

Whatever the class of cable or the method of laying it, however, the conductor should not be in continuous lengths of more
than about 500 yards, in order that a fault may be readily localised. At the junctions of sections cast-iron boxes (see Fig. 183,
p. 312) should be placed, with screw terminals on porcelain bases,
through which the necessary connections can be quickly made or

unmade. On no account should cut-outs be placed at these points unless special provision be made to enclose them in flame-tight boxes ; they should be confined as far as possible to the generating station and pit bottom.

At the far ends of the levels where the distributing mains are joined to the feeders, it is good practice to place cast-iron junction boxes in the "gates," so that any or all of the distributors may be coupled up according to requirements. It is imperative, if cut-outs be used here, as is sometimes necessary, that they be placed in flame-tight boxes, for this position is close to the coal faces, from which gas may be given off at any time.

The "distributors" are necessarily of a temporary character and will be shifted as the work progresses. They must therefore be light and flexible. If convenient they may be carried upon the props, and in the absence of timber may be laid beside the roads, unless they are hung on rough iron spikes driven into the rock. In any case they will often be subjected to rough usage. For attachment to movable machines, such as coal cutters and rock drills, it is usual to provide either concentric or twin cables, insulated with vulcanised rubber and braided with steel wires. Either of these can be made sufficiently flexible, and they are also strong enough to stand being hauled across rough shale and coal.

If a distributor feed a dip pump it should be erected in a semi-permanent manner ; since such pumps are liable to be drowned-out, and the plant may have to be withdrawn several hundred feet or yards at short notice. In such cases the conductors should be brought back with the motor, and not cut. They should be coiled and hung up at the pumping station, being uncoiled again as the water recedes. The best insulation for this work is vulcanised rubber well braided with hemp and ozokerited over all.

For a polyphase current system it is necessary to run three or four conductors in place of the two required for a continuous-current or single-phase alternate system. For permanent work three conductors may be built into a concentric cable, which

may be lead-covered and armoured.   With a four-wire system
two concentric cables of two conductors each are preferable.

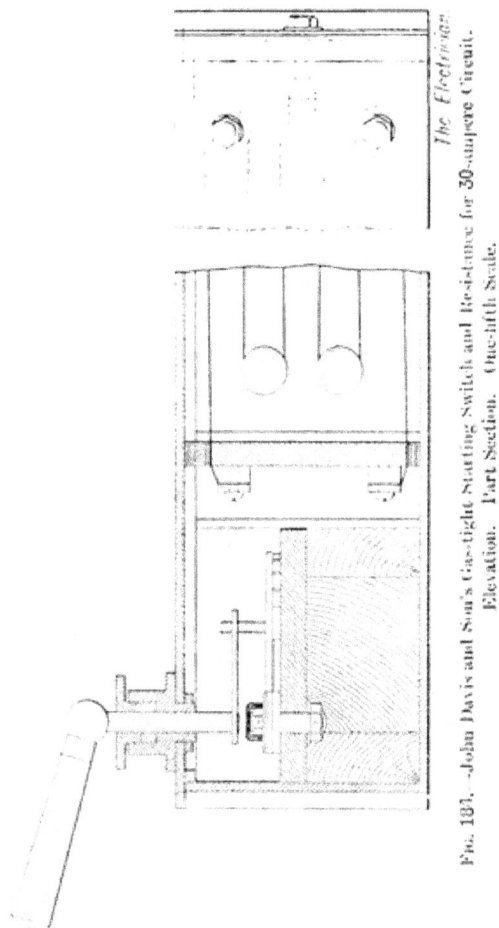

FIG. 181.—John Davis and Son's Gas-tight Starting Switch and Resistance for 30-ampere Circuit.
Elevation. Part Section. One-fifth Scale.

The method of erection is in no respect different from that
adopted with continuous-current working.   The three wires are
carried practically as readily as two.

The use of switches and fuses below ground in places likely to contain explosive gas requires careful regulation, and is likely to become the subject of special legislation in the future. The

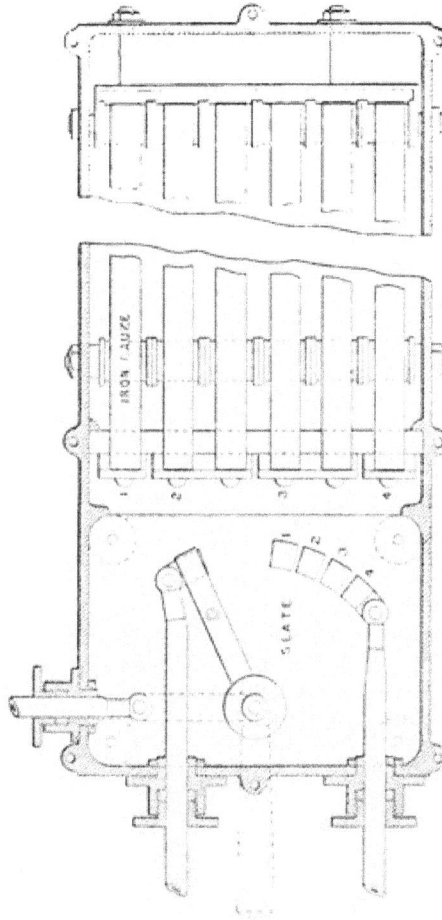

Fig. 184a.—J. Davis and Son's Gas-tight Starting Switch and Resistance for 30-ampere Circuit. Plan. Part Section. Lid Removed. One-fifth Scale.

The Electrician.

author states emphatically that, in his opinion, no switch, cut-out, junction box, or resistance frame that is not absolutely flame and dust-tight should be allowed in a coal mine where naked

lights are prohibited.   It is madness to forbid the use of a naked
light and yet to allow the types of switches and cut-outs sometimes found in pits.

To ensure absolute freedom from firing gas with a continuous-
current motor is perhaps impossible, but no such difficulty
should be found with small stationary devices, such as switches,
cut-outs, and junction boxes.   It is merely a matter of expense,
and proper precautions should be insisted upon in all permanent
plants.

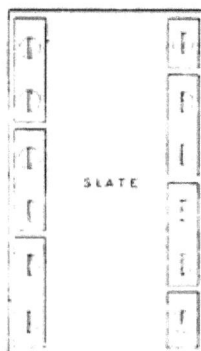

FIG. 184B.—Showing Plan of Slate Base and Attachments to the Resistance
Coils in Figs. 184 and 184A.

In Figs. 184, 184A and 184B are shown various views of Davis's
Mining Switch with resistance coils attached.   It is used for
starting and regulating motors.   The whole is enclosed in an
iron case, and the switch contact pieces break circuit in a gas-
tight compartment.

§ 65. UNINSULATED RETURNS; SAFETY CABLES.

It has been the custom at some pits to use old iron ropes
as returns.   These are simply laid along the roads or hung up
by hooks to the side props.   The practice, however, is not to be
recommended in pits where there is any chance of gas, and in
a large plant, with a pressure of 500 volts, it would in any case
be impossible.   The author coupled a return in this manner at
Andrew's House Pit, Durham, in 1887, the arrangement being

adopted simply from considerations of first cost. The pressure
at the dynamo terminals averaged 225 volts, and the power in
this part of the circuit never exceeded 8 E.H.P. The rope lay
for a considerable part of its total length of 1·25 miles in
water. The leakage of current, inevitable in a pit, decomposed
water at various places, and the iron corroded rapidly at the
junction to the copper return near the motor, and small
sparks could often be obtained between the rope and the tram
metals near the pit bottom, thus showing that the iron rope,
although lying on the wet ground, did not make good "earth."
The danger was recognised and caused no difficulty ; but if the
pit had been gaseous, it is probable that an explosion would
have resulted.

A bare separate return, if justifiable at all, can only be so on
account of cheapness : but the apparent saving is in most cases
more than counterbalanced by after troubles.

If, however, a system of concentric mains be used with the
outer conductor in metallic contact with iron or steel wire
armouring,* the conditions are altogether different ; and probably as great an immunity from the danger due to a broken
conductor can be obtained by this arrangement as by any
of the patented safety cables.

There are two classes of bare return concentric cables in use.
They are shown in Figs. 185 and 186. The former has a lead
covering, which is squirted around a copper conductor, the two
together forming the outer or return circuit. Armouring can
be applied outside, and this adds to the conductivity while protecting the soft lead, whose chief function is to keep water
from the insulation between the centre and outer conductors.

The second class is cheaper, and the return circuit is formed
simply by the armouring, which is placed directly upon the
insulation. It is only suitable for use in dry places, where it
will give good results.

The chief advantages offered by these cables, in connection with mining work, is the security they give against

* For example. the concentric cable system of Messrs. Mavor and Coulson.

external sparking, and their mechanical strength, owing to
which they may be stapled to rough walls, like gas piping.

There are also conductors which may be collectively classed
as "safety cables." In all of them the object is entirely to

Fig. 185.
Concentric Cable with uninsulated outer conductor consisting of a spiral of
sheet copper under the lead, the lead, and the steel armouring.

obviate sparking at the point of rupture when an active conductor is cut in two. The general method is to provide a
small auxiliary cable in parallel to the main conductor. This,
in the event of a fall cutting the cables, is arranged to break

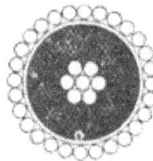

Fig. 186.
Concentric Cable with uninsulated outer conductor, consisting of a spiral
of sheet copper under the steel armouring—no lead.

either *before* or *after* the main conductor is broken; and in each
case actuates a magnetic cut-out which opens the main circuit.
The idea will be best understood by reference to some of the
better known solutions of the safety-cable problem.

Mr. Ll. Atkinson, of Messrs. Easton, Anderson and Goolden, was the first to patent a cable of this class. It was not designed for concentric wiring, and hence two wires are required for each circuit. The cable (*see* Fig. 187) consists of two concentric conductors in parallel, insulated from each other, excepting at the main terminals of the circuit to be protected. The current therefore is in the same direction in each conductor. The outer, which is designed to carry the larger part of the current, is of any ordinary make. The inner is constructed of fine wire in the form of a continuous cylindrical coil, the outer diameter of which is covered by insulation, and forms the core around which the main conductor is laid. It is assumed that when a stone falls on the cable the outer con-

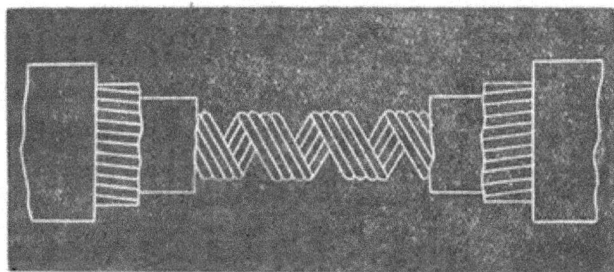

Fig. 187.—Atkinson Safety Mining Cable, showing the outer conductor cut through and the inner one drawn out.

ductor, offering much resistance, may be cut through; but the loosely-coiled centre conductor will be simply drawn out of the insulation. The entire current will then pass through the fine wire, and blow a magnetic cut-out in the engine-room or other convenient place; the arc being thus transferred from a possibly dangerous spot to a safe one. The efficient action of this cable depends upon two conditions :—

(1.) Perfect insulation between the inner and outer conductors.

(2.) The continuity of the fine wire circuit being preserved during the fall.

In order to prevent an accidental contact between the two
conductors interfering with the working of the cable, the inventor provides a resistance in series with the fine wire at
its negative end, so that the fall of pressure is less along the
inside than along the outside.   The result is that there is
always a difference of pressure between the inner and outer
conductors at each part of the cable.   And a short circuit
between the two at any point causes an increase of current
in the fine wire which operates the magnetic cut-out.

Generally it is found that cables hung lightly, with plenty
of slack, are dragged down and buried without suffering any
damage other than slight surface abrasion ; but if they are
firmly fixed they may be cut through as if with a knife.

It, may, therefore be reasonably questioned whether a fall which
is sufficiently heavy to cut a stout cable in two will not sever
the fine conductor before the spirals have time to draw out.

In recent installations a small separated insulated wire has
been hung (beside the main conductor) with a few spirals
twisted at intervals so as to allow a ready increase of length in
the event of a fall or of any undue strain.   This is found to
be cheaper than the compound cable, and to give equally good
results.   It should be mentioned that Mr. Atkinson regards
the cable as of more importance as a preventive against firing
timber from arcs than as an absolute means of preventing
sparks.   The author is in accord with this view.   For the time
taken to move the detaching mechanism on the surface will
chiefly determine the spark at the break in the cable, though
its effect is limited by the resistance of the unbroken fine wire
in parallel with the main wire.

The Charleton cable, patented by Messrs. R. J. Charleton, of
Newcastle-on-Tyne, is of similar make, but the inner conductor
is merely a small wire, about No. 18 S.W.G.   The currents in
the two conductors are in the same direction.   The action
depends upon the fine wire breaking *before* the large one, when
a magnetic cut-out is designed to open the main circuit.   Successful working depends upon the maintenance of the insulation between the two circuits, and also upon the time required

to move the magnetic cut-out levers. If this is not considerably less than the period of time elapsing between the cutting of the two cables it is doubtful whether sparks will not occur almost simultaneously at the break and the cut-out. The problem is complicated by self-induction and capacity, which under suitable conditions may seriously affect the *time* required to open the cut-outs after the breakage of the main conductor. It will be easy, however, to experiment for any particular case.

Fig. 188.—Sectional View of Nolet Safety Cable.
A, Armouring ; C, Main Cable ; C¹, Auxiliary Conductor.

Another interesting safety cable is the invention of M. Nolet, engineer of the Cockerill Company, Seraing, in conjunction with M. Jasper, of Liége, and has been applied to several coal mines in Belgium. It provides for a small wire insulated and built into the cable, as in the Charleton cable, but the currents in the two circuits are in opposite directions.

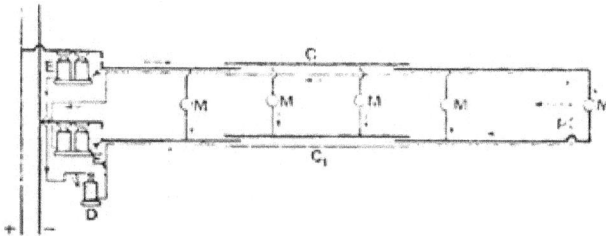

Fig. 189.—Diagram of Connections of Nolet Safety Cable.

C, Main Cable ; C_1, Auxiliary Conductor ; E and E_1, Magnetic Cut-outs ; M, Motors. Large arrows show direction of main current ; small arrows that of the shunt current.

A sectional view showing the method of construction is given in Fig. 188.

The fine wire is coupled up as shown in Fig. 189. The function of the small wire circuit is to close the main circuit through the two magnetic cut-outs (one to each of the mains).

Therefore, in the event of a breaking of the fine wire, the cut-out will operate and open the principal circuit.

M. Nolet has recognised that repairs to cables, as generally constructed, cannot be properly carried out without the use of soldering irons, which are inadmissible in a mine. To provide for this contingency his cable is made in short lengths, each section being provided with metal couplings for both of the conductors, as shown in Fig. 190. The rings for the subsidiary wire are shown at C′, and those for the main conductor at C. The couplings are made to fit together with easy friction, and are so arranged that a sliding movement of about one quarter of an inch is sufficient to open the auxiliary circuit, while a considerable displacement is required to uncouple the outer conductor. The joints are protected by india-rubber tubes.

The construction appears to admit of a *certain* time interval between the opening of the main circuit and the lifting of the

Fig. 190.—Sectional View of Nolet Safety Cable, showing the Method of Coupling-up.

magnetic cut-out, and also provides a ready means of repairing a damaged section. But the cable must be costly, and a slow movement of the measures, throwing weight on the cable, might interrupt the auxiliary circuit unnecessarily. It may be questioned whether the device would have time to act in the event of a fall shearing the cable instantaneously, as sometimes occurs.

In the author's judgment, none of these safety conductors gives more security than that obtainable with an armoured concentric one buried in the floor, or suspended loosely with plenty of slack between the supports. If a fall occurs, it will be brought down and buried. If the armour and outer conductor are cut through, the incision must first take place on one side, and then most probably the severed strands of the outer conductor will be pressed against the inner conductor, and so blow the magnetic cut-out without opening the circuit; because the bottom half of the outer conductor cannot be cut through until the top half is brought into contact with the inner one. The truth of this can be easily demonstrated by cutting the cable through with a cold chisel and hammer.

Success lies in hanging the cable so loosely that it cannot be cut through instantly. If this occur, it is doubtful whether the two conductors would always be jammed into good contact by a sudden blow, and so a spark might occur coincidently with the blowing of the magnetic cut-out.

Another solution of the safety cable problem has been proposed by Mr. Frederick Hurd. The conductors are run in iron pipes, through which a current of fresh air is forced by a blower, the motors and regulating devices being protected in a similar manner. The plan is good, and, if properly carried out, would seem to afford safety against a large variety of risks. But there is no provision for opening the electric circuit when the pipes are broken.

Messrs. R. B. Pownall & Son have carried the method a step farther. In a patent dated December 2, 1893, they provide means by which the maintenance of the electric circuit is made to depend directly upon the air pressure in the pipes; so that, in the event of damage to the pipes, resulting in a leak or in total breakage, the escape of air lowers the plenum and opens the magnetic cut-out controlling the particular circuit. The details of the system have not yet been made public; but the author is in a position to assert confidently that the scheme is a very practical one, and meets the varied requirements of colliery work in a most complete manner.

The danger of firing gas through a broken cable is, however, very remote. There is, so far as the author knows, no authentic instance of this having occurred, although the possibility is sufficiently evident.

The fact is that, for obvious reasons, cables are not usually run in roads with "weight" on them, and, where this is unavoidable, sufficient slack is allowed to prevent the nuisance

z

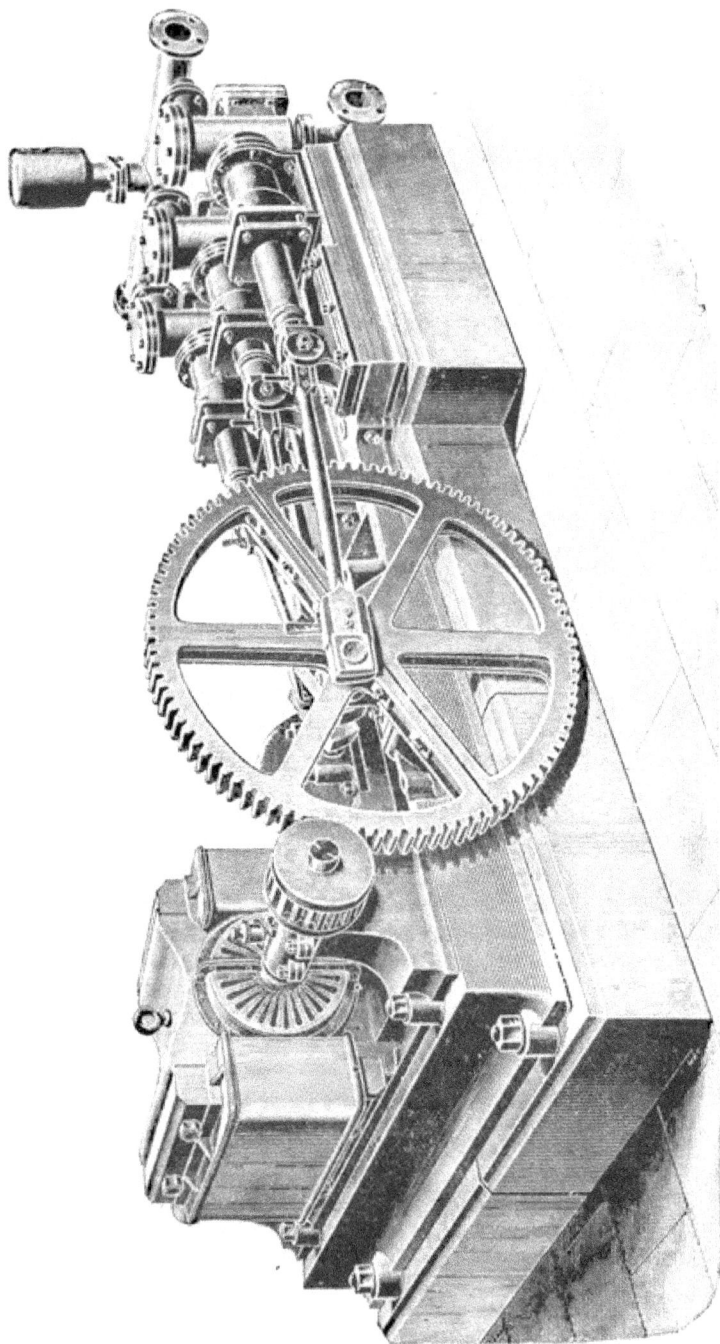

FIG. 191.—General Electric Power and Traction Company's Three-throw Main Pumps, Geared to Motor.

of frequent repairs, quite irrespective of the risk of an explosion, and, therefore, broken cables are rare.

With the gradual extension of electric work in coal mines for driving portable as well as stationary machines, the risks from this cause will be increased, and it is wise to recognise the danger and to adopt the simple and efficacious remedies suggested by experience. This is specially desirable with new plants.

§ 66. MOTORS AND DRIVEN MACHINES; SELECTION OF TYPE OF MOTOR; CONTINUOUS AND POLYPHASE CURRENTS.

The selection of the type of motor for any special work is generally not difficult. An approximate estimate of the power is easily arrived at, and a small error in its determination is of little importance; the crux lies in estimating the average and maximum torque. The cogency of this will be apparent on reference to the equations for torque of motor (§11, p. 43), and to the mechanical characteristics of motors (§ 14, p. 59).

For example: Suppose it is required to drive a main pump, as shown in Fig. 191, on a continuous-current circuit at 500 volts pressure at the motor terminals. The theoretical power in the water is known, and the losses in rising main, pumps, gearing and motor, are easily determined. The maximum torque will occur at starting, when the motor has to overcome the statical friction of the pumps and set the column of water in motion. The average torque will be simply that due to the constant load with full delivery of water. To meet these conditions, it is necessary to have a motor with a magnetic field sufficiently strong to give the necessary starting torque without undue sparking or an excessive quantity of current; and to give the working load at the designed speed with the highest possible efficiency consistent with a fair margin of power.

The motor winding should, preferably, be in series (*see* § 14, p. 60; also § 23, pp. 116 and 121).

The chief constructional point is sufficient strength in the armature shaft and bearings to start the pumps safely. If this

z2

Fig. 192.—Endless Rope Haulage, with Goulden Safety Motor.

be attended to the motor will be mechanically strong enough for the running load.

Again, assume it is required to erect an endless rope hauling plant—as illustrated in Fig. 192—on the same circuit. The various losses must first be determined, the weight of tubs, loaded and unloaded, debited and credited according to the grades, and a suitable allowance made for tub-wheel and rope friction. From these data the starting and the average torque can be determined. If the running load is fairly constant—as is generally the case with an endless rope—the series-wound motor is the most suitable, and the suggestions made with reference to the main pumping motor are equally applicable. The margin of power, however, requires to be greater than with pumps, as the load may be largely augmented at times by tubs leaving the rails. There is not much fear of overloading during busy periods, because an increase of torque will slow the motor and so decrease the actual power; but, since the current will be temporarily increased, a margin for the heating effect in the windings must be provided.

If, however, the system of haulage be a main-and-tail one, as indicated in Fig. 193, where the motor has to be frequently stopped, and to deal with a continually varying load, good speed regulation and perfect control of the motor become of more importance than efficiency. In this case a specially designed cumulative compound motor (*see* § 14, p. 65) may be used, with a switch and starting resistance designed to keep the shunt coils always in circuit during work time, and to introduce a variable resistance in series with the armature and series windings. Reversals of motion should be made by means of friction clutches and gearing, and *not* by the motor, as this would introduce complications which are better avoided in pit work. The maximum torque in this case is generally that at starting, but not necessarily, as the load may start on the level and afterwards have to climb a steep grade.

The best arrangement for working main-and-tail rope systems is to run the motor continuously, and to give forward or backward motion to the rope drums through friction clutches

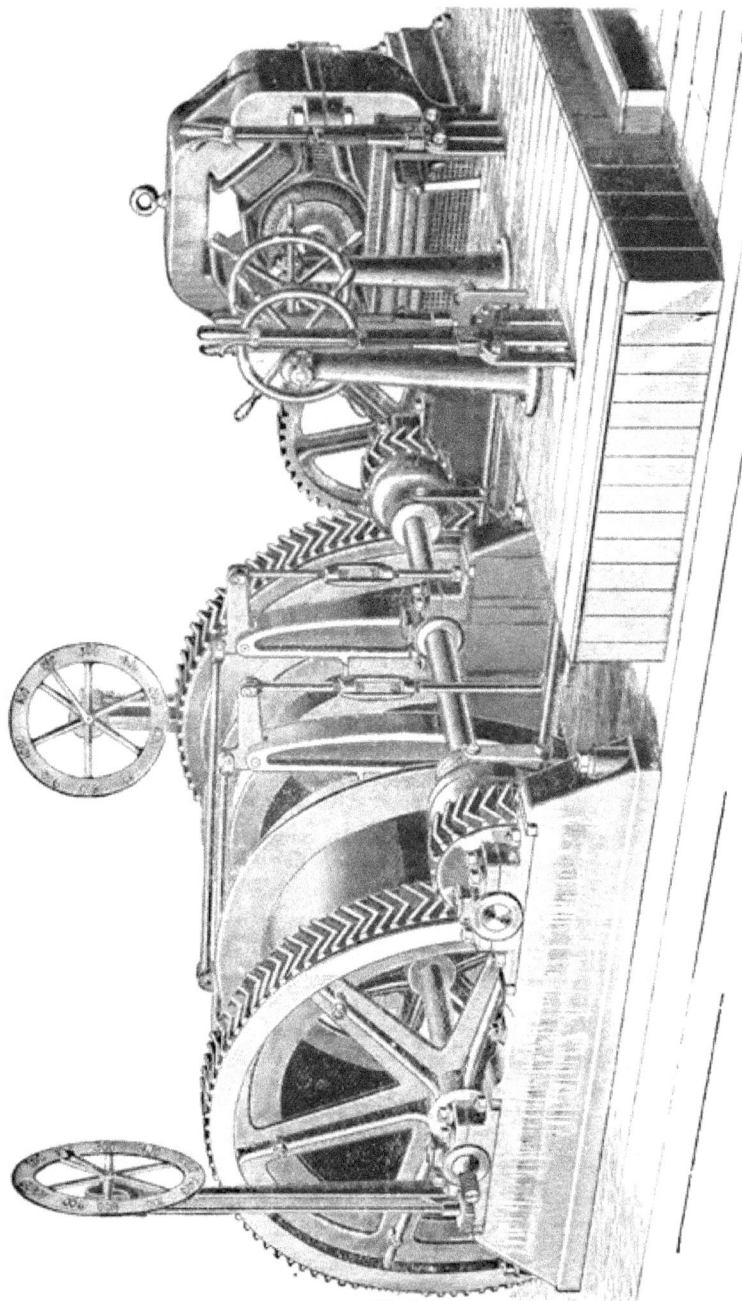

Fig. 193.—Thomson-Houston Main-and-Tail Rope Haulage.

(330)

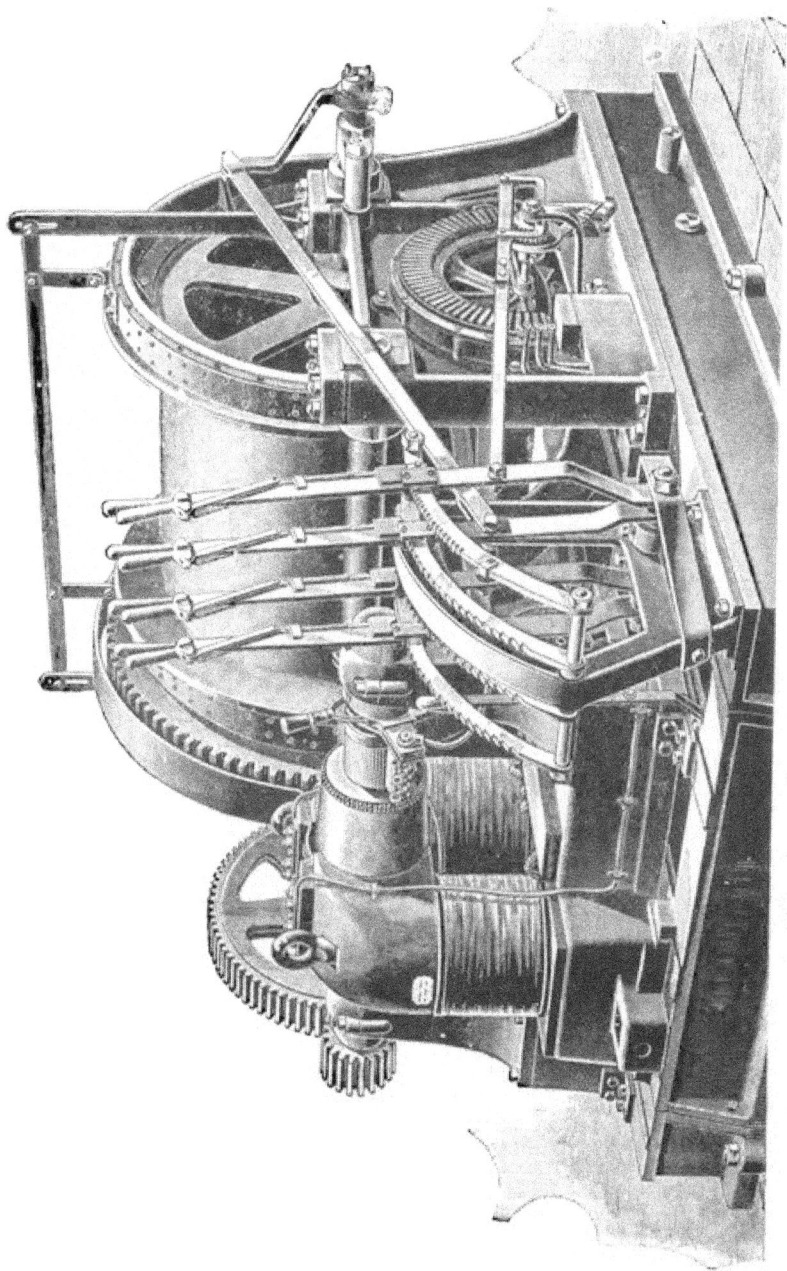

Fig. 194.—Thomson-Houston Single Rope Haulage or Winding Engine.

mounted on an intermediate shaft. If the motor be designed
in accordance with the above suggestions, no trouble will be
experienced.

The preceding remarks are also generally applicable to
single-rope haulage, although in some cases, where the load is
small and the pressure low, a series motor with a large starting
resistance may prove satisfactory. A suitable arrangement for
single rope haulage or winding is shown in Fig. 194.

For main winding gear, which is worked practically continuously during a shift, the author prefers a cumulative
compound motor having a large regulating resistance, the
motor running in one direction only, and the reversals being
made through friction clutches as already described for main-
and-tail hauling. But if the winding be used only occasionally, as frequently occurs with staple-pits, then a series
motor with a large regulating resistance is preferable. The
time occupied in raising the load is so short that the question
of speed regulation does not enter into the problem. What is
required is a large power at starting, gradually yet quickly
applied, and perfect control over the cage at all parts of the
run, but especially when stopping. It is also necessary to be
able to raise or lower the cage a few inches. For such work
the author has found series motors give excellent results.
He finds it convenient to couple the brake lever to the friction
clutch, so that when the brake is applied the circuit is opened,
and the motor is then ready for raising or lowering as required.

Dip pumps should invariably be driven by series-wound
motors* if continuous currents be used.

Several typical designs are shown in Figs. 195 and 195A.
The first is a view of the Goolden-Atkinson pump, which has
been largely used for draining dip areas. The second design
is that of the pump made by the Jeffrey Company in America.
It is permanently mounted on a trolley with removable wheels,
and is a light and efficient pump.

---

* See remarks on main pumps, p. 326.

Centrifugal pumps have generally a limited use in mines on account of their requiring to be " primed " before they will suck water, and also because there is always a risk of the pump ceasing to act if air gets into the suction pipe; in fact, they require frequent attention. These pumps are not suitable for

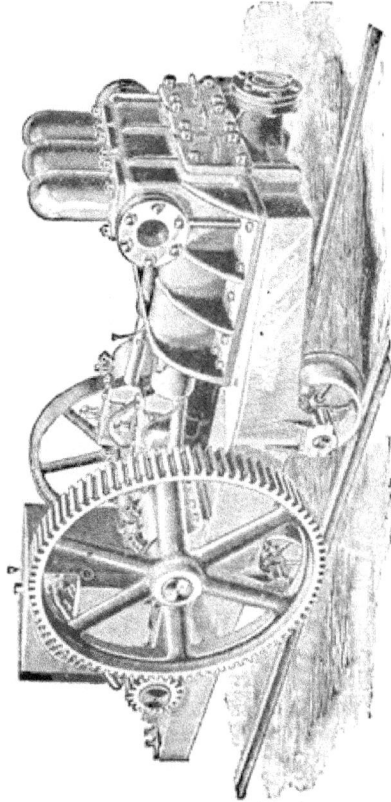

Fig. 195. Jeffrey Three-Throw Dip Pump, mounted on a Trolley.

working against higher heads than 28ft. or 30ft. of water unless special precautions are provided for starting. They are suit-able for a low head and a large volume, and will be found useful in cases where the water is too " dirty " for a force pump. A usual arrangement is shown in Fig. 196.

Fig. 196A refers to an American double rotary force pump, having a capacity of from 50 gallons to 75 gallons per minute. The design forms a very compact combination of pump and motor, which can be easily wheeled about a pit by one man. The

FIG. 196.—General Electric Power and Traction Company's Centrifugal Pip Pump, suitable for mounting on a Trolley.

rotary type of pump does not find much favour with English mining engineers.   It has the disadvantage of requiring to be primed like a centrifugal pump, and is not adapted for working with dirty water.   It has, however, practically no slip, and can be worked at varying speeds without seriously affecting the efficiency.   Both centrifugal and rotary pumps give the best results for dip working when the level of the suction pipe is so arranged as to charge the pump.

Fans, either main or auxiliary, should also be driven by series motors, if run off a continuous-current circuit.   The electric motor is peculiarly adapted for running fans, and it is somewhat surprising that this has not been recognised more fully.   The author designed a 25-H.P. main fan plant for a tin

Fig. 196A. — Jeffrey Double Rotary Dip Pump, mounted on a Trolley.

mine at Eger, in Bohemia, in 1889, the dynamo being driven by water power at a distance of about half a mile from the motor.

Stamp batteries and ore crushers should be driven by shunt or compound-wound motors, if continuous-currents are used.

Coal cutters and rock drills have hitherto, with few exceptions, been run on continuous-current circuits.   In order to avoid "racing" when running light, both differential compound and shunt windings have been tried with indifferent success—comparative regularity of speed has been gained at the expense of torque. After numerous experiments, the author has found series-wound motors to give the best results for driving coal cutters and drills. He prefers to control the speed by an external resistance. Other workers in this field have arrived at the same conclusions. But the success attained has not been sufficient to warrant the hope that coal will be cut largely in the future by machinery driven by continuous-current motors.

The future appears to lie with the polyphase systems. The polyphase motor, having neither collector nor brushes, obviates the danger from sparking at the brushes—a defect always present with continuous-current motors. And, since the speed can never exceed that of synchronism with the generator, racing is impossible. The regulation of polyphase currents can generally be accomplished in a small space by means of impedance coils *without any possibility of sparking*, for no part of the circuit need be opened or shunted (*see* Impedance Coil, p. 241), as is necessary with a continuous-current resistance regulator. (*See* § 59, p. 293, for method of obtaining a large starting torque.)

The sole point on which continuous-current motors are superior to polyphase ones, and then only with series winding, is as regards the *starting* torque. From this point of view they are undoubtedly the best and most efficient motors that can be built. The three-phase motor at starting is roughly comparable to a shunt motor ; it requires a starting resistance. (*See* Tables B B and C C, p. 289 ; also curves in Figs. 172 and 173, p. 294.)

It is always possible, however, to start a motor before coupling it to the load if the initial torque be too great for the maximum motive effort. With drilling machines the drill car be slackened if necessary, and with coal-cutters the initial torque of the motor can be made sufficiently great to start the cutter against the friction of the débris, which is all that is necessary. A "fall" should be cleared away by hand.

In driving large pumps and hauling gear, it will generally be advisable to use friction clutches, through which the load can be gradually applied after the motor has acquired it proper speed. The clutch does not entail extra expense in most cases, since its use is generally advisable, and often absolutely necessary even with continuous-current motors.

§ 67. SAFETY MOTORS; THE MINING MOTOR OF THE FUTURE.

Attempts have been made to design a continuous-current motor perfectly free from the risk of firing gas from sparks at the commutator or from short circuits in the coils. It is admitted on all sides that so long as brushes and commutators (sliding contacts) are used it is impossible to work without sparks, which, under certain conditions, may fire explosive gas. It is, indeed, found experimentally that the small blue sparks occasionally seen on even the best continuous-current motors are able to fire gas if the metal of the commutator becomes heated, as it will at times. It is, therefore, necessary to confine the explosion area to as small a space as convenient, in order to limit the effects and prevent communication to the outer air.

The first attempts were made in the direction of enclosing the entire motor in a metal case. It was soon found, however, that the quantity of air enclosed was too great to be safely fired, and this method has been abandoned by the majority of motor builders.

Messrs. Mavor and Coulson have, however, devised a completely enclosed motor (*see* Fig. 197) of admirable mechanical design, which is by far the best illustration of this class. Whether it may be classed among the "Safety" motors is perhaps a matter for experiment; but there can be no doubt about the security it affords from accidental damage from any outside cause, and it is thoroughly dust-proof.

The next obvious course was to enclose only the armature, commutator, and brushes. By suitable arrangements this can be done in a very effective manner, and the space so enclosed can be made small enough to limit an explosion to the protected area. But other difficulties are introduced. The design of the

FIG. 197.—Mavor and Coulson's Steel-Clad Mining Motor.

brush gear is complicated, and the adjustment of the brushes
is rendered exceedingly difficult, especially to unskilled hands.
The author has built motors for coal cutting in this manner,
and Messrs. Easton, Anderson and Goolden have patented
various devices with a similar object.

FIG. 198.—Goolden Safety Mining Motor.

In Fig. 198 is shown a perspective view of the excellent
safety mining motor made by Messrs. Easton, Anderson and

Goulden. The general arrangement is easily understood. The armature, commutator, and brushes are encased in a metal compartment, the free space of which is small. The makers say that an explosion of gas within it cannot be communicated to the outer atmosphere. The author has had no opportunity of testing this important statement, but from his knowledge of the motor it appears to be correct, always assuming that the

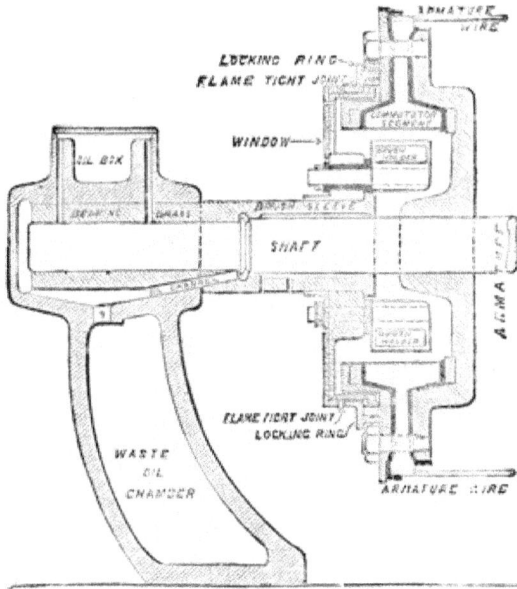

Fig. 199.—Sectional view of Davis and Stokes' Inverted Commutator as used on their Safety Mining Motor.

casing is in good condition and the inspection doors properly fastened. (*See* remarks on this subject on page 342.)

Messrs. Davis and Stokes, of Derby, have also invented a "safety motor," a sectional view of part of which is shown in Fig. 199. It will be seen that the commutator is inverted, and the brushes placed inside of it, instead of on the outer perimeter, as is usual. The brush arms are attached

A A

to a movable ring, sliding upon the bearing parallel to the shaft, and arranged to lock in the proper running position through a screw collar. The brushes are thus placed in a small circular space inside the commutator, the cubic contents of which in small motors is about the same as that of a safety lamp; and which, even in large machines, can always be made sufficiently small to render an explosion within it quite harmless. It should be noticed that the brushes can only be got at by the attendant when the motor is standing, and, therefore, when there is no chance of danger. Motors of this make have been tested in an explosive mixture of air and gas by Mr. John Rhodes, at Aldwark Main Colliery, who found that, although the gas continually flashed inside the commutator, it did not communicate flame to the outside. This is, perhaps, the safest form of continuous-current motor yet invented, and has been successfully used for the past four years in connection with a variety of underground work.

It is questionable, however, whether all these designs are not chiefly important as dust protectors. (It is absolutely necessary that a motor worked on a coal face should be dust-proof.) Though the cases may fit well when new, they probably soon get loose and defective from the rough usage inseparable from the class of work, and thus become useless as safety devices; indeed, they may then be a positive source of danger from the apparent security attached to them.

The only real safeguard against explosions from sparking at brushes and commutators (sliding contacts) is to do away with them altogether, and the only way in which this is practicable at present is by the use of polyphase motors. The mechanical simplicity and excellent speed regulation of these machines have already been urged. These advantages, in combination with their safety, must gradually tell. They will necessarily take the place of all other types of motors for driving coal cutters, rock drills, and other machines which have to be driven in a dusty atmosphere liable at any moment to be charged with gas.

In the author's judgment the mining motor of the future will probably present the appearance of a circular iron box.

The revolving part will have no electrical connection with the supply mains, and will resemble an iron cylinder with closed ends and steel shaft. A breakdown of the insulation inside the cylinder will simply affect the efficiency ; *it will not cause sparking.* The windings in connection with the mains will be few in number, laid in recesses, and be entirely protected from external damage by means of the outer framework. The conductors will be of the concentric type ; and in permanent plants it is probable that the Pownall Safety Mining System, or some development of it, will be largely adopted. The junction boxes, switch, cut-outs, and regulating devices will be all flame-tight, and be protected by oil or by air under pressure. An explosion from an inadvertent spark will then be a more remote contingency than one from a safety lamp.

CHAPTER X.

COAL-CUTTING AND ROCK-DRILLING BY ELECTRIC POWER.

§ 68. INTRODUCTORY.

THERE has always been a certain fascination about the application of electric power to coal-cutting and rock-drilling. It is easy to appreciate this feeling when one has had some experience of coal-getting. Indeed, it has been felt for a long time that a mechanical means of hewing coal is desirable from all points of view. Yet, strange to say, the problem has presented so many difficulties that its successful solution on a large scale still belongs to the future, although coal has been got by mechanical means in isolated cases for a long time past.

The difficulties are well known, and may be regarded under two heads, each of considerable importance. The first refers to the mechanical side of the question, and hence properly belongs to the present discussion. The second relates to the system of working and the general management of mines, and cannot be treated here at length. But reference must be made to some of the obstacles thrown in the way of mechanical coal-getting by the men and the masters. The use of a coal-cutting machine may seem to be simply a question of the price of coal hewn by it in comparison with the cost of hand-hewn coal; and this is undoubtedly the case. But the cost of machine-won coal is at present necessarily based on figures resulting from the use of one or two machines, instead of on figures from a dozen worked on "faces" specially laid out for the purpose. It should be noted that the cost of the prime power for two or three machines is practically as great as that required for a dozen, and that the system of mains need not be proportionately increased for the larger number, because the power does not 'ncrease in direct proportion to the number of machines, as it is not possible to keep them all running at the same time. It is important, too, that machinery should be run as nearly continuously as possible, which suggests the employment of several shifts of men. And this is only practicable, even if the men's union permit it, when the conditions are such that the coal can be carried away sufficiently quickly. In the majority of faces, as arranged to-day, this is impossible, and a machine cannot be kept at work continuously for even a shift of eight hours; so that, from causes which are quite independent of the working capabilities of the machine, its theoretical output is never reached. This fact points to the necessity for a coal-cutter of smaller capacity than those now in use—say a machine which would hole from 5 to 10 yards an hour, working under normal conditions and allowing the necessary time for clearing away the coal. Hitherto the aim has been rather to produce machines which, under favourable circumstances, will cut from 20 to 40 yards of face per hour. But a very casual inspection of a coal mine will demonstrate that such a rate cannot be maintained for any length of time, simply from the difficulties of clearing away through gates two chains apart, even neglecting for the moment the difficulty of keeping the face clear from falls of roof during the time necessary for removing the coal.

If mechanical coal-getting is to be carried out on a large scale with the long-wall system, it will be necessary to modify the number and position of the gates in order to increase the facilities for clearing away. The question belongs to mining managers, and not to electrical engineers; and this is, perhaps, one of the chief reasons why mechanically-driven coal-cutters have not made as much progress as their undoubted utility appears to warrant. A colliery is being opened out (October, 1894) near Eckington, in which the whole of the work will be done electrically, even to the driving of the fan. The coal will be worked on the pillar and stall system by Jeffrey coal-cutters. This undertaking is due to American enterprise, and will be watched with the highest interest.

It does not lie in the province of the present work to discuss the question of machines *versus* hand coal-getting. But it may be remarked that one of the best electric coal-cutters, with two men to work it, will easily hole as much as from 20 to 30 hewers; thus largely reducing the number of men required below ground. And since the number of accidents necessarily bears some direct proportion to the number of men employed, the introduction of machinery should have the effect of largely reducing the loss of life.

The number of mechanical principles applied to coal-getting machines is astonishing, and the ingenuity displayed by many inventors, who evidently have not been in close touch with the principles of mining, is equally surprising. All conceivable methods of cutting, drilling, boring, shearing, and picking coal have been designed, on paper at least, and many of them have been applied to more or less practicable machines. The earliest attempts were made in the direction of a mechanical pick. As might be expected, the pick with the man at the back of it directing each blow did much the better work, and this method was soon dropped. The drill and the circular saw have given the most successful models upon which to work out a cutter, hence a variety of machines embodying these principles.

The machines which have survived the initial stages and give some hope for future success are divisible into three main classes :—

(*a*) Bar cutters.

(*b*) Rotary wheel or disc cutters.

(*c*) Drills.

§ 69. ROTARY BAR COAL-CUTTERS: GOOLDEN-ATKINSON, JEFFREY, Etc.

Under the first head may be included all those machines which have a cutter-bar revolving either at right angles or parallel to the face of the coal. This type has so far proved the most successful in this country and in America, the best known being the Goolden-Atkinson bar cutter of English invention, and the Jeffrey machine of American origin.

The essential feature of the Goolden-Atkinson machine is a tapered rotary bar, which is fitted with a series of cutters driven into tapered holes. They are arranged in a spiral form, the object being to clear out the cuttings and prevent the jamming of the bar. The shape, size, and general arrangement are the result of long and careful experiment by Messrs. Llewellyn and Claude Atkinson, the staff of Messrs. Easton, Anderson and Goolden, and the Electrical Coal Cutting Corporation. The main idea is shown in Figs. 200 and 200A.

The cutter-bar (see Fig. 201) is usually geared to the armature shaft by a pair of cast steel double helical wheels running in oil, and makes from 300 to 500 revolutions per minute, according to the material being cut. The motors are always wound in series, and hence the speed varies inversely as the resistance ; which is exactly the best condition for work of this kind, since it prevents undue shocks to the motor, and tends to keep the power absorbed within safe limits. It will be seen that the motor is thoroughly protected from mechanical injury either from falls of roof or sides, and that the armature and commutator are fitted into a dust-tight and flame-tight compartment of dimensions as small as are consistent with the safe and effective handling of the brushes. The cutting mechanism as a whole is made to swing through an arc of rather more than 90deg. on the turntable which carries it (see Fig. 201A). This gives facilities for shifting the machine from point to point in the mine, and also enables the bar to cut its way into the coal until it is at right angles to the face and ready for holing. Fig. 201A gives a view of the machine as arranged for work. In the early machines of this

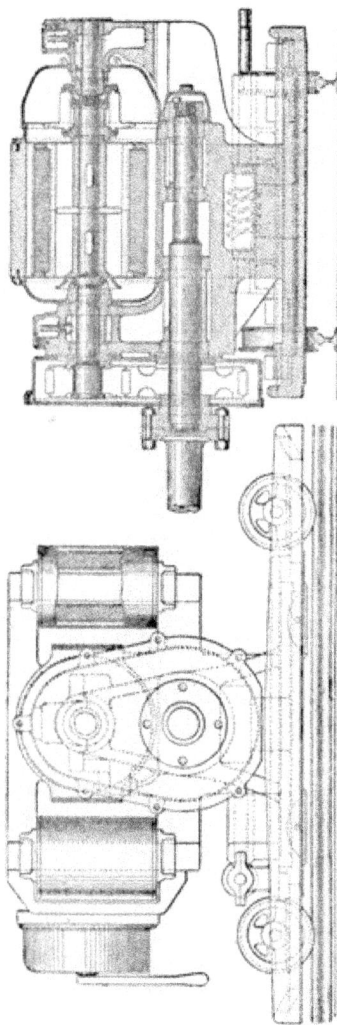

FIG. 200.—Goolden-Atkinson Coal-Cutter ; Elevation facing Cutter-Bar.

FIG. 200A.—Cross Section, showing details of safety motor and gearing.

class the coal was always picked out to give the bar a start.
Since the bar has to be carried by bearings it may be thought

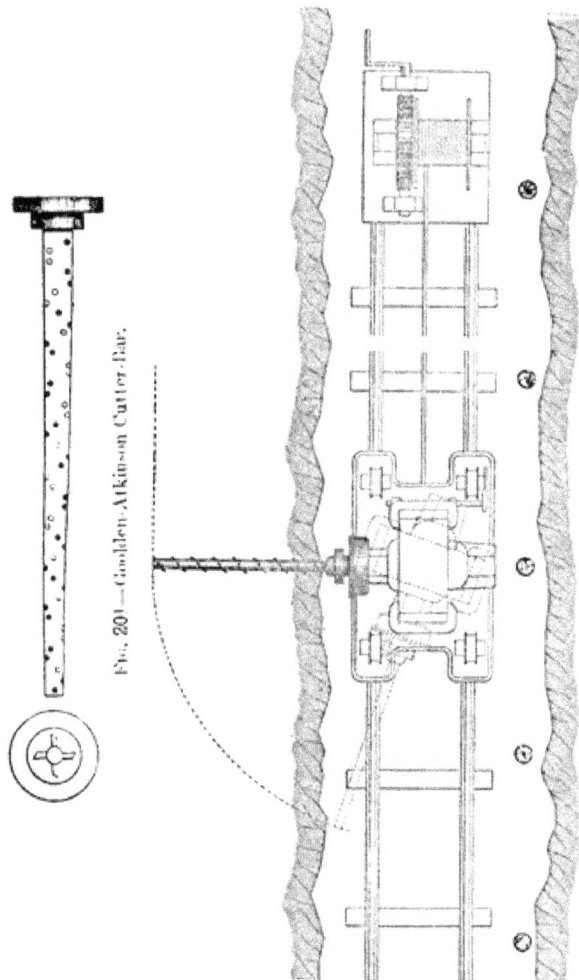

Fig. 201.—Gadden-Atkinson Cutter-Bar.

Fig. 201A.—Gadden-Atkinson Coal-Cutter.   Method of Working.

that coal cannot be cut level with the floor by the bar type of
machine ; but this difficulty is successfully met.   By arranging

the seats of the turntable at a proper angle, and causing it to revolve in an inclined plane, the makers contrive to keep the bar nearly horizontal, and below the level of the rails when at right angles to the face in the cutting position, and yet to lift over the rails when turned parallel to the face. In this way a flooring of regular slope is made at every cut, and forms the floor for the rails during the next cut. In the majority of cases it is not found necessary to level the floor, although when cutting in coal it may pay to do so. The feed is made by a hand winch as in Fig. 201A, or is worked automatically from the motor shaft. The author believes that hand feeding is found to be the more satisfactory method with this machine.

The Electrical Coal Cutting Corporation have for some years been working coal at a contract price per ton hewn or per linear yard undercut to a specified depth. They supply the whole of the plant and the men to work the machine, while the colliery proprietors find the labour for clearing away the coal from the faces, leaving them ready for the machines. This system, on the whole, has worked well, and, were it not for the difficulties already referred to, the machines would turn out a much larger tonnage and, of course, return a larger interest on the capital outlay.

The experience of the Corporation, whilst it has demonstrated that nearly every mine, working long-wall, can be successfully operated by electric coal-cutting machines, has also brought into further prominence the essential difference between hand and machine working, and the great difficulties of getting the owners and managers to appreciate the difference, and to lay themselves out to apply to mining the discipline and regularity obtaining in manufacturing industries.

A modification of the Goolden-Atkinson machine has lately been introduced by Mr. Frederick Hurd, who for some years was on the staff of Messrs. Goolden and Co. No particulars of its performance have yet been published. A general idea of the machine may be gained from Fig. 202. There is a scoop built at the back of the bar to assist the clearing of the cuttings from the hole. This may or may not be an improvement; experience alone can determine; but the author believes its use has been abandoned after a few trials. It should be noted that the Goolden bar is claimed to clear itself by

FIG. 202.—Hurd's Coal-Cutter.

virtue of the spiral arrangement of the cutters. There are also some interesting devices for ventilating the motor and safeguarding it against the risk of exploding gas. These consist of an elaborate system for circulating fresh air under pressure along the conductors, which are encased in pipes for the purpose, and through the interior of the motor, the starting resistance and switch-board boxes, &c. The object in view is,

FIG. 203.—Birtley Rotary Bar Coal-Cutter, with case removed, showing gearing and chain for clearing the cuttings at back of Cutter-bar.

of course, most desirable ; but the expense necessary to execute the system described in Mr. Hurd's patent specification is probably too great for ordinary requirements.

Another bar cutter has been patented quite recently by Messrs. T. Heppell and Patterson, which embodies some novel features. One of these machines, built by Messrs. Ernest Scott

Fig. 204.—Birtley Rotary Bar Coal-Cutter, with case removed, showing the propelling gear and Cutter-Bar with teeth removed.

and Mountain, is shown in Figs. 203 and 204. The case is removed to show the motor and gearing. One of the special points is the endless chain used to clear away the cuttings. This is

shielded by a strong guard. The gearing consists of cast steel
bevelled wheels and pinions. The machine is propelled by
means of a steel rope which passes round a drum geared to the
armature shaft by worm gearing, an eccentric, rod, pawl and
ratchet. The design is very compact, and appears to be
capable of being worked into a less height than the type of bar-
cutter adopted by other makers ; but this remains to be proved,
for many alterations may be required before the machine gets
into regular work. The author believes that at present one of
these machines is at work in the Cannock Chase district. Mr.
Heppell says that one of these cutters at the Pilner Main
Collieries is doing good work, cutting at the rate of about 20
yards per hour, 3ft. under. A sectional view of the cutter-bar
is shown in Fig. 205. It is arranged to swing about the centre
of supports, which permits of the cutters being readily examined,

Fig. 205.—Section of Birtley Cutter-Bar.

and also facilitates the moving of the machine about a coal
pit. It has three dovetailed grooves cut along its length.
The cutters have corresponding dovetails, and are distanced
by suitable pieces of steel. The cutters can obviously be
arranged spirally or otherwise, as may be required.

The Jeffrey coal-cutter is of different construction from any
machine of English origin, and was, no doubt, first designed to
meet the conditions of the American coal seams. Special
machines are built to suit the conditions obtaining in British
mines. The method of working will be readily grasped on an
inspection of Fig. 206. The details of construction are shown
in Fig. 207.

It will be seen that the bar lies parallel to the coal face when
in the cutting position. It is designed to hole as deep as 7ft.

Deep undercutting is an especial advantage in pillar and stall working with an average roof. This machine consists of a bed-frame, occupying a space 3ft. wide by 7ft. 2in. long, built up of

Fig. 206.—Jeffrey Electric Coal-Cutter, showing method of jacking it and position of attendants.

two steel channel bars firmly braced, the top plates on each carrying racks with the teeth downwards. Mounted upon and engaging with this bed-frame is the sliding frame, consisting

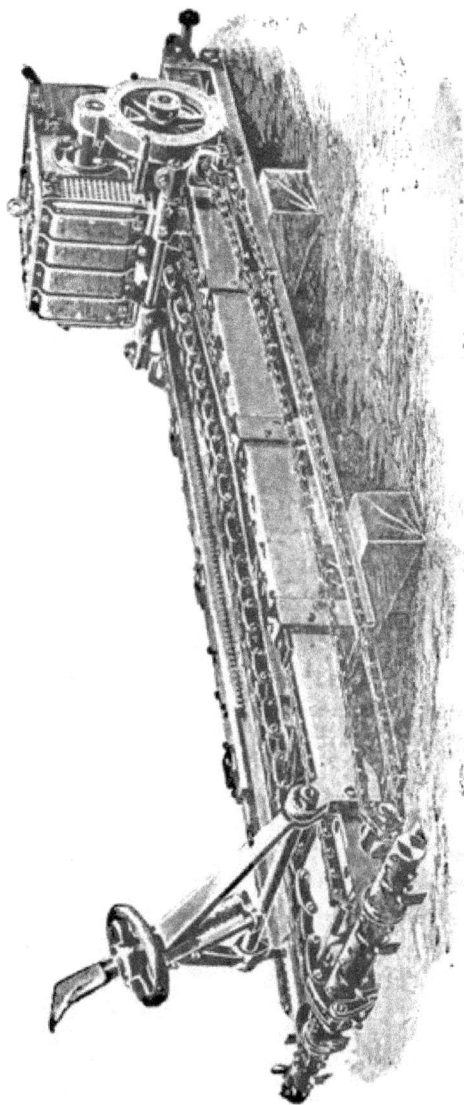

Fig. 207.—Jeffrey Electric Coal-Cutter, showing Cutter-Bar and Jack for fixing the machine.

mainly of two steel bars, well braced, upon which is mounted
at the rear end an electric motor. Upon the front end of this
sliding frame is mounted the cutter-bar, held firmly by two
solid steel shoes with brass bearings. The cutter-bar receives
motion through an endless steel or bronze chain from the
driving shaft, and as it revolves is fed forward by means of
straight-cut gear and a worm and wheel. The bar is usually
36in. wide. In America the seams are generally much thicker
than those in this country, and the roofs are sufficiently good
to enable the face to be kept open for the distance necessary
with a 6ft. or 7ft. holing, and pillar and stall working is chiefly
practised. But in England the conditions of long-wall working
rarely permit of such deep undercutting, and from 4ft. to 5ft.
is, as a rule, the deepest cut advisable. At the Cannock and
Rugeley Collieries, where two of the Jeffrey machines are at
work, one holes 4ft. and the other 7ft.

The electric motor occupies a space of about 20 inches square.
It is designed to absorb a maximum of 15 E.H.P., and in some
veins of coal does not require more than 7·5 H.P. The American
standard pressure is 220 volts, but this can be altered to suit
circumstances. The armature is designed to run at 1,000
revolutions per minute, and the cutter-bar at 200. The
momentum of the moving parts is such that the machine over-
comes ordinary obstacles without difficulty, and, since the
motor is not coupled rigidly to the bar, the vibrations are
not transmitted to it, and the machine runs steadily and
comparatively quietly. Two men are required to work it—
one at the switch at the back and the other as helper to clear
the *debris* and to assist in shifting. The machine is taken
from stall to stall on a special truck, shown in Fig. 208. When
at work it is slid on to two boards or bars of iron in front of
the coal, and fastened firmly by means of the front and rear
jacks against the face and roof of the coal, which is thus supported to some extent during the under-cutting. When the
full depth of the cut is reached the feed is thrown off, and the
cutter bar is returned to the starting position by means of
reversing gear. The machine is then moved sideways for a
distance equal to the length of the cutter bar, and another cut
is made in the same manner. Each cut is made in ordinary

FIG. 208.—Jeffrey Coal-Cutter, on truck.

working in from three to six minutes. The output depends to a great extent upon the skill of the men in moving and jacking the machine, and also upon the hardness of the coal, and whether the cut is made in coal or "dirt." In some seams these machines are said to cut from 40 to 66 linear yards of face to a depth of 6ft. in a shift of 10 hours. Since the depth of the undercut is roughly twice as great as that usually made with a bar machine of the Goolden-Atkinson type, this is equal to a length of face of from 80 to 130 yards, in comparison with the performance of the latter. Assuming these figures to be correct, it would appear that although the Jeffrey machine has to be shifted after every cut, yet it may cut a greater area of coal than the Goolden-Atkinson machine, which, the author understands, does not average more than from 80 to 100 yards of face in this time.

In comparing the two, stress must be laid on their relative suitability for different kinds of work. The Goolden-Atkinson machine requires a space between the face and the props at the goaf of not more than four feet at the outside, whereas the Jeffrey machine cannot be worked in a smaller space than six feet; but with the latter the depth of undercut may be nearly double that with the former, though, of course, the total length of roof unsupported will be then nearly twice as great. It is, therefore, obvious that the side cutter-bar can be used in places where the front cutter-bar cannot possibly be. It must be remembered also that the Jeffrey coal-cutter makes a capital heading machine, while the side cutter cannot be used for this purpose. The two machines are not, therefore, rivals for the same work in all cases. They cost in the first instance about the same, the motors are of the same rated capacity, and the actual power required in practice appears to be about the same. Hence there does not seem to be much to choose between them as regards the question of cost, but it is rather a matter of suitability for any particular case.

The English agents for the Jeffrey mining machines are Messrs. John Davis and Son, who are placing these useful coal-cutters in various mines in England, Wales, and Scotland.

Mr. Williamson, of Cannock and Rugeley Collieries, who is using one of the Jeffrey machines with a 36in. bar, holing 4ft. in a 5ft. seam, says that the output of round coal is increased 15 per cent.—that is to say, hand cutting makes 75 per cent. of round coal and the Jeffrey cutter 90. And that one machine, working single shifts, has increased the output by at least 60 tons per day.

In Table D D are given a few results from a series of exhaustive trials of electric coal-cutters in mines in Ohio, U.S.A., made in 1890 by Mr. R. M. Haseltine, Chief Inspector of Mines. This gentleman found that the only electric coal-cutter in general use was the Jeffrey machine, and, therefore, most of his tests refer to this type. The Leichner and Thomson-Van Depoele machines (see the end of the series) are of comparatively small output, and do not seem to have made much headway since the date of the report, so far as the author is aware; while, on the other hand, the use of Jeffrey machines has steadily increased in American mines, and is now gaining a hold on mining engineers in this country.

The figures are not directly comparable with those now obtaining in England, for the conditions are very different, and the machines have been since improved in many respects; but the results, taken collectively, may prove instructive to future users of electric coal-cutting plant, as they show, on the authority of an independent official and expert, what was being done in this direction as far back as 1890.

The General Electric Company of America have built a modification of the front-cutter type machine, which is of interest. In it the cutter-bar is made to revolve in such a direction that the coal is cut upwards. It is claimed for this method that it assists the holding down, and hence there is less need for careful fastening by jacks and braces. In this machine there are four chains instead of the single one used with the Jeffrey machine. The Stine-Smith coal-cutter is also similar in design, and appears to be a soundly built machine.

The author has no information about performances of either of the two last-mentioned machines.

Table DD.—Showing Results of Experiments made with Electrical Coal-Cutters in Ohio, 1890, by R. M. Haseltine, Chief Inspector of Mines.

Name of Mine where test was made.	Number of cuts made while testing.	Number of machine.	Width of cut. Ft.	In.	Depth of cut. Ft.	In.	Square feet undercut.	Time in cutting, including fixing. Min.	Sec.	Time in making the cut only. Min.	Sec.	Electric horse-power used in cutting. Maximum.	Minimum.	Average.	Average h.p. used in running back.	Electric horse-power required to overcome frictional load. With feed off.	With feed on.	To feed.	Net electric h.p. required in cutting.	H.P. required to undercut 1 sq. ft. coal in 1 min.	Speed of armature.	Voltage.
Jan. 7, 1891. Brush Fork, No. 2 Hocking Co., Morris Coal Co.	1st	Jeffrey No. 11	3	3	5	8	18.4	8	45	6	45	17.1	15.9	16.2	7.6	5.4	6.7	1.3	9.5	10.2	...	...
	2nd	"	3	3	5	6	18.4	7	30	7	00	16.0	12.1	15.7	7.7	5.4	6.7	1.3	9.0	10.1	...	...
	3rd	"	3	3	5	8	18.4	7	37	7	00	17.0	13.7	15.6	7.9	5.4	6.7	1.3	8.9	10.0	...	...
	1st	Jeffrey No. 10	3	3	5	8	18.4	6	54	5	30	18.7	16.2	17.3	9.0	5.3	5.7	0.4	11.6	8.8	...	...
	2nd	"	3	3	5	8	18.4	6	00	5	00	20.6	16.2	18.8	11.1	5.3	5.7	0.4	12.7	9.5	...	...
	3rd	"	3	3	5	8	18.4	6	25	5	43	23.2	19.1	22.0	9.9	5.3	5.7	0.4	16.3	10.8	...	...
	4th	"	3	3	5	8	18.4	6	30	7	00	20.0	18.3	21.9	10.7	5.3	5.7	0.4	14.0	11.9	...	...
	5th	"	3	3	5	8	18.4	7	34	7	15	21.7	14.8	19.7	10.2	5.3	5.7	0.4	13.7	11.2	...	...
	6th	"	3	3	5	8	18.4	9	00	6	30	20.9	15.6	19.1	10.4	5.3	5.7	0.4	12.3	9.8	1,420	225
	7th	"	3	3	5	6	18.4	7	28	7	15	20.9	15.4	18.6	8.9	5.3	5.7	0.5	...	...	...	...
Jan. 8, 1891.	1st	Jeffrey No. 9	3	3	5	8	18.4	8	55	4	45	22.8	18.2	20.8	9.0	7.5	7.6	0.1	13.2	10.9	...	...
	2nd	"	3	3	5	8	18.4	6	05	4	15	20.7	17.1	18.8	10.8	7.5	7.6	0.1	11.2	10.4	...	...
	3rd	"	3	3	5	8	18.4	6	32	4	00	23.7	22.9	23.4	11.7	7.5	7.6	0.1	15.8	11.0	...	...
	4th	"	3	3	5	8	18.4	7	38	5	30	24.0	23.1	23.6	10.7	7.5	8.0	0.5	16.0	11.7	...	...
	5th	"	3	3	5	8	18.4	8	00	4	00	24.0	21.9	22.9	11.6	7.5	8.0	0.5	14.9	11.7	...	...
	6th	"	3	3	5	8	18.4	9	49	5	15	24.5	21.9	23.5	9.3	7.5	8.0	0.5	15.5	12.4	...	...
	1st	Jeffrey No. 12	3	3	5	8	18.4	8	30	5	45	16.8	12.3	15.3	8.4	6.0	6.4	0.4	8.9	9.2	1,233	245
	2nd	"	3	3	5	8	18.4	6	41	5	15	16.7	14.7	15.7	8.4	6.0	6.4	0.4	9.3	9.1	...	...
	3rd	"	3	3	5	8	18.4	6	45	5	15	16.0	14.6	15.4	9.1	6.0	6.4	0.4	9.0	9.0	...	...
	4th	"	3	3	5	5	18.4	7	33	5	00	16.1	12.1	15.0	8.8	6.0	6.4	0.4	8.6	8.7	...	...

Mine	Cut																		
Sand Run Mine ... Morris Coal Co. ...	1st 2nd	*Jeffrey* No. 1,021	3 3	3 3	6 6	7 7	17·8 17·8	1 6	25 15	4 5	00 45	16·3 15·5	11·7 14·4	13·9 11·9	6·5 6·8					: :
Upson's Mine, Shawnee, O.	1st 2nd 3rd 1st	*Jeffrey* No. 1,023 " " No. 1,025	3 3 3 3	3 3 3 3	7 7 7 7	18·1 18·1 18·1 18·1	5 5 5 5	59 49 44 29	5 5 5 5	00 00 15 00	12·0 12·1 13·2 11·5	9·6 9·6 9·6 9·7	11·2 11·2 11·2 10·5	6·3 6·7 7·1 7·1	5·6 5·6 5·6 4·4	5·8 5·8 5·8 4·6	0·2 0·2 0·2 0·2	5·4 5·4 5·4 5·9	7·3 7·3 7·4 6·3	: : : : 1,230 220 235
Prospect Slope, East Palestine, Colbia County, Jan. 24, 1891.	1st 2nd	*Jeffrey New* No. 1,027	3 3	3 3	2 2	13·5 13·5	5 5	30 33	6 6	00 30	16·8 19·2	10·1 14·3	15·0 17·6	7·7 8·3	5·8 5·8	6·0 6·0	0·2 0·2	9·0 11·6	9·3 10·8	: : :
	1st 2nd 3rd	*Jeffrey New* No. 1,026 " "	3 3 3	3 3 3	6 6 6	14·6 14·6 14·6	6 6 6	35 45 30	6 6 6	00 15 00	27·0 28·9 26·5	17·1 17·5 16·6	22·3 25·3 22·7	8·9 12·0 10·4	5·8 5·8 5·8	5·9 5·9 5·9	0·1 0·1 0·1	16·4 19·4 16·8	13·9 14·2 12·9	: : :
	1st	*Jeffrey Old* No. 1,030	3	3	6	14·6	6	53	5	30	29·6	18·5	24·5	11·6						:
Shawnee and Iron Point	1st 2nd	...	3 3	3 3	7½	17·8 17·8	4 4	00 47	3 4	30 00	24·9 23·5	11·8 15·0	20·2 19·9		6·2 6·2	7·4 7·4	1·2 1·2	12·8 12·5	9·9 10·2	1,455 :
Shawmut, Elk County, Pa.	1st 2nd 3rd 4th	*Leichner*	3 3 3 3	0 0 0 0	4 4 4 4	12·7 12·6 12·5 13·9	4 4 4 4	30 33 14 10	2 2 2 2	30 33 45 42	17·8 18·0 18·7 19·3	15·2 14·2 14·6 13·4	16·4 16·3 16·2 16·8	10·1 9·7 9·8 10·7	4·96 4·96 4·96 4·96	4·98 4·98 4·98 4·98	0·02 0·02 0·02 0·02	11·4 11·3 11·2 11·8	7·3 7·2 7·4 7·3	: : : :
Shawmut, Elk County, Pa.	1st	*Thomson Vanhoevele*	3	0	7½	10·9	4	21	2	45	11·5	5·6	9·0	5·0	2·70	3·06	0·36	6·0	7·0	: :
	2nd 3rd	... ...	3 3	0 0	4½ 2½	10·0 9·6	3 3	55 20	2 3	10 00	9·3 15·4	7·5 8·6	8·3 12·2	4·0 6·0	2·70 2·70	3·06 3·06	0·36 0·36	5·3 9·2	4·2 6·0	: :

(a) The sixth cut with machine No. 10 made with new knives. (b) The first cut with machine No. 1,023 made with new knives. (c) The first cut with machine No. 1,027 made with new knives. (d) The knives in machine No. 1,026 had made three cuts before test. (e) The knives in machine No. 1,025 had made one cut before test. (f) The knives in machine No. 1,030 had mine second cut made in coal. (f) At the Shawnee and Iron Point mine first cut made in the fireclay which forms the floor. (g) At the Shawnee and Iron Point

§ 70. ROTARY SIDE-WHEEL COAL-CUTTERS. GILLOTT
AND   COPLEY,   YORKSHIRE   ENGINE   CO., AND
SNELL-WATERHOUSE COAL-CUTTERS.

A number of patents have been taken out for machines with
side wheels.   In a few the wheels have been designed to cut
vertically, or even at any angle, but in the majority the aim
has been simply to make a horizontal cut as near the bottom
as possible.   The diameter of the wheel determines the depth
of the undercut, which is limited in practice to from 39 to
55 inches, the latter being found to be as deep a cut as is
desirable.   If this depth be exceeded there is a strong probability of the coal shearing at the back and fouling the wheel
before it is clear of the cut, and if the cut be shallower the coal
will not break readily by its own weight, but requires wedging
down.

Passing by the early inventors of these machines and coming
to modern history, the firm of Messrs. Gillott and Copley
stands in the front rank ; but from some cause, which is not
very apparent, they have not shown a disposition to apply
electric motors to their machines, preferring to continue the
use of compressed air.

The Yorkshire Engine Company have also built some
excellent machines of this type, embodying several improvements on those of the former firm.   They also have confined
themselves to compressed air, although they have built to
order a few electrically-driven machines.   An illustration of
one of their machines is given in Fig. 209, in order to show
the general requirements of this type of plant, and also because
the author has found from actual experience that the machines
do excellent work.

Fig. 210 is an enlarged view of the cutter-wheel, showing the
method of supporting it in bearings.   The design is arranged
to hole at a few inches above the rail tops.

Owing to the success which this class of coal-cutter has
obtained when driven by compressed air, no radical change has

Fig. 209.—Yorkshire Engine Company's Side-Wheel Coal-Cutter.

been attempted by inventors when applying electricity. They have accepted the main principle, and have simply changed the motive power.

On referring to Fig. 209 it will be seen that the side wheel is fitted with two kinds of cutters. Every alternate one has Y-shaped teeth, while the remainder are straight. The function of the latter is to tear out the centre of the cut, while the former clears it out to the desired width, usually about 3in. In air-driven machines the speed of the wheel varies from 15 to 30 revolutions per minute, according to the kind of dirt and its resistance. This speed is found to be too slow for the electric motor, since the reducing gear is costly and cumbersome ; therefore all who have tried to apply electricity have

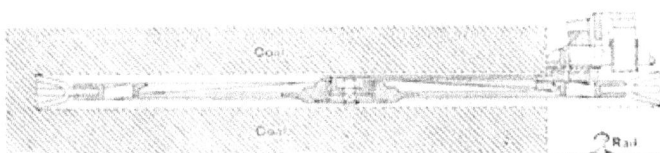

Fig. 210.—Cutter-Wheel : Yorkshire Engine Company.

endeavoured to raise the cutter-wheel speed as much as possible. This has necessitated strengthening the machine as a whole, and has increased the dead weight from about 25cwt. to about 35cwt. The increased speed, however, has caused no trouble in working. In fact, the momentum of the moving parts is so great that obstacles which caused serious shocks at the slower speed are scarcely felt at the higher speed of from 70 to 100 revolutions per minute. And the increased weight tends to prevent derailment.

While upon the subject of weight, it may be remarked that if it is desirable to limit the weight of a coal-cutter to, say, 10cwt., which is about as much as any one man can shift about a pit with the aid of crowbars, it may be well worth while to pay special regard to the making of a light machine. But if it be necessary, from any cause, such as the need for a machine of greater capacity, to increase the weight much beyond the limit just mentioned, then it appears to the author to be a matter of little importance whether the machine weighs 20cwt., 30cwt., or even 40cwt. In each case it is too heavy to be handled without special tools, and a screw jack is as much a part of its equipment as it is of a locomotive's. In fact, with the larger type of electric machines the increase of weight is by no means an unmixed evil, for it gives increased stability, and, no matter how rough the road, or uneven the dirt, the machine rides steadily through it all.

The advantage of the wheel machine compared with the rotary bar lies in the low head room required, and perhaps in the enormous capacity of the wheel on long walls where the conditions are favourable.

The author has, however, never seen a machine at work which had anything like a chance of working continuously, and so reaching the theoretical output. The majority of pits are so laid out that continuous working is practically impossible, and the most useful machine is that which, under the given conditions of normal working, will give the largest average output, and not a maximum output for a short time under the best conditions.

A number of patents have been taken out for side-wheel coal-cutters; but perhaps the first practical machine was designed by the author and Mr. Major Waterhouse, of Glass Houghton Collieries, about four years ago. The general construction of this machine will be readily seen on an inspection of Figs. 211, 212, and 213, which are plan, longitudinal, and end views. Much detail is left out in order to show up the main features. It will be noticed that the design roughly resembles that of the Yorkshire Engine Company's compressed air machine. But, in fact, the whole of the details have been modified to suit the increased rate of speed and to accommodate the gearing. The machine was built to hole 3ft. 6in. in a seam 24in. in height, on a *level* with the floor, this being the first machine so designed. The maximum height above the rails was 22in. The cutter made from 80 to 100 revolutions per minute in various kinds of shale and coal, no difficulty being

experienced either from their hardness or from iron pyrites.
The weight was about 35cwt. At least two men were necessary to work the machine, and in soft coal with a rapid feed
a third was required to keep the wheel from being fouled by

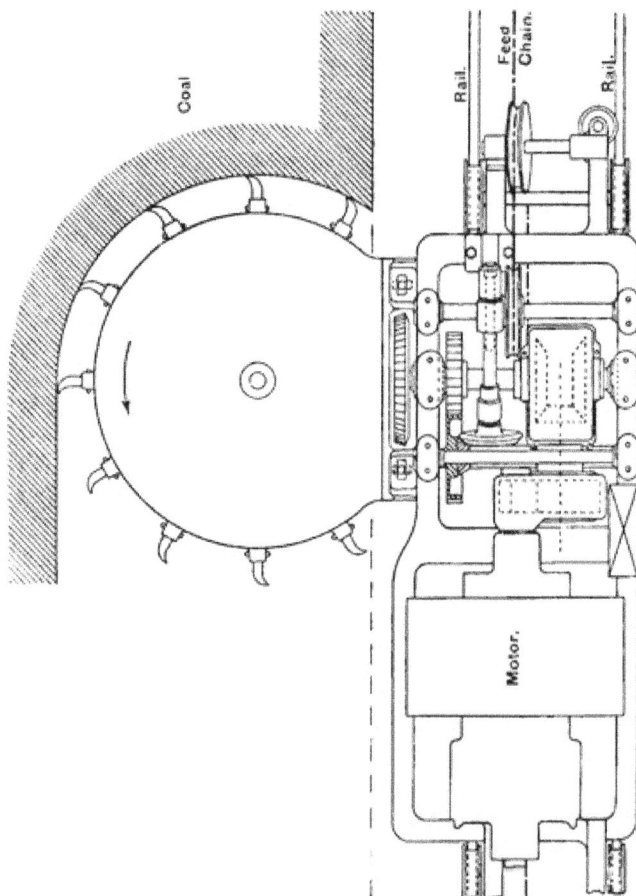

FIG. 211.—Snell-Waterhouse Coal-Cutter, Plan.

the coal. The wheel was designed to run in either direction,
but the inventors always arranged the cutters to feed out the
cuttings at the back of the machine, instead of at the front, as
is frequently done in air machines. It was found, however, that

the wheel did not sufficiently clear itself, and that a man was required to shovel out the loose coal at the front from time to time. The power absorbed did not vary so much as was

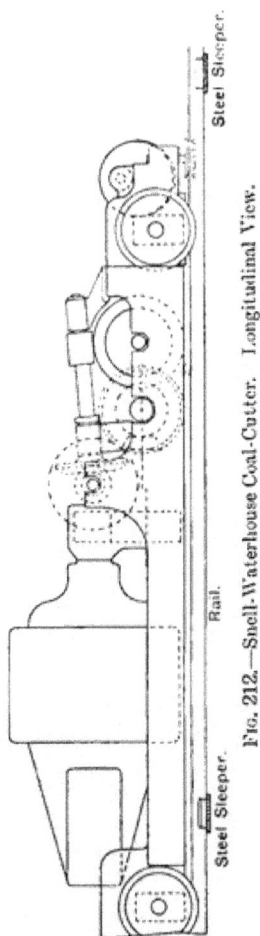

Fig. 212.—Snell-Waterhouse Coal-Cutter. Longitudinal View.

expected. The automatic feed, actuated by a worm and wheel from the first cross shaft, nicely adjusted the speed of the machine to the resistance of the cut. The motor was wound

in simple series, and thus the current varied in proportion to
the torque. And, since the pressure decreased slightly as the
load increased, the tendency was to equalise the power absorbed
and to vary the rate of feed with the degree of resistance to
the cutters.

The chief difficulties met with at Glass Houghton were due
to the peculiar formation of the seam, which is 22in. in height
with a thin parting near the middle. The holing was made on
the level of the floor in the fireclay, and presented no serious difficulty. But the lower coal broke away from the parting, leaving the tops in position, and frequently fouled the wheel before
a cut could be made sufficiently long for the introduction of
wedges. Under these conditions the time wasted in clearing

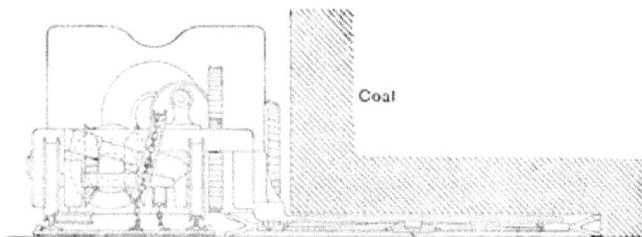

FIG. 213.—Snell-Waterhouse Coal-Cutter. End View.

the wheel more than counterbalanced any gain from the high
rate of speed at which the machine holed when the coal
remained firmly in position. A bar-cutter might have suc-
ceeded here if more head room had been permissible, for it
is obvious that the fall of coal behind the bar, or even on the
top of it, would not be so serious as with a side wheel. But in
this case the large number of iron balls in the clay were held
to be an obstacle in the way of the bar-cutter, and, in addition,
the bar could not be made to cut *perfectly* level with the flooring, as was desirable. The power absorbed varied from 10 to
17 E.H.P. The motor was designed to give 15 B.H.P. at about
800 revolutions, and was of the armoured type adopted by
Messrs. Lahmeyer and other Continental firms, but not much
known in England. The design is well adapted for use on a

coal-cutting machine, as the massive field-magnets protect the field windings and the armature from damage by falls of roof or accidental injury during working.

At present there are running in Yorkshire collieries a few electric side-wheel machines which have been developed from the Glass Houghton one, but no particulars of their working have been published. Under suitable conditions, however, they must give economic results.

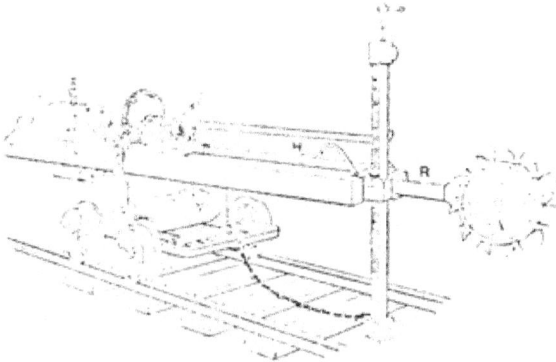

Fig. 214.—Wantling-Johnson Electric Coal-Cutter and Heading Machine.

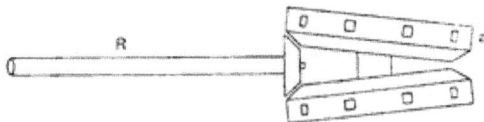

Fig. 215.—Plan of Cutter-Wheels of Wantling-Johnson Electric Coal-Cutter.

A smaller type of wheel machine of American origin is shown in Fig. 214. It is the invention of Messrs. Wantling and Johnson, and was shown at the World's Fair, Chicago, 1893. The author has no particulars about its performance, but it seems to be a practical machine. It is adapted for making vertical or horizontal cuts, but cannot be set to cut level with the floor. It appears to be especially suitable for heading work. The cutter-wheels are shown in plan in Fig. 215. There are two wheels revolving in opposite directions.

FIG. 216.—Thomson-Houston "Mowing Machine" Coal-Cutter.

Their diameter can be made to suit requirements. The sketch in Fig. 214 shows the machine arranged for heading, in which case it is always parallel with the rails. For long-wall or pillar and stall working the motor and cutter would be at right angles to the rails and the coal. The method of steadying the cutter is clearly indicated in the figure.

The Thomson-Houston "Mowing Machine" coal-cutter, shown in Fig. 216, presents several features entirely different from those of the preceding machines. It may be regarded as a modification of the Baird coal-cutter. The cutters are carried on a chain which passes round an overhanging arm capable of rotation through an arc of nearly 180deg. The illustration shows most of the working details. It appears to be suitable for long-wall working. There does not seem to be any definite information published with respect to its performances.

Messrs. Mavor and Coulson have adopted Baird's design, but with several modifications; the principal one being the use of worm gear for driving the cutter-chain. Their machine is designed to cut at the rate of one foot per minute through ordinary soft Scotch coal, the chain-wheel revolving at 60 revolutions per minute.

The Jeffrey Company are also trying a cutter-chain on one of their special English machines, and expect it will effect a saving in power, since the width of the cut is reduced to 2·5in.

§71. ELECTRIC POWER DRILLS: ROTARY AND PERCUSSIVE TYPES; JEFFREY, MARVIN, AND VAN DEPOELE DRILL.

In spite of the obvious advantages to be derived from the application of electric power to rock and coal drills, much progress does not seem to have been made in this direction. The reasons for this are probably partly the same as those, already referred to, which have hindered the introduction of mechanical coal-getting, and, in addition, the cost of the conductors, which, no doubt, has been in many cases prohibitive. But there are also some difficulties peculiar to the work.

Driving stone drifts is about as rough work as can be imagined, and the men employed are as rough as the work. But even more important than the rough treatment is the question of ventilation. Obviously, when driving a heading, there is a tendency for the air to stagnate, and there is difficulty in removing the fumes of powder after shots are fired unless fresh air can be brought up to the face. Now the compressed air drill exactly meets this requirement. The exhaust from the drill causes a steady draught, which cools and purifies the air, and powder fumes can always be cleared away by turning on a little air. The electric drill, on the contrary, tends, if anything, to increase the difficulty, for the power wasted in the drill raises the temperature of the air at the face of the heading. This disadvantage is probably sufficient, in many cases, to render the use of electric drills impracticable, the mere question of cost of boring (gauged, usually, by work done in well-ventilated spots) being altogether subordinated to other considerations. Of course, for a variety of purposes, such as quarrying, working in well-ventilated spots, &c., these objections have no weight. But even under the most favourable circumstances the electric drill does not seem to have made much progress.

There are two main principles upon which rock drills are designed : the rotary and the percussive.

The rotary type is usually driven through gearing from a small motor. The percussive type is sometimes worked by a motor winding up a coiled spring, which at the proper moment uncoils and spends its energy upon the drill shaft. In other types the principle of magnetic attraction is used, the drill shaft being attached directly to the iron core of a solenoid.

Hitherto the motor-driven drills have proved to be the more successful in general work. Several firms build these machines, each having a different way of gearing the motor to the drill shaft, but the general principle is the same. The best known in this country, as well as in America, is the Jeffrey drill. It is chiefly used for working in hard coal, and is found to give satisfactory results. In general the makers design the motors to work at a pressure of 220 volts; but this can be varied to suit special requirements. The rated output of the drills for 2in. holes is 2 H.P., but a factor of safety of two is allowed. They are stated to drill, in the hard "boney" found in anthracite mines, a 2in. hole, 6ft. deep, in from 2min. to 2½min., while in the coal they will cut a similar hole in less than a minute. One of these

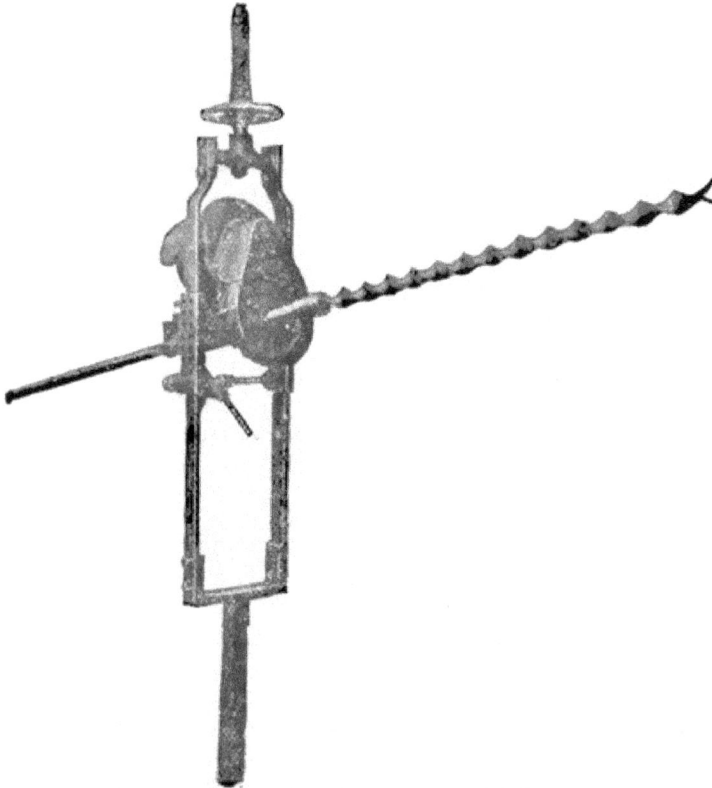

Fig. 217.—Jeffrey Rotary Electric Rock and Coal Drill.

drills is in use at Newbattle Colliery, Dalkeith. The manager, Mr. John Morrison, reports that it makes, in the hard splint coal, a 2in. hole, 4ft. deep, in the space of 5min., including the time taken to set the drill. ¡This, he adds, is equal to the work of four men, while the Jeffrey drill requires only one man.

c c 2

A good idea of the general shape of the drill, and of the method of fixing it in position, will be gathered from Fig. 217. It will be seen that the motor is protected from mechanical injury by an iron shield. Any type of motor can be fitted to the drill, and probably a polyphase machine will ultimately be adopted.

Various English firms have built electric drills for mining work, but, judging from the number in use, none of these has been so successful as the Jeffrey drill. The designs are usually heavier than that shown in Fig. 217.

The reciprocating action of a solenoid is apparently well suited for actuating percussive drills. Yet the solenoid, energised by a continuous current, is the least effective means of using magnetic power through comparatively long distances. Its proper function is to attract an armature through a very short range, or to hold it firmly in position.*

However, in spite of this manifest disadvantage, solenoidal action offers so many advantages in the design of percussive drills that it appears highly probable that this type of drill will ultimately come into more general use than the motor type. Therefore, a brief reference to the subject may be useful.

The earliest attempts to utilise the principle were made in accordance with the diagram shown in Fig. 218. The current was supplied to the centre of two coils forming the solenoid, and was passed through them successively; the changing being controlled by means of a two-way switch, indicated at a and c. The iron shaft of the drill formed the core, and was attracted first by one coil and then by the other. This is a very wasteful design, for the power absorbed during the instroke is the

---

* Prof. S. P. Thompson has recently demonstrated that an alternate-current magnet possesses very different properties from those of a continuous-current one. He shows that by means of an alternate current the pull of an electro-magnet on its armature may be extended through comparatively long distances, and may be even greater at, say, three inches than when closer. This important fact probably opens up a new field, and may, therefore, assist in the development of percussive drills.

same as that during the outstroke.   An additional objection is
the trouble given by the contacts at the switch points.

An improvement is possible by the use of a single solenoid
drawing in a plunger against a strong coiled spring.   The blow
given by the recoil of the spring is very rapid, and resembles
the impact of a hammer if the bit be at the right striking distance ; and, further, the instroke is made comparatively slowly,
and does not shake the drill carriage.   In case the stroke is
too short to strike the work a spring cushion is provided to
receive the blow.   A two-way switch is required, however,
and this practically renders the design useless.

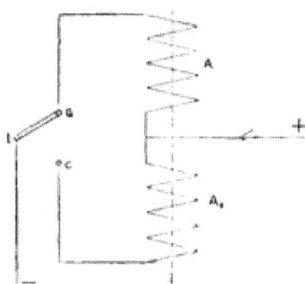

FIG. 218.—Diagram of Electric Percussive Drill, with two-way
switch at a c.

A distinct advance is made by the use of pulsating or
alternate currents, or a combination of them.   There are
many methods, but they all embody the following conditions :
No switch is necessary, three or more wires are required, and
also, in some cases, a special dynamo.   The first is an absolute
gain, for there is no trouble from sparking, and the last two
are not serious difficulties when it is considered that these
drills are only likely to be used in mines and places where it
will pay to provide a suitable dynamo if the use of electric
drills is deemed advisable.   The simplest drill of this class is
one coupled to a two-phase circuit, as shown in Fig. 219; but
this has the disadvantage that the power used on the instroke
is nearly as great as that on the outstroke.   The number of
blows given by drills of these types corresponds with the frequency of the alternator, and they are therefore adapted for

delivering many quick light blows per minute rather than a few heavy blows. A frequency of from four to ten per second is found to give the best results.

A still better device is that applied in the Marvin drill. This is of the double solenoid type, and is one of the most

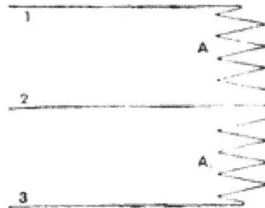

Fig. 219.—Diagram of Electric Percussive Drill, for use on a two-phase circuit. No Switch required.

successful in America. It is run off a special dynamo giving pulsating currents in two circuits. The connections are shown in Fig. 220. The dynamo armature, in its simplest form, has only one coil. One end of this is fixed to a continuous metal ring d_1, on which rubs the brush B_1; the other is joined to a metal segment, d_2, extending for 180 deg. and concentric with

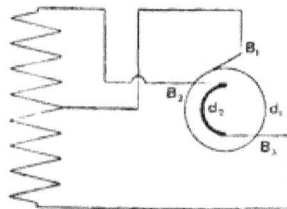

Fig. 220.— Diagram of Marvin Electric Percussive Drill.
d_1, d_2, rings on dynamo ; B_1, B_2, B_3, fixed brushes.

d_1. Two brushes, B_2 and B_3, placed diametrically opposite to each other, rub against it. Three conductors are required ; one from the middle point of the solenoid is coupled to B_1, and those from the ends of the coils are respectively joined to B_2 and B_3. The two coils successively carry a current, and so give reciprocating motion to the drill.

The best results with reciprocating drills are likely to
be obtained by the use of a combination of pulsating and
alternating currents; for the instroke can then be made
without much waste of power, and yet a very effective blow
can be delivered at each outstroke. This principle has been
applied in the Van Depoele electric percussive drill. One form
of the arrangement is shown diagrammatically in Fig. 221.
The dynamo may be of any continuous current make, with an
ordinary commutator, and the two usual fixed brushes at F_1
and F_2; but it requires the addition of two revolving brushes,
R_1 and R_2, separated by 180 deg. The coils in the solenoid
are three in number, 1 and 3, coupled in opposition to each

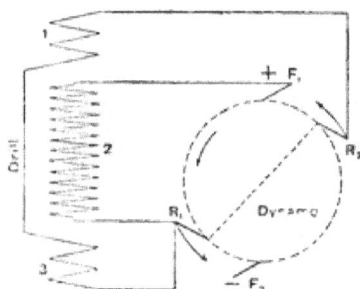

Fig. 221.—Diagram of Van Depoele Electric Percussive Drill.
R_1, R_2, revolving brushes; F_1, F_2, fixed brushes; 1, 2, 3, coils of drill.

other, having few turns relatively to the main coil 2. The connections are made as follows:—Coils 1 and 3 are in series
with each other, and are coupled to the revolving brushes, R_1
and R_2; these coils are, therefore, traversed by an alternate
current. Coil 2 is coupled to the fixed brush F_1 at one end,
and at the other to the junction of coil 3 with the rotating brush R_1; it is, therefore, energised by a pulsating current,
and exerts the chief effect on the plunger with a polarity
always in the same direction. The blow is made mainly by the
action of the centre coil, the end coils practically neutralising
each other. The return stroke is chiefly due to the resultant
pull of the two end coils, as the centre one varies between zero
and a maximum. At a first glance there seems to be in this
design a departure from simplicity without any compensating

gain, but on a closer examination it will be seen that it marks a considerable advance, and that the chief difficulties referred to in the early part of this section are to a great extent overcome.

The instroke can be made as light and the outstroke as heavy as required, both of these desirable ends being attained without the use of any switch or complicated gear in the drill. Also only three small conductors are required, and the dynamo gives a continuous current, and can be used for lighting or motor work.

The chief difficulty likely to be experienced with the Van Depoele drill appears to lie in the heating of the iron cores from hysteresis and eddy currents. This, however, will be gradually overcome as experience suggests better methods of building the core and framing.

APPENDIX.

LONG-DISTANCE TRANSMISSION OF POWER.

The distance through which power may be economically transmitted by electricity depends upon so many varying conditions that it is not possible to give a definite limit, although one can be found for any particular case.* The theoretical considerations affecting the losses in the line are discussed fully in § 16, pp. 70-79, and the practical difficulties in § 17 to § 20. The data given there are sufficient to enable a complete investigation to be made for any case likely to occur in practice. It may be useful, however, to have a table of the relative costs of continuous-current power plants for the usual pressures, and for such distances as are likely to commend themselves for ordinary practice. It will be obvious, however, that the figures must necessarily be but approximations, and in some cases may be even misleading, unless the requirements of the problem are determined by one who has a practical knowledge of the work.

The author has prepared Table E.E, in which the costs are all expressed in terms of £ per H.P. delivered by the motor. The line is supposed to be of bare sicilium bronze, erected on wooden posts, with oil insulators. The dynamo cost includes a proportional share of station instruments, and that of the motor includes switches and starting frames, whilst the line cost is inclusive of posts, insulators, lightning arresters, and other details.

The prices, which allow a slight margin for market fluctuations, include packing. The erection cost is a variable quantity, and must be determined for each case.

* *See* page 132.

Table E E.—*Cost of Continuous-Current Plants for Transmission of Power.*

Distance in feet between Dynamo and Motor.	E.M.F. at Motor Terminals.	Current in Amperes per H.P. delivered by Motor.	Commercial Efficiency.	H.P. required at Dynamo per H.P. delivered by Motor.	Per H.P. delivered by Motor. Cost in £.			
					Dynamo.	Wire.	Motor.	Total Electrical equipment.
5,000	500	1·62	75	1·33	14·595	4·331	9·975	28·901
10,000	500	1·62	65	1·54	16·891	7·218	9·975	34·084
15,000	500	1·62	50	2·00	21·945	7·166	9·975	39·086
5,000	800	1·013	80	1·25	15·763	3·715	10·972	30·450
10,000	800	1·013	75	1·33	16·773	6·825	10·972	34·570
15,000	800	1·013	70	1·43	18·033	9·187	10·972	38·192
20,000	800	1·013	65	1·54	19·411	10·828	10·972	41·211
25,000	800	1·013	60	1·67	21·052	13·072	10·972	45·096
30,000	800	1·013	55	1·82	22·955	14·437	10·972	48·364
35,000	800	1·013	50	2·00	25·226	15·2 5	10·972	51·423
10,000	1,000	0·81	80	1·25	16·405	9·515	11·471	37·392
15,000	1,000	0·81	75	1·33	17·456	9·896	11·471	38·823
20,000	1,000	0·81	70	1·43	18·768	10·500	11·471	40·739
25,000	1,000	0·81	65	1·54	20·212	11·366	11·471	43·049
30,000	1,000	0·81	60	1·67	21·918	12·022	11·471	45·411
35,000	1,000	0·81	55	1·82	23·887	12·521	11·471	47·879
40,000	1,000	0·81	50	2·00	25·250	12·678	11·471	50·399
45,000	1,000	0·81	45	2·25	29·531	11·312	11·471	52·814
50,000	1,000	0·81	45	2·25	29·531	14·765	11·471	55·767

In many of the problems brought before the electrical engineer the distances are far greater than those assumed in Table E E, and much higher pressures are necessary. The author is of opinion that for *extra* long distances the best way to deal with power transmission is by polyphase currents (*see* Chapter VIII., especially § 54, p. 267). If the pressure at the alternators can be kept below 7,000 volts, it will not be necessary to use step-up transformers; but if the pressure exceeds this it will generally be advisable to use them.

The cost of the alternators and transformers rapidly decreases as the output is increased. The cost of alternators of, say, 100 kilowatts capacity is about £7 per kilowatt, including station apparatus; while it has been estimated in connection with the Niagara plant that, with outputs of 4,000

kilowatts, the cost will not exceed £2. Transformers of 100
kilowatt capacity average £5 per kilowatt, and in very large
sizes are offered by manufacturers at £1 per kilowatt. These
figures are so far apart that it is not possible to give any
general data for the generating plant, or even for the motors;
but with the line a fair approximation is possible, since the
cost of copper may be taken as fairly constant within the
limits of the quantities required. Table F F has been pre-

Table F F.—*Cost of Line and of H.P. delivered, with Three-phase
Circuits, for various Distances and Pressures.*

Distance in Feet.	Area of each conductor in square inches.	Effective pressure between mains in volts.	Current in each main in amperes.	Efficiency of conductors in per cent.	Total cost of line at 1s. per lb. of copper.	Cost per H.P. delivered by motor.
For 100 brake horse-power.						
					£	£ s.
25.000	0·0067	5.000	8·75	97·0	200	2 0
50.000	0·0067	7.000	6·25	97·0	400	4 0
100.000	0·0067	10.000	4·375	97·0	800	8 0
150.000	0·0067	10.000	4·375	95·5	1.200	12 0
250.000	0·0067	10.000	4·375	92·5	2.000	20 0
375.000	0·0067	10.000	4·375	89·0	3.000	30 0
500.000	0·0067	10.000	4·375	86·0	4.000	40 0
For 200 brake horse-power.						
					£	£ s.
25.000	0·0067	5.000	17·50	94·0	200	1 0
50.000	0·0067	7.000	13·5	92·5	400	2 0
100.000	0·0067	10.000	8·75	94·0	800	4 0
150.000	0·0067	10.000	8·75	91·0	1.200	6 0
250.000	0·0067	12.500	7·0	90·0	2.000	10 0
375.000	0·0067	15.000	5·83	90·0	3.000	15 0
500.000	0·0067	20.000	4·375	92·5	4.000	20 0
For 500 brake horse power.						
					£	£ s.
25.000	0·0134	10.000	26·25	97·5	400	0 16
50.000	0·0134	10.000	26·25	94·5	800	1 12
100.000	0·0134	10.000	26·25	92·0	1.600	3 4
150.000	0·0134	10.000	26·25	84·5	2.400	4 16
250.000	0·0134	12.500	21·00	84·0	4.000	8 0
375.000	0·0134	15.000	17·50	83·5	6.000	12 0
500.000	0·0134	20.000	13·15	87·0	8.000	16 0

pared to give a rough idea of the cost of copper for overhead
lines for powers of 100, 200, and 500 H.P., delivered by motors
with three-phase currents at various efficiencies and pressures
up to 20,000 volts, and transmitted through various distances.

In these calculations an allowance of 5 per cent. has been
made for the sagging of the wires, so that a distance of
5,000ft. requires 1 mile, or 5,280ft., of conductor.   For first
approximations each 5,000ft. in the first column may be taken
as a mile.   The size of conductor chosen in the first two series
is No. 13 S.B.W., which is, perhaps, the smallest consistent with
the requirements of mechanical stability.   The line losses are
chosen arbitrarily.   The most instructive thing to notice is
the cost per horse-power delivered by the motor.   The necessity
for high pressures is very apparent.

Tables EE and FF should be compared with Table X,
p. 268.

FIG. 208A.—The New Rotary-Bar Coal-Cutter of the General Electric Company, U.S.A.

(*This figure illustrates text on page 354 ante.*)

[To face page 384.

GENERAL ALPHABETICAL INDEX.

D D

	PAGE
Steam Dynamos	10, 304
——— Engines	6, 303
Steel, Cast	29
—— Induction Curves of	35
—— Mild, Use of, in Dynamo Design	31
Steinmetz's Law	173
Step-up Transformers	208
Stokers, Mechanical	304
Stoneware Conduits	100
Sub-stations for Power Distribution	139, 217
Sumpner, Dr. W. E.	179, 259
Symbols, List of	15, 197
Synchroniser	239
Synchronism	213
Synchronous Motors	149, 213, 282
——— Polyphase Motors	282
Systems of Transmitting Power, Weight of Copper required, etc.	257, 381
Tabulated Form for Designing Dynamos and Motors...	56
Temperature, Rise of, in Armature	20
——— in Field Coils	31, 32
Tests of American Electric Coal-Cutters	362
——— American Polyphase Motors	289
——— Oerlikon Polyphase Motors	286, 289
——— Oerlikon Single-Phase Asynchronous Motors...	183
Thompson, Prof. S. P., on Alternate-Current Electro-Magnets ...	376
——— Alternate-Current Motors	178
——— Phase Conversion	278
Thomson-Houston Air-Blast Lightning Discharger	108
——— Alternator	240
——— Arc Dynamo	232
——— Coal-Cutter	372
——— Compensator ... ...-	236
——— Transformers	195, 196, 200
——— Winding Engine, Electric .-	331
Three-Phase Circuits, Power in	261
——— Connections of...	276
——— Transmission of Power by	276
Three-Phase Alternators	256, 273
Three-Wire Systems of Distribution ... 123, 132, 228, 237, 254, 272, 276
——— Size of Middle Wire	133
Toothed Armatures	42, 292
Torque 43, 54, 118, 175, 263, 293, 337
Total E.M.F. of Armature	18
Traction, Motor	28, 41
——— Series	120
Transformer with Sub-divided Secondary Coil	237
Transformers, Brown, C. E. L.	194
——— Brush	192, 196
——— Change Ratio of	229, 246
——— Constant Current	199
——— Continuous Current (see Dynamotor)	
——— Copper Loss in	201, 205
——— Curves of 198, 202, 204
——— Design of Alternate Current... ... ... 190, 197, 200, 205
——— Dobrowolski Three-Phase	274
——— Drop of Pressure	197

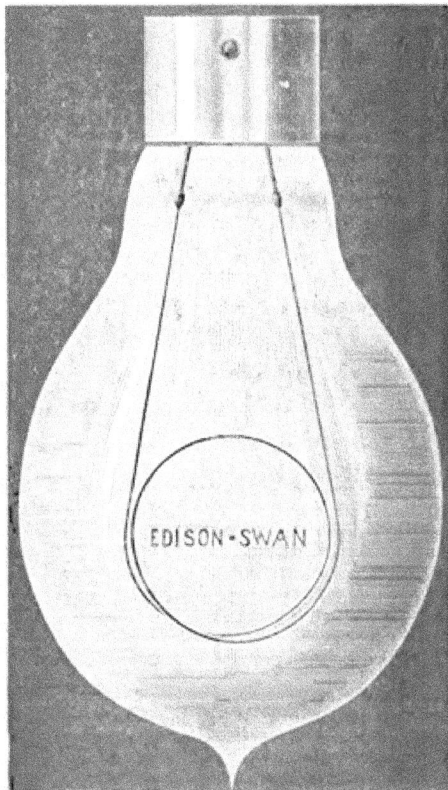

INDEX TO ILLUSTRATIONS.

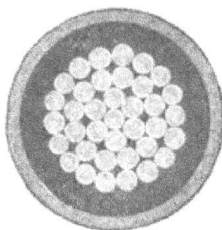

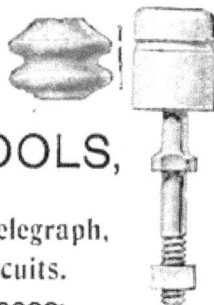

LIST OF TABLES OF DATA AND TESTS.

E E

LIST OF FORMULÆ.

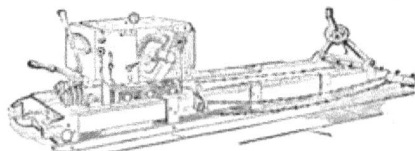

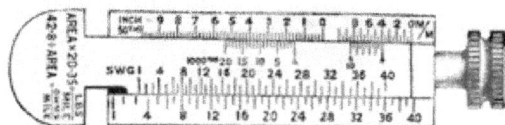

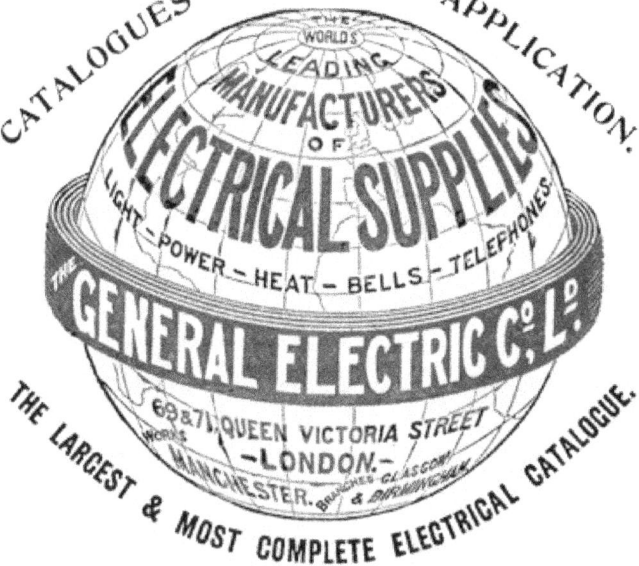

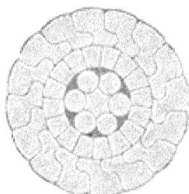

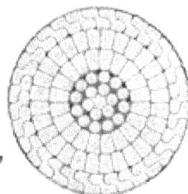

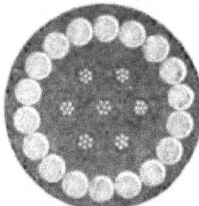

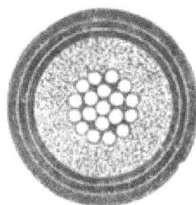

𝔅ooks for 𝔈lectricians, 𝔈lectrical 𝔈ngineers, and 𝔈lectrical 𝔖tudents.

February, 1895.

THE Publisher of "THE ELECTRICIAN" has compiled the following List of Books on Electrical and Allied Subjects, all of which can be obtained direct from "THE ELECTRICIAN" PRINTING AND PUBLISHING COMPANY, LIMITED, Salisbury Court, Fleet Street, London.

Any of the Works mentioned in this List will be forwarded to any address in the United Kingdom, on receipt of the quoted price. Five per cent. must be added for postage on foreign orders.

Copies of any known Work on the Theory or Practice of Electricity, or relating to any particular application of Electricity, or of an Electrical or Engineering character (either British or Foreign), can be ordered through this Company. Inquiries respecting the Latest and Best Books on Electrical Subjects will be answered by post, Free of Charge, and all possible assistance rendered to inquirers.

Daily, Weekly, or Monthly parcels of Miscellaneous Books and Publications, for London, the Provinces, or Abroad, made up on the best possible terms.

Cheques and Postal Orders to be made payable to "The Electrician Printing and Publishing Company, Limited," and to be crossed "Coutts and Co."

CONTENTS.

"𝔗he 𝔈lectrician" 𝔖eries.

"THE ELECTRICIAN" PRINTING & PUBLISHING CO., LIMITED.
I, 2, and 3, Salisbury Court, Fleet Street. London, E.C.

"THE ELECTRICIAN" SERIES.

466 pages, price 12s. 6d., *post free,* 13s.

ELECTROMAGNETIC THEORY.

VOL. I.

By OLIVER HEAVISIDE.

EXTRACT FROM THE PREFACE.

This work was originally meant to be a continuation of the series "Electromagnetic Induction an its Propagation," published in *The Electrician* in 1885-6-7, but left unfinished. Owing, however, to the necessity of much introductory repetition, this plan was at once found to be impracticable, and was, by request, greatly modified. The result is something approaching a connected treatise on electrical theory, though without the strict formality usually associated with a treatise. The following are some of the leading points in this volume. The first chapter is introductory. The second consists of an outline scheme of the fundamentals of electromagnetic theory from the Faraday-Maxwell point of view, with some small modifications and extensions upon Maxwell's equations. The third chapter is devoted to vector algebra and analysis, in the form used by me in my former papers. The fourth chapter is devoted to the theory of plane electromagnetic waves, and, being mainly descriptive, may perhaps be read with profit by many who are unable to tackle the mathematical theory comprehensively. It may be also useful to have results of mathematical reasoning expanded into ordinary language for the benefit of mathematicians themselves, who are sometimes too apt to work out results without a sufficient statement of their meaning and effect. But it is only introductory to plane waves. I have included in the present volume the application of the theory (in duplex form) to straight wires, and also an account of the effects of self-induction and leakage, which are of some significance in present practice as well as in possible future developments. There have been some very queer views promulgated officially in this country concerning the speed of the current, the impotence of self-induction, and other material points concerned.

Vol. II. is in preparation.

SECOND ISSUE. *370 pages, 150 illustrations. Price* 10s. 6d., *post free.*

MAGNETIC INDUCTION IN IRON AND OTHER METALS.

By J. A. EWING, M.A., B.Sc.,

Professor of Mechanism and Applied Mechanics in the University of Cambridge.

SYNOPSIS OF CONTENTS.

After an introductory chapter, which attempts to explain the fundamental ideas and the terminology, an account is given of the methods which are usually employed to measure the magnetic quality of metals. Examples are then quoted, showing the results of such measurements for various specimens of iron, steel, nickel and cobalt. A chapter on Magnetic Hysteresis follows, and then the distinctive features of induction by very weak and by very strong magnetic forces are separately described, with further description of experimental methods, and with additional numerical results. The influence of Temperature and the influence of Stress are next discussed. The conception of the Magnetic Circuit is then explained, and some account is given of experiments which are best elucidated by making use of this essentially modern method of treatment. The book concludes with a chapter on the Molecular Theory of Magnetic Induction, and the opportunity is taken to refer to a number of miscellaneous experimental facts on which the molecular theory has an evident bearing.

A German Edition is also published, price 8s. 6d.

I, 2, and 3, Salisbury Court, Fleet Street, London, E.C.

"THE ELECTRICIAN" SERIES—*continued.*

• • • • • • • • • • • • •

Now Ready. *Very fully illustrated; handsomely bound, on good paper, price* 7s. 6d.

ELECTRIC LAMPS AND ELECTRIC LIGHTING.

Being a Course of Four Lectures delivered at the Royal Institution, April-May., 1894,

By PROF. J. A. FLEMING, M.A., D.Sc., F.R.S., M.R.I.,

Professor of Electrical Engineering in University College, London.

SYNOPSIS OF CONTENTS.

I.—A Retrospect of Twelve Years—Factors in the Development of Electric Illumination—The Historical Starting Point—Davy's Researches on the Electric Arc—The Evolution of Incandescent Electric Lighting—Definition of Fundamental Terms—Units of Measurement of Current, Pressure, Work, and Power—Board of Trade Units—Conditions of Public Supply under Acts of Parliament—The Heating Effect of an Electric Current—Joule's Law—Experimental Proofs—Radiation from Incandescent Bodies—Temperature of Radiant Bodies—Surface Efficiency and Specific Radiant Qualities of Various Materials—Peculiar Properties of Carbon—Brightness of Various Lights—Methods of Photometric Comparison—Standards of Light—Sun, Moon, Electric Arc, Incandescent Lamp as Illuminants—Quality and Intensity of Light—Luminosity and Candle-power—The Physiological Question of Vision.

II.—The Physics of an Incandescent Lamp—Characteristic Curves—Relation of Candle-power to Current and Pressure—Effects of Position on Candle-power—Age of Incandescent Lamps—Lamp Mortality—Causes of it—Judicious and Injudicious Arrangement of Lamps—Sockets—Switches—Fuses—Decorative Employment of Incandescent Lamps—House Wiring—Fire Office Rules—Good and Bad Work—Causes of Destruction of the Carbon Filament—Molecular Physics of the Glow Lamp—Edison Effect—Large Incandescent Lamps—Electric Meters—Methods of Testing and Comparing Glow Lamps—Advantageous Utilisation of Current.

III.—The Electric Arc Lamp—Method of Production of the Arc—Study of the Arc by Projection—Laws of the Electric Arc—The Convection of Carbon in the Arc—The Crater—The Distribution of Electric Pressure in the Arc—Arc Lamp Mechanism—Recent Improvements—Distribution of Light from the Arc—Luminous Efficiency of the Arc—Comparison with Incandescent Lamps—Street and Interior Arc Lighting—Proper Distribution of Light—Arc Light Photometry—The Alternating and Continuous Arc—The Inverted Arc—The Use of the Arc in Metallurgy—Electrical Reduction of Metals by the Arc—Temperature of Arc Light Crater—The Solar Temperature.

IV.—The Production of Current for Electric Lighting—Generating Stations—Systems of Supply—Low Pressure Continuous and High Pressure Alternating—Structure of a Dynamo and Transformer—Views of Generating Stations at Home and Abroad—Underground Conducting Mains—Networks of Conductors—House Services—Long Distance Transmission—Electric Lighting of Rome—Tivoli-Rome Transmission—Utilisation of Water-power—Load Diagrams of Stations—Supply of Currents for Purposes other than Light.

PRESS NOTICES.

"Treats the whole subject with the lucidity for which Prof. Fleming's expositions are remarkable, and in language which has been, as far as possible, divested of technicalities. . . . Those who may contemplate the electric lighting of their houses will find in the book many valuable hints and practical suggestions."—*Times.*

"Prof. Fleming possesses the rare gift of being able to interest almost any audience in his subject. . . . The work under notice consists of a transcript of four lectures delivered at the Royal Institution, and though much of the charm which the lecturer was able to throw round the subject is now absent, still enough remains to render a perusal of these pages a recreation rather than a study. . . . Besides supplying the non-technical student with such information as he is sadly in need of, he has given to the serious student the best account which we have yet seen of the physical properties of arc and glow lamps. . . . The volume contains some 220 pages of large type, and is well illustrated throughout. It is the only one we have ever seen that we can thoroughly recommend to the non-technical reader. On the other hand, for the electrical student we cannot too forcibly recommend a careful study of the second and third lectures."—*Daily Chronicle.*

'We have no hesitation in recommending all who may be interested in the subject to buy a copy of Prof. Fleming's book, which is well worth the price asked for it; being fresh as regards matter, and abundantly provided with illustrations which are more or less original. . . . As we commenced this article by stating, the book is worth buying for a guide to a fair general knowledge of the principles which underlie the industrial uses of electricity. We can also compliment the publishers upon the way in which the volume is turned out."—*Journal of Gas Lighting.*

"When one comes across a reprint of a series of afternoon Royal Institution lectures he expects an elementary book, with its contents strung together to connect a lot of experiments that the author was anxious to show to an audience already jaded with a variety of subjects no one human brain could take in. Dr. Fleming has given us nothing of this sort, but a concise account of incandescent and arc lamps, followed by a short sketch of electric lighting generally. It is by no means a mere popular book, however, and few electricians, no matter how well up in their subject, will find they can afford to pass it by. . . . The binding, printing, and general get up is admirable."—J.S. in *The Electrician.*

"Dr. Fleming is known for his high abilities in setting forth scientific facts in a popular lecture in clear and most attractive language, and anyone familiar with the elements of physical science will read the book with great pleasure and derive much sound instruction. . . . The book is very handsomely got up, and would make an exceedingly suitable presentation volume to anyone interested in, and who already knows at least a little of, the subject."—*Glasgow Herald.*

DIGEST POST FREE ON APPLICATION.

1, 2, and 3, Salisbury Court, Fleet Street, London, E.C.

"THE ELECTRICIAN" SERIES—*continued.*

FOURTH ISSUE. *500 pages, 157 illustrations. Price* 7s. 6d., *post free* 8s.

THE ALTERNATE CURRENT TRANSFORMER
IN THEORY AND PRACTICE.

By J. A. FLEMING, M.A., D.Sc., F.R.S., M.R.I., &c.,

Professor of Electrical Engineering in University College, London.

VOL. I.—THE INDUCTION OF ELECTRIC CURRENTS.

SYNOPSIS OF CONTENTS.

CHAPTER I.—Introductory.

Faraday's Electrical Researches—Early Experiments on Current Induction—Electro-Dynamic Induction—First Induction Coil—Electro-Magnetic Induction—Lines of Force—Methods of Ampère, Arago, and Faraday—Physical Nature of Lines of Magnetic Force.

CHAPTER II.—Electro-Magnetic Induction.

Magnetic Force and Magnetic Induction—Tubes of Magnetic Induction—Rate of Change of Magnetic Induction through a Circuit—Inductance—Electromotive Force of Induction—Electro-Magnetic Momentum—Electro-Magnetic Energy—Dimensions of the Co-efficient of Self-Induction or Inductance—Unit of Self-Induction or Inductance—Constant and Variable Inductance—Curves of Magnetisation—Graphical Representation of Variation of Co-efficient of Induction—Magnetic Hysteresis.

CHAPTER III.—The Theory of Simple Periodic Currents.

Variable and Steady Flow—Current Curves—Simple and Complex Harmonic Motion—Fourier's Theorem—Simple Periodic and Sine Curves—Current Growth in Inductive Circuits—Logarithmic Curves—Geometrical Illustrations—Graphic Representation of Periodic Currents—Mean Value of the Power of a Periodic Current—Power Curves—Experimental Measurement of Periodic Current and Electromotive Force—Wattmeter Method of Periodic Power—Mutual Induction of Two Circuits of Constant Inductance.

CHAPTER IV.—Mutual and Self-Induction.

Prof. Joseph Henry's Researches in Electro-Magnetism—Mutual Induction—Elementary Theory of the Induction Coil—Comparison of Theory and Experiment—Magnetic Screening and the Action of Metallic Masses in Induction Coils—Transmission of Rapidly Intermittent, Alternate, or very Brief Currents through Conductors—Effects of Saturation and Magnetic Hysteresis—Characteristic Curves of the Series and Parallel Transformer—Efficiency of Transformers—Ferrari's Experiments.

CHAPTER V.—Dynamical Theory of Current Induction.

Electric Displacement—Maxwell's Theory of Molecular Vortices—Velocity of Propagation of an Electro-Magnetic Disturbance—Electrical Oscillations—Function of the Condenser in an Induction Coil—Impulsive Discharges—Alternative Paths—Impulsive Impedance—Relation of Impedance to Periodicity—Dr. Hertz's Researches on Electrical Oscillatory Induction—Resonance Phenomena—Interference Phenomena at Various Distances—Recent Experiments—Poynting's Views on the Propagation of Electro-Magnetic Energy—Possible Direction of Future Research.

PRESS NOTICES.

"It would be very difficult to pick out from amongst the electrical literature of the past ten years any work which marks, as emphatically as does Dr. Fleming's book, the manner in which the practical problems of the day ave compelled electrical engineers to advance in their knowledge of theoretical science. . . . . It is a book which the electrical engineer of the present and of the future alike will read—he of the present, if he can ; he of the future, because he must."—*Prof. Silvanus P. Thompson in "The Electrician."*

"The practical importance and interest of the subject treated is so great that there should be little need to urge students and electrical engineers to make themselves acquainted with this book, but I do urge them nevertheless ; and they may think it fortunate that Dr. Fleming has managed to find time to issue so instructive and readable and well-timed a volume."—*Dr. Oliver J. Lodge in "Nature."*

"Dr. Fleming's book contains an enormous amount of valuable matter . . . . which cannot be got anywhere else in the plain and concise way it is given by Dr. Fleming. It is one of those books every electrician should have."—*Electrical Review.*

"A most important, timely, and valuable book. . . . . The author has earned the thanks of everyone interested in this great branch of electrical investigation and practice."—*Electrical World* (New York).

"If anyone wants this difficult subject treated in the clearest way, he cannot do better than read this book." *Industries.*

1, 2, and 3, Salisbury Court, Fleet Street, London, E.C.

"THE ELECTRICIAN" SERIES—*continued.*

SECOND ISSUE. *More than 600 pages and over 300 illustrations. Price* **12s. 6d.**, *post free.*

THE ALTERNATE CURRENT TRANSFORMER
IN THEORY AND PRACTICE.

By J. A. FLEMING, M.A., D.Sc., F.R.S., M.R.I., &c.,

Professor of Electrical Engineering in University College, London.

VOL. II.—THE UTILISATION OF INDUCED CURRENTS.

SYNOPSIS OF CONTENTS.

CHAP. I.—Historical Development of Induction Coil and Transformer.

The Evolution of the Induction Coil—Page's Researches—Callan's Induction Apparatus—Sturgeon's Induction Coil—Bachhoffner's Researches—Callan's Further Researches—Callan's Great Induction Coil—Page's Induction Coil—Abbot's Coil—Automatic Contact Breakers—Ruhmkorff's Coils — Poggendorff's Experiments — Stöhrer's, Hearder's, Ritchie's Induction Apparatus—Grove's Experiments—Apps' Large Induction Coils—Jablochkoff's Patent—Fuller's Transformer—Early Pioneers—Gaulard and Gibbs—Zipernowsky's Transformers—Improvements of Rankin Kennedy, Hopkinson, Ferranti, and others—The Modern Transformer since 1885.

CHAP. II.—Distribution of Electrical Energy by Transformers.

Detailed Descriptions of Large Alternate-Current Electric Stations using Transformers in Italy, England, and United States—Descriptions of the Systems of Zipernowsky-Déri-Bláthy, Westinghouse, Thomson-Houston, Mordey, Lowrie-Hall, Ferranti, and others—Plans, Sections, and Details of Central Stations using Transformers—Illustrations of Alternators and Transformers in Practical Use in all the chief British, Continental, and American Transformer Stations.

CHAP. III.—Alternate-Current Electric Stations.

General Design of Alternating-Current Stations, Engines, Dynamos, Boilers—Proper Choice of Units—Water Power—Parallel Working of Alternators—Underground Conductors—Various Systems—Concentric Cables—Capacity Effects dependent on Use of Concentric Cables—Phenomena of Ferranti Tubular Mains—Safety Devices—Regulation of Pressure—Choice of Frequency—Methods of Transformer Distribution—Sub-Stations—Automatic Switches.

CHAP. IV.—The Construction and Action of Transformers.

Transformer Indicator Diagrams—Ryan's Curves—Curves of Current—Electromotive Force and Induction-Analysis of Transformer Diagrams—Predetermination of Eddy Current and Hysteresis Loss in Iron Cores—Calculation and Design of Transformers—Practical Predetermination of Constants—Practical Construction of Transformers—Experimental Tests of Transformers—Measurement of Efficiency of Transformers—Calometric Dynamometer and Wattmeter Methods—Reduction of Results.

CHAP. V.—Further Practical Application of Transformers.

Electrical Welding and Heating Transformers for producing Large Currents of Low Electromotive Force—Theory of Electric Welding—Other Practical Applications—Conclusion.

PRESS NOTICES.

"In reviewing the first volume of this work we found much to admire and praise, much to raise high expectations for the volume which was to follow. These expectations have by no means been disappointed. The new volume is in many ways of even greater interest than its predecessor."
Professor Silvanus P. Thompson in "The Electrician."

"The book is really a valuable addition to technical literature."—*Industries.*

"A valuable addition to the somewhat meagre literature on a subject which is sure to grow in importance, and we congratulate Dr. Fleming on his work."—*The Engineer.*

"Le sujet traité par le Dr. Fleming est un de ceux qui, pour le moment, attirent l'attention générale ; son ouvrage est certainement un des plus importants de la littérature électrique. Tous les problèmes relatifs à l'application des courants alternatifs y sont traités avec une très grande compétence et de plus avec une clarté et avec une précision sans égales. Nous ne pouvons donc que recommander vivement cet ouvrage à l'attention de tous les électriciens."—*La Lumière Électrique.*

"L'ouvrage de M. Fleming est une œuvre vraiment pratique qui doit rendre à l'industrie de grands services par l'amas de renseignements qu'elle contient."—*L'Industrie Électrique.*

"Das Fleming'sche Werk füllt entschieden eine Lücke in der Literatur aus und kann durchaus empfohlen verden."—*Elektrotechnische Zeitschrift.*

"THE ELECTRICIAN" SERIES—*continued.*

In Two Volumes.—Price: stout paper, **2s.,** *post free 2s. 2d. each; strong cloth covers,* **2s. 6d.,** *post free 2s. 9d. each. Single Primers, 3d., post free 3½d.*

"THE ELECTRICIAN" PRIMERS.

(FULLY ILLUSTRATED.)

A Series of Helpful Primers on Electrical Subjects for the use of Colleges, Schools, and other Educational and Training Institutions, and for Young Men desirous of entering the Electrical professions.

TABLE OF CONTENTS.

The object of "The Electrician" Primers is to briefly describe in simple and correct language the present state of electrical knowledge. Each Primer is short and complete in itself, and is devoted to the elucidation of some special point or the description of some special application. Theoretical discussion is as far as possible avoided, the principal facts being stated and made clear by reference to the uses to which they have been put. Both volumes are suited to readers having little previous acquaintance with the subject. The matter is brought up to date, and the illustrations refer to instruments and machinery in actual use at the present time. It is hoped that the Primers will be found of use in Schools, Colleges, and other Educational and Training Establishments, where the want of a somewhat popularly written work on electricity and its industrial applications, published at a popular price, has long been felt; while artisans will find the Primers of great service in enabling them to obtain clear notions of the essential principles underlying the apparatus of which they may be called upon to take charge.

"The articles are generally so well written, and the subject matter so judiciously condensed, that there is but very little to criticise, though much to praise."—*Electrical Review.*

"The books are well printed, and we can heartily commend them as stepping stones to more advanced works."—*Electrical Plant.*

"Clearly written, and all that can be desired in the form of enunciation and explanation."—*Work.*

"The contents of each one of these volumes is of that quality and description which at once constitute a book a welcome addition to the library of the student or of the artisan."—*Amateur Work*

"These Primers are admirably adapted for teaching purposes; they are calculated to be exceedingly useful in connection with the preparation of object lessons; they are very suitable for presents to boys of a mechanical turn, and they might well find a place in school libraries."—*School Board Chronicle.*

Issued annually, price **3s.,** *post free.*

A DIGEST OF THE LAW OF ELECTRIC LIGHTING,

AND OTHER SUBJECTS.

(Revised to January in each year.)

By A. C. CURTIS-HAYWARD, B.A., M.I.E.E.

An abstract of the Electric Lighting Acts, 1882 and 1889, and of the various documents emanating from the Board of Trade dealing with electric lighting. The digest treats first of the manner in which persons desirous of supplying electricity must set to work, and then of their rights and obligations after obtaining Parliamentary powers; and gives in a succinct from information of great value to Local Authorities, Electric Light Contractors, &c., up to date. The Board of Trade Regulations, the London County Council Regulations as to Theatre Lighting, British and Foreign Rules and Regulations for the Prevention of Fire Risks arising from Electric Lighting, and the Installation Regulations of the Electric Supply Companies of the Metropolis are also given.

1, 2, and 3, Salisbury Court, Fleet Street, London, E.C.

"THE ELECTRICIAN" SERIES—*continued.*

Fully Illustrated. Price 7s. 6d., *post free*

THE

INCANDESCENT LAMP AND ITS MANUFACTURE.

By GILBERT S. RAM.

EXTRACT FROM THE PREFACE.

With the expiration of Edison's master-patent for the carbon incandescent electric lamp, the attention of electric light engineers, as well as of all those who use the light, is once more directed to the consideration of the lamp itself, to the possibility of obtaining better lamps, and to the probable reduction in price which will naturally follow. Owing to the long prevailing monopoly in the sale and manufacture, there has been little inducement for those interested to experiment and to study the problems connected with the incandescent lamp. As a result of this, the literature of the lamp is very scanty, and is entirely confined to the pages of the leading technical journals. While dynamos, alternators, transformers, arc lamps, and almost every piece of apparatus connected with electrical engineering and lighting, have been written on at length and discussed at meetings of scientific societies, the incandescent electric lamp, which has been the chief cause of the very existence of these machines and apparatus, has been comparatively neglected. With the exception of the valuable series of articles by Mr. Swinburne which appeared in *The Electrician* some years ago, no comprehensive or detailed account of lamp manufacture has appeared. The manufacture of the incandescent lamp and the principles underlying it are, consequently, but little known, except to those actually engaged in the work.

The Author has attempted to impart such information as he has acquired in the course of a considerable experience in lamp-making, and to give this information with as little mathematical embellishment as, under the circumstances, is possible.

Fully Illustrated. Price 7s. 6d., *post free.*

DRUM ARMATURES AND COMMUTATORS

(THEORY AND PRACTICE).

By F. MARTEN WEYMOUTH.

A complete treatise of the theory and the different modes of construction of Drum Winding and also of close-coiled continuous-current Commutators, together with a full résumé of some of the principal points of consideration that are involved in their design.

The first chapter relates entirely to the theory of the Drum Winding. An explanation, it i hoped with sufficient fulness, is there given of the generation of electromotive force and current within that form of winding ; and questions of magnetism connected therewith are also considered Chapters II. to IX. are devoted to the description of various methods in which Drum Winding ha been carried out in practice, with special reference to what is termed the "end-winding." Chapters X. to XIII. touch upon the mechanical construction of Commutators ; and Chapter XIV. to XXVI. deal with Commutator Sparking, with which has become necessarily involved th whole subject of what is known as "armature reactions." The book closes with a chapter on th Taper of Commutator Segments.

The various subjects are treated without, or almost without, the use of mathematics. But th author ventures to think that, in this case, the absence of mathematics is far from unjustifiable.

"THE ELECTRICIAN" SERIES—*continued.*

Over 300 pages, 100 illustrations. Price **10s. 6d.**, *post free.*

The ART of ELECTROLYTIC SEPARATION of METALS.
(THEORETICAL AND PRACTICAL.)

By GEORGE GORE, LL.D., F.R.S.

THE ONLY BOOK ON THIS IMPORTANT SUBJECT IN ANY LANGUAGE.

SYNOPSIS OF CONTENTS.

HISTORICAL SKETCH.
Discovery of Voltaic and Magneto-Electricity—First Application of Electrolysis to the Refining of Copper—List of Electrolytic Refineries.

THEORETICAL DIVISION.
Section A.: *Chief Electrical Facts and Principles of the Subject.*—Electric Polarity and Induction, Quantity, Capacity, Potential—Electromotive Force—Electric Current—Conduction and Insulation—Electric Conduction Resistance.

Section B.: *Chief Thermal Phenomena.*—Heat of Conduction Resistance—Thermal Units, Symbols, and Formulæ.

Section C.: *Chief Chemical Facts and Principles of the Subject.*—Explanation of Chemical Terms—Symbols and Atomic Weights—Chemical Formulæ and Molecular Weights—Relation of Heat to Chemical Action.

Section D.: *Chief Facts of Chemico-Electric or Voltaic Action.*—Electrical Theory of Chemistry—Relation of Chemical Heat to Volta Motive Force—Volta-Electric Relations of Metals in Electrolytes—Voltaic Batteries—Relative Amounts of Voltaic Current produced by Different Metals.

Section E.: *Chief Facts of Electro-Chemical Action.*—Definition of Electrolysis—Arrangements for Producing Electrolysis—Modes of Preparing Solutions—Nomenclature—Physical Structure of Electro-Deposited Metals—Incidental Phenomena attending Electrolysis—Decomposability of Electrolytes—Electro-Chemical Equivalents of Substances—Consumption of Electric Energy in Electrolysis.

Section F.: *The Generation of Electric Currents by Dynamo Machines.*—Definition of a Dynamo and of a Magnetic Field—Electro-Magnetic Induction—Lines of Magnetic Force.

PRACTICAL DIVISION.
Section G.: *Establishing and Working an Electrolytic Copper Refinery.*—Planning a Refinery—Kinds of Dynamos Employed—Choice and Care of Dynamo—The Depositing Room—The Vats—The Electrodes—The Main Conductors—Expenditure of Mechanical Power and Electric Energy—Cost of Electrolytic Refining.

Section H.: *Other Applications of Electrolysis in Separating and Refining Metals.*—Electrolytic Refining of Copper by other Methods—Extraction of Copper from Minerals and Mineral Waters—Electrolytic Refining of Silver Bullion and of Lead—Separation of Antimony, of Tin, of Aluminium, of Zinc, of Magnesium, of Sodium and Potassium, of Gold—Electrolytic Refining of Nickel—Electric Smelting.

Appendix.—Useful Tables and Data.

Second Edition, price **2s.**, *post free.*

ELECTRO-CHEMISTRY.
By GEORGE GORE, LL.D., F.R.S.

This book contains, in systematic order, the chief principles and facts of electro-chemistry, and is intended to supply to the student of electro-plating and electro-metallurgy a scientific basis upon which to build the additional practical knowledge and experience of his trade. A scientific foundation, such as is here given, of the art of electro-metallurgy is indispensable to the electro-depositor who wishes to excel in his calling, and should be studied previously to and simultaneously with practical working. As the study of electro-chemistry includes a knowledge not only of the conditions under which a given substance is electrolytically separated, but also of the electrolytic effect of a current on individual compounds, both are described, and the series of substances are treated in systematic order. *An indispensable book to Electro-Metallurgists.*

1, 2, and 3, Salisbury Court, Fleet Street, London, E.C.

"THE ELECTRICIAN" SERIES—*continued.*

Electrical Laboratory Notes & Forms.

ARRANGED AND PREPARED BY

Dr. J. A. FLEMING, M.A., F.R.S.

Professor of Electrical Engineering in University College, London.

These " Laboratory Notes and Forms " have been prepared to assist Teachers, Demonstrators and Students in Electrical Laboratories, and to enable the Teacher to economise time. They consist of a series of (about) Twenty Elementary and (about) Twenty Advanced Exercises in Practical Electrical Measurements and Testing. For each of these Exercises a four-page Report Sheet has been prepared, two pages of which are occupied with a condensed account of the theory and practical instructions for performing the particular Experiment, the other two pages being ruled up in lettered columns, to be filled in by the Student with the observed and calculated quantities. Where simple diagrams will assist the Student, these have been supplied. These Exercises are for the most part based on the methods in use in the Electrical Engineering Laboratories of University College, London ; but they are perfectly general, and can be put into practice in any Electrical Laboratory.

Each Form is supplied either singly at **4d.** nett, or at **3s. 6d.** per dozen nett (assorted or otherwise as required) ; in sets of any three at **1s.** nett ; or the set of (about) Twenty Elementary (or Advanced) Exercises can be obtained, price **5s. 6d.** nett. The complete set of Elementary and Advanced Exercises are price **10s. 6d.** nett, or in a handy Portfolio, **12s.** nett, or bound in strong cloth case, price **12s. 6d.** nett.

Spare Tabulated Sheets for Observations, price **1d.** each nett.

Strong Portfolios, price **1s.** each.

The very best quality foolscap sectional paper (16in. by 13in.) can be supplied, price **1s.** per dozen sheets nett.

ELEMENTARY SERIES.—(*Now ready.*)

1. The Exploration of Magnetic Fields.
2. The Magnetic Field of a Circular Current.
3. The Standardization of a Tangent Galvanometer by the Water Voltameter.
4. The Measurement of Electrical Resistance by the Divided Wire Bridge.
5. The Calibration of the Ballistic Galvanometer.
6. The Determination of Magnetic Field Strength.
7. Experiments with Standard Magnetic Fields.
8. The Determination of the Interpolar Field of an Electromagnet with Varying Lengths of Air Gap.
9. The Determination of Resistance and Temperature Coefficients with the Post Office Pattern of Wheatstone's [Bridge.
10. The Determination of Electromotive Force by the Potentiometer.
11. The Determination of Current Strength by the Potentiometer.
12. A Complete Test of a Primary Battery.
13. The Calibration of a Voltmeter by the Potentiometer.
14. A Photometric Examination of an Incandescent Lamp.
15. The Determination of the Absorptive Powers of Semi-Transparent Screens.
16. The Determination of the Reflective Powers of Various Surfaces.
17. The Determination of the Electrical Efficiency of an Electromotor by the Cradle Method.
18. The Determination of the Efficiency of an Electromotor by the Brake Method.
19. The Efficiency Test of a Combined Motor Generator Plant.
20. Test of a Gas Engine and Dynamo Plant.

ADVANCED SERIES.—(*Ready shortly.*)

21. The Determination of the Specific Electrical Resistance of a Sample of Wire.
22. The Measurement of Low Resistances by the Potentiometer.
23. The Measurement of Armature Resistances.
24. The Standardization of an Ampere-meter by Copper Deposit.
25. The Standardization of a Voltmeter by the Potentiometer.
26. The Standardization of an Ammeter by the Potentiometer.
27. The Determination of the Magnetic Permeability of a Sample of Iron.
28. The Standardization of a High Tension Voltmeter.
29. The Efficiency Test of a Transformer.
30. The Delineation of the Curves of Current and Electromotive Force of a Transformer
31. The Photometric Examination of an Arc Lamp.
32. The Measurement of Insulation and High Resistance.
33. The Examination of a Secondary Cell by the Potentiometer.
34. The Efficiency Test of an Alternator.
35. The Complete Efficiency Test of a Secondary Battery.
36. The Calibration of Electric Meters.
37. The Determination of the Hysteresis Curve of Iron by the Magnetometer.
38. The Determination of Hysteresis Loss by the Wattmeter.
39. The Determination of the Capacity of a Concentric Cable.
40. The Complete Hopkinson Test of a Pair of Dynamos.

1, 2, and 3, Salisbury Court, Fleet Street, London, E.C.

"THE ELECTRICIAN" SERIES—*continued.*

320 pages, 155 illustrations.   Price **6s. 6d.**, *post free.*

PRACTICAL NOTES FOR ELECTRICAL STUDENTS.

LAWS, UNITS, AND SIMPLE MEASURING INSTRUMENTS.

By A. E. KENNELLY and H. D. WILKINSON, M.I.E.E.

SYNOPSIS OF CONTENTS.

CHAPTER I.—Introductory.
Early Ideas—Electricity produced by Chemical Energy—Requirements in a good Cell—Chemical Action.

CHAPTER II.—Batteries.
Daniell, Minotto, Thompson Tray, Leclanché, Fuller, De la Rue and Standard Cells.

CHAPTER III.—Electromotive Force and Potential.
Connecting Cells in Series—Distribution of Potential in a Battery.

CHAPTER IV.—Resistance.
Relative Resistance of Metals—Relation between Length, Diameter, and Weight of Telegraph Conductors—Resistances in Series and in Multiple Arc.

CHAPTER V.—Current.
Effect of "Opening" or "Closing" a Circuit—Velocity of Current—Retardation—Period of Constant Flow—The Ampère—The Coulomb—The Milliampère—Ohm's Law.

CHAPTER VI.—Current Indicators.
Detectors or Indicators—Directions for Making—Detectors for Telegraph and for Telephone Work—Indicators for Large Current.

CHAPTER VII.—Simple Tests with Indicators.
Tests for "Continuity," for Fault in Telegraph Apparatus, for Identity of Wires, for Insulation—Overhead Line Insulators—G.P.O. Standard Indicator.

CHAPTER VIII.—Calibration of Current Indicators.
Calibration by Low-Resistance Cells—Calibration Curves—Simultaneous Calibration of Instruments of Similar Sensitiveness and of Differing Sensitiveness—Use of the "Shunt"—Comparison by Tangent Galvanometer.

CHAPTER IX.—Magnetic Fields and their Measurements.
Permanent Magnetic Fields—Electro-Magnetic Fields—Magnetic Fields of Coils and Solenoids.

TABLE OF NATURAL TANGENTS.

---

190 pages, 116 illustrations.   Price **3s. 6d.**, *post free.*

THE STEAM-ENGINE INDICATOR & INDICATOR DIAGRAMS.

A PRACTICAL TREATISE ON.

Edited by W. W. BEAUMONT, M.I.C.E., M.I.M.E., &c.

This useful book considers the object of an Indicator Diagram, or what it is desired that the Diagram shall show; describes the construction for the Indicator in its various forms; describes the apparatus necessary for the attachment of the Indicator to the engine, and how to use the instrument; gives examples of diagrams from all kinds of engines most in use, comparing these diagrams and showing how far they agree with theoretical diagrams; and shows the most simple methods of calculating and constructing theoretical curves of expansion, and of comparing the actual with the theoretical performance of steam in the steam engine cylinder.

Fully illustrated.   Price **1s. 6d.**, *post free 1s. 9d.*

THE MANUFACTURE OF ELECTRIC LIGHT CARBONS.

A Practical Guide to the Establishment of a Carbon Manufactory.

Contains the results of several years' experiments and experience in carbon candle-making, and gives full particulars, with many illustrations, of the whole process.

1, 2, and 3, Salisbury Court, Fleet Street, London, E.C.

"THE ELECTRICIAN" SERIES—*continued.*

Over 400 pages, nearly 250 illustrations. Price **10s. 6d.**

ELECTRIC MOTIVE POWER.

By ALBION T. SNELL, Assoc.M.Inst.C.E., M.I.E.E.

The rapid spread of electrical work in collieries, mines, and elsewhere has created a demand for a practical book on the subject of transmission of power. Though much had been written, there was no single work dealing with the question in a sufficiently comprehensive and yet practical manner to be of real use to the mechanical or mining engineer; either the treatment was adapted for specialists, or it was fragmentary, and power work was regarded as subservient to the question of lighting. The Author has felt the want of such a book in dealing with his clients and others, and in "ELECTRIC MOTIVE POWER" has endeavoured to supply it.

In the introduction the limiting conditions and essentials of a power plant are analysed, and in the subsequent chapters the power plant is treated synthetically. The dynamo, motor, line, and details are discussed both as to function and design. The various systems of transmitting and distributing power by continuous and alternate currents are fully enlarged upon, and much practical information, gathered from actual experience is distributed under the various divisions. The last two chapters deal exhaustively with the applications of electricity to mining work in Great Britain, the Continent, and America, particularly with reference to collieries and coal-getting, and the results of the extensive experience gained in this field are embodied.

In general, the Author's aim has been to give a sound digest of the theory and practice of the electrical transmission of power, which will be of real use to the practical engineer, and to avoid controversial points which lie in the province of the specialist, and elementary proofs which properly belong to text-books on electricity and magnetism.

A LARGE=SHEET TABLE,

Giving full particulars of the Electricity Supply Stations throughout Great Britain up to January, 1895, can be obtained mounted on stout board, with cord for hanging. Price: Varnished, **3s. 6d.**, Unvarnished, **3s.**— each post free. A Map, showing positions of Supply Stations, is mounted on the back of the Table. A Coloured Map, showing the Streets of London in which Mains for Private Lighting are laid up to January, 1895, together with the areas allotted to the different Supply Companies, is also mounted on some copies of the above Table; and the price of these, complete, post free, is **5s.**

NEW VOLUMES IN PREPARATION.

SUBMARINE CABLE-LAYING AND REPAIRING.

By H. D. WILKINSON, M.I.E.E., &c., &c.

This work will describe the procedure on board ship when removing a fault or break in a submerged cable and the mechanical gear used in different vessels for this purpose; and considers the best and most recent practice as regards the electrical tests in use for the detection and localisation of faults, and the various difficulties that occur to the beginner.

MOTIVE POWER AND GEARING
FOR ELECTRICAL MACHINERY.

By E. TREMLETT CARTER, C.E.

(COPIOUSLY ILLUSTRATED WITH SCALE DRAWINGS & NUMEROUS PLATES.)

The purpose of this work is the explanation of the principles and practice of modern mechanical motive power and gearing, especially in their application to electrical machinery. Electrical engineering is as much a matter of engines and gearing as of dynamos and cables; but the conditions of electric light and power distribution are such that a special study of the mechanical plant is necessary. Just as marine or locomotive steam practice is treated in a special manner in works on the subject; so the Author has endeavoured to hold in view the special requirements of electrical practice, and to produce a work on steam and other motive power which shall be solely devoted to these requirements.

"MOTIVE POWER AND GEARING" is adapted equally to the needs of the practical engineer and of the student, and the treatment is such as may be easily understood without special mathematical training. Besides steam plant, as used in electric power stations, the work treats of gas, oil, and water-power engines, and the chapters on these, as well as the section on Gearing, are written on the lines of the latest practice in electric power stations. The best points in the development of motive power for electrical engineering on the Continent and in the United States have also been considered, and are fully treated, and compared with English practice. This work constitutes the only existing treatise on the Economics of Motive Power and Gearing for Electrical Machinery.

1, 2, and 3, Salisbury Court, Fleet Street, London, E.C.

ELECTRICITY AND MAGNETISM.

AN ELEMENTARY TREATISE ON FOURIER'S SERIES, and Spherical
Cylindrical and Ellipsoidal Harmonics, with Applications to Problems on Mathematical Physics. By
Prof. Byerly, Harvard University. 12s. 6d.

MODERN VIEWS OF ELECTRICITY. By Oliver J. Lodge, F.R.S., Professor o
Physics in University College, Liverpool. Illustrated. 6s. 6d.

THE ELECTRO-MAGNET AND ELECTROMAGNETIC MECHANISM. By
Silvanus P. Thompson, D.Sc., F.R.S. 450 pages, 213 illustrations. 15s.

ELECTRICITY: ITS THEORY, SOURCES AND APPLICATIONS. By
John T. Sprague. Third Edition. Revised and enlarged. 15s.

MODERN APPLICATIONS OF ELECTRICITY. By E. Hospitalier. Trans
lated by Julius Maier, Ph.D. Second Edition. 28s.

ELECTRICAL PAPERS. For Advanced Students in Electricity. By Olive
Heaviside. 2 vols. 31s. 6d. net.

PROCEEDINGS OF THE INTERNATIONAL ELECTRICAL CONGRESS
HELD IN THE CITY OF CHICAGO, August 21st to 25th, 1893. 12s. 6d.; post free, 13s.

A COURSE OF LECTURES ON ELECTRICITY, DELIVERED BEFORE THE
SOCIETY OF ARTS. By George Forbes, M.A., F.R.S. (L. & E.) With 17 illustrations, crown 8vo, 5s.

SHORT LECTURES TO ELECTRICAL ARTIZANS. By Dr. J. A. Fleming
M.A., F.R.S., &c. Fourth Edition. 4s.

LECTURES IN ELECTRICITY AT THE ROYAL INSTITUTION, 1875-76
By John Tyndall, D.C.L. 2s. 6d.

NOTES OF A COURSE OF SEVEN LECTURES ON ELECTRICAL
PHENOMENA. By John Tyndall. 1s. 6d.

ELECTRIC WAVES: Being Researches on the Propagation of Electric Action
with Finite Velocity through Space. By Dr. Heinrich Hertz. Translated by D. E. Jones. 10s. 6d. net

LECTURES ON SOME RECENT ADVANCES IN PHYSICAL SCIENCE
By Prof. P. G. Tait. Third Edition. 9s.

ELEMENTARY LESSONS IN ELECTRICITY AND MAGNETISM. By Si
vanus P. Thompson, Principal and Professor of Physics in the Technical College, Finsbury. With illu
trations. Fcap. 8vo, 4s. 6d.

PRACTICAL ELECTRICITY. For First Year Students of Electrical Engineering
By W. E. Ayrton, F.R.S., Professor of Electrical Engineering in the City and Guilds of London Centr
Institution. Seventh Edition. Illustrated throughout, 7s. 6d.

A TREATISE ON ELECTRICITY AND MAGNETISM. By J. Clerk Max
well, M.A., F.R.S. Third Edition. 2 vols., demy 8vo, cloth, £1. 12s.

RECENT RESEARCHES IN ELECTRICITY AND MAGNETISM. By
Prof. J. J. Thomson, M.A., F.R.S. 18s. 6d.

MAGNETISM AND ELECTRICITY, AN ELEMENTARY MANUAL OF
With Examination Questions and many illustrations. By Prof. Jamieson. Third Edition. Crown 8vo, 3s. 6d

ELECTRICITY AND MAGNETISM. By Prof. Balfour Stewart, F.R.S., an
W. W. Haldane Gee, Demonstrator and Assistant Lecturer in Owens College, Manchester. Crown 8vc
7s. 6d.; School Course, 2s. 6d.

AN ELEMENTARY TREATISE ON ELECTRICITY. By J. Clerk Maxwell
M.A., F.R.S. Edited by William Garnett, M.A. Demy 8vo, cloth, 7s. 6d.

INVENTIONS, RESEARCHES, AND WRITINGS OF NIKOLA TESLA
Edited by T. Commerford Martin. 16s. 6d.

MAGNETISM AND ELECTRICITY. By A. W. Poyser, M.A. Cloth, 2s. 6d.

FIRST BOOK OF ELECTRICITY AND MAGNETISM. By W. Perren May
cock. 84 illustrations. 2s. 6d.

MANUAL OF ELECTRICAL SCIENCE. By George J. Burch, B.A. 3s.

ELECTROMAGNETIC THEORY—*see page 4.*

THE ALTERNATE-CURRENT TRANSFORMER IN THEORY AND PRACTICE—*see pages 6 and 7.*

A.B.C. OF ELECTRICITY. By W. H. Meadowcroft. 2s.

MAGNETISM AND ELECTRICITY. For the use of students in schools and science classes. By H. C. Tarn. F.S.Sc. With numerous diagrams. Cloth, 2s.

ELEMENTS OF STATIC ELECTRICITY. By P. Atkinson, Ph.D. 7s. 6d.

MAGNETISM AND ELECTRICITY. By R. Wallace Stewart. 160 Illustrations, 5s. 6d.

MAGNETISM AND ELECTRICITY. By Edward Aveling, D.Sc. With numerous woodcuts. 6s.

ELECTRICITY TREATED EXPERIMENTALLY, for the use of schools and students. By Linnæus Cumming, M.A. 4s. 6d.

ELECTRICITY AND MAGNETISM. By S. R. Bottone. 3s. 6d.

Book E. ARITHMETICAL PHYSICS. Part IIA.—MAGNETISM AND ELECTRICITY, ELEMENTARY AND ADVANCED. With Supplement on Lines of Force. By C. J. Woodward, B.Sc. 2s.

Book F. ARITHMETICAL PHYSICS. Part IIB.—MAGNETISM AND ELECTRICITY, DEGREE AND HONOURS STAGES. By C. J. Woodward, B.Sc. New Edition. 3s. 6d.

MAGNETIC INDUCTION IN IRON AND OTHER METALS—*see page 4.*

ELEMENTS OF EXPERIMENTAL AND NATURAL PHILOSOPHY. By Jabez Hogg. More than 400 woodcuts. 5s.

MAGNETISM AND ELECTRICITY. By W. Jerome Harrison and Charles A. White. 2s.

MAGNETISM AND ELECTRICITY FOR BEGINNERS. By W. G. Baker, M.A. 1s.

ELECTRICITY. By Dr. Ferguson; revised and extended by Prof. James Blyth, M.A. 3s. 6d.

MANUAL OF MAGNETISM AND ELECTRICITY. By John Cook, M.A 1s.

MAGNETISM AND ELECTRICITY. By John Angell. 1s. 6d.

MAGNETISM AND ELECTRICITY. By F. Guthrie, B.A., Ph.D. 3s. 6d.

ELECTRICITY AND MAGNETISM. By Fleeming Jenkin. 3s. 6d.

ELECTRICITY FOR PUBLIC SCHOOLS AND COLLEGES. By W. Larden, M.A. 6s.

ELEMENTARY TREATISE ON PHYSICS, Experimental and Applied. Translated from Ganot's "Elements de Physique," by E. Atkinson, Ph.D. Twelfth Edition. 15s.

PAPERS ON ELECTRO-STATICS AND MAGNETISM. By Lord Kelvin. Second Edition. 18s.

PHYSICS. Advanced Course. By G. F. Barker. 21s.

A TEXT-BOOK OF THE PRINCIPLES OF PHYSICS. By Alfred Daniel. Second Edition. 21s.

ELECTRICITY AND MAGNETISM. A Popular Treatise. By Amédée Guillemin. Translated by Silvanus P. Thompson. 31s. 6d.

QUESTIONS AND EXAMPLES IN EXPERIMENTAL PHYSICS, SOUND, LIGHT, HEAT, ELECTRICITY, AND MAGNETISM. By B. Loewy. 2s.

LESSONS IN ELEMENTARY PHYSICS. By Prof. Balfour Stewart. 4s. 6d.

ELECTRICITY AND MAGNETISM. By L. Cumming, M.A. 2s. 6d.

MAGNETISM AND ELECTRICITY. By J. Spencer, B.Sc. 1s. 6d.

STUDENT'S TEXT-BOOK OF ELECTRICITY. By Henry M. Noad, Ph.D., F.R.S. New Edition, with introduction and additional chapters by W. H. Preece, C.B., F.R.S. 12s. 6d.

HANDBOOK OF ELECTRICITY, MAGNETISM, AND ACOUSTICS. By Dr. Lardner. Edited by Geo. Carey Foster, B.A. 5s.

PRACTICAL NOTES FOR ELECTRICAL STUDENTS—*see page 13.*

ELECTRICAL ENGINEERING FORMULÆ—*see page 9.*

INTRODUCTION TO THE THEORY OF ELECTRICITY. By Linnæus
Cumming, M.B. 8s. 6d.

PHYSICAL TREATISE ON ELECTRICITY AND MAGNETISM. By J. E
H. Gordon. £2 2s.

THE PRACTICAL MEASUREMENT OF ELECTRICAL RESISTANCE. By W
A. Price. 11s.

MATHEMATICAL THEORY OF ELECTRICITY AND MAGNETISM. Vol.
I., Electrostatics; Vol. II., Magnetism and Electrodynmics. By H. W. Watson, D.Sc., and S. H.
Burbury, M A. 10s. 6d. each.

MATHEMATICAL THEORY OF ELECTRICITY AND MAGNETISM, AN
INTRODUCTION TO. By W. T. A. Emtage. M.A. 7s. 6d.

ELECTRICITY : A Sketch for General Readers. By E. M. Caillard. 7s. 6d.

ELEMENTS OF DYNAMIC ELECTRICITY AND MAGNETISM. By
Philip Atkinson, A.M., Ph.D. 10s. 6d.

THE ELECTRIC CURRENT ; How Produced and How Used. By R. Mullineux
Walmsley, D.Sc. 10s. 6d.

ALTERNATING CURRENTS OF ELECTRICITY. By T. H. Blakesley
Third Edition. 5s.

ALTERNATING CURRENTS OF ELECTRICITY : Their Generation, Measure
ment, Distribution, and Application. By Gisbert Kapp. 4s. 6d.

POLYPHASED ALTERNATING CURRENTS. By E. Hospitalier. 3s. 6d.

EXPERIMENTS WITH ALTERNATING CURRENTS OF HIGH POTENTIAL
AND HIGH FREQUENCY. By Nikola Tesla. 5s.

ALTERNATING CURRENTS. An Analytical and Graphical Treatment for
Students and Engineers. By Dr. F. Bedall and Dr. A. C. Crehore. Second Edition. 11s.

THE ARITHMETIC OF ELECTRICAL MEASUREMENTS. By W. R. P
Hobbs, Head Master of the Torpedo School, H.M.S. "Vernon.' New Edition. 1s.

THEORY AND PRACTICE OF ABSOLUTE MEASUREMENTS IN ELEC
TRICITY AND MAGNETISM. By Andrew Gray, M.A., F.R.S.E., Professor of Physics in the Universit
College of North Wales. In 2 vols., crown 8vo. Vol. I., 12s. 6d. Vol. II., in 2 parts, 25s.

ABSOLUTE MEASUREMENTS IN ELECTRICITY AND MAGNETISM
By Prof. Andrew Gray. Second Edition. 5s. 6d.

PRACTICAL ELECTRICAL MEASUREMENT. By James Swinburne
Cloth, 4s. 6d.

EXERCISES IN ELECTRICAL AND MAGNETIC MEASUREMENTS, with
Answers. By R. E. Day. 3s. 6d.

"THE ELECTRICIAN" PRIMERS—*see page 8.*

ARITHMETIC OF MAGNETISM AND ELECTRICITY. By Robert Gunn
2s. 6d.

AN INTRODUCTION TO PHYSICAL MEASUREMENTS. By Dr. F
Kohlrausch. Translated by T. H. Waller and H. R. Procter. 15s.

PHYSICAL ARITHMETIC. By Alexander Macfarlane. 7s. 6d.

REPORTS OF THE COMMITTEE ON ELECTRICAL STANDARDS
Appointed by the British Association, with a Report to the Royal Society on Units of Electrica
Resistance. 9s.

ELECTRICAL MEASUREMENT AND THE GALVANOMETER, ITS CON
STRUCTION AND USES. By T. D. Lockwood. Second Edition. 6s.

THERMO-ELECTRICITY, Theoretically and Practically Considered. By Arthu
Rust. 2s.

ELECTRICITY IN THEORY AND PRACTICE. By Lieut. Bradly A. Fiske
10s. 6d.

ELECTRICITY FOR ENGINEERS. By Charles Desmond. Revised Edition
10s. 6d.

POTENTIAL : Its Application to the Explanation of Electrical Phenomena Popu
larly Treated. By Dr. Tumlirz. Translated by D. Robertson, M.A. 3s. 6d.

KIRCHOFF'S LAWS AND THEIR APPLICATION. By E. C. Rimington. Cloth, 1s. 6d.

ELECTRICAL INFLUENCE MACHINES: containing a full account of their Historical Development, their Modern Forms, and their Practical Construction. By John Gray, B.Sc. 89 illustrations. 4s. 6d.

ELECTRICITY IN THE SERVICE OF MAN. A Popular and Practical Treatise on the Applications of Electricity in Modern Life. Revised by R. Mullineux Walmsley, D.Sc. (Lond.), F.R.S.E. Medium 8vo., with nearly 850 illustrations. 10s. 6d.

ELECTRICITY IN MODERN LIFE. By G. W. de Tunzelmann. With 88 illustrations, 3s. 6d.

DOMESTIC ELECTRICITY FOR AMATEURS. Translated from the French of E. Hospitalier, with additions By C. J. Wharton, Assoc. Soc. Tel. Engineers. Numerous illustrations, demy 8vo, cloth, 6s.

PRACTICAL ELECTRICS: A Universal Handy Book on Every-Day Electric Matters. Third Edition. 3s. 6d.

ELECTRICITY: A Hundred Years Ago and To-Day. By Prof. E. J. Houston. 4s. 6d.

ELECTRICITY IN DAILY LIFE. A Popular Account of its Application to Every-day Uses. 125 illustrations. 9s.

ELECTRICITY IN MINING. By Silvanus P. Thompson, D.Sc., F.R.S. 2s.

ELECTRICITY FOR SCHOOLS. By J. E. H. Gordon. 5s.

A CENTURY OF ELECTRICITY. By T. C. Mendenhall. 4s. 6d.

STANDARD METHODS IN PHYSICS & ELECTRICITY CRITICISED, AND A TEST FOR ELECTRIC METERS PROPOSED. By H. A. Naber. Demy 8vo., cloth gilt, 5s., post free.

ELECTRICAL ENGINEERING LEAFLETS (HOUSTON AND KENNELLY'S). In three Grades—Elementary, Intermediate, and Advanced—of 35 Leaflets each. Price of single Leaflet, 6d., post free. Subscription price for any one Grade, 12s. 6d. ; for any two Grades, 23s. ; for all three Grades, 33s. 6d. Further particulars on application.

A HANDBOOK FOR OPERATORS IN MEDICAL ELECTRICITY AND Massage. By H. Newman Lawrence, M.I.E.E. 1s.

A PRACTICAL TREATISE ON THE MEDICAL AND SURGICAL USES OF ELECTRICITY. By Drs. Beard and Rockwell. 200 illustrations. Royal 8vo. 28s.

PRACTICAL APPLICATION OF ELECTRICITY IN MEDICINE AND SURGERY. By Drs. Liebig and Rohé. Royal 8vo, 400 pages, profusely illustrated, 11s. 6d.

ELECTRICITY IN THE DISEASES OF WOMEN. By G. Betton Massey, M.D. Second Edition. 12mo, 8s. 6d.

THE WORK OF HERTZ.—*see page 9.*

A TEXT-BOOK OF ELECTRICITY IN MEDICINE AND SURGERY. By George Vivian Poore, M.D. Crown 8vo, 8s. 6d.

A MANUAL OF ELECTRO-THERAPEUTICS. By W. Erb, M.D., translated by A. de Watteville, M.D., &c. Demy 8vo, 18s.

ELECTRICITY AND ITS MANNER OF WORKING IN THE TREATMENT OF DISEASE. A Thesis for the M.D. Cantab Degree, 1884. By the late William E. Steavenson, M.D., M.R.C.P., Casualty Physician and Electrician to St. Bartholomew's Hospital. To which is appended an Inaugural Medical Dissertation on Electricity for the M.D., Edin. Degree, written in Latin by Robert Steavenson, M.D. in 1778, with a Translation by the Rev. F. R. Steavenson, M.A. 4s. 6d.

INTERNATIONAL SYSTEM OF ELECTRO-THERAPEUTICS. For Students, General Practitioners, and Specialists. 32s.

TRAITÉ ÉLÉMENTAIRE D'ÉLECTRICITÉ. By J. Joubert. With 321 illustrations. 6s.

SUR LA PROPAGATION DU COURANT ÉLECTRIQUE. By A. Bandsept. 1s.

ÉLECTRICITÉ INDUSTRIELLE. By D. Monnier. With 338 illustrations. 16s.

LEÇONS SUR L'ÉLECTRICITÉ. By Eric Gerard. 2 vols., fully illustrated. 19s.

FORMULAIRE DE L'ÉLECTRICITÉ. By E. Hospitalier. 10th year: 1892. 4s. 6d.

TRAITÉ PRATIQUE DE L'ÉLECTRICITÉ. By Felix Lucas. 13s. 6d.

ELECTRIC LIGHTING & TRANSMISSION OF POWER.

THE ALTERNATE-CURRENT TRANSFORMER IN THEORY AND PRACTICE—*see pages 6 and 7.*

DYNAMO ELECTRICITY : Its Generation, Application, Transmission, Storage and Measurement. By G. B. Prescott. 545 illustrations. £1. 1s.

"THE ELECTRICIAN" PRIMERS—*see page 8.*

DYNAMO MACHINERY AND ALLIED SUBJECTS (ORIGINAL PAPERS ON). By Dr. John Hopkinson. 5s.

HOW TO BUILD DYNAMO-ELECTRIC MACHINERY. By Edward Trevert. 10s. 6d.

PRACTICAL NOTES FOR ELECTRICAL STUDENTS—*see page 13.*

CONTINUOUS-CURRENT DYNAMOS AND MOTORS. An Elementary Treatise for Students. By Frank P. Cox. 9s.

THE INCANDESCENT LAMP AND ITS MANUFACTURE—*see page 10.*

THE DYNAMO : Its Theory, Design and Manufacture. By C. C. Hawkins, A.M.Inst.C.E., and F. Wall's. 10s. 6d.

ARMATURE WINDINGS OF ELECTRIC MACHINES. By H. F. Parshall and H. M. Hobart. *In preparation.*

DRUM ARMATURES AND COMMUTATORS—*see page 10.*

PRACTICAL MANAGEMENT OF DYNAMOS AND MOTORS. By F. B. Crocker and S. S. Wheeler. Second Edition. 4s. 6d.

DYNAMO ATTENDANTS AND THEIR DYNAMOS. By Alfred H. Gibbings. 1s.

ELECTRICAL ENGINEERING FORMULÆ—*see page 9.*

DYNAMO AND MOTOR BUILDING FOR AMATEURS. By C. D. Parkhurst. 4s. 6d.

A DIGEST OF THE LAW OF ELECTRIC LIGHTING, &c.—*see page 8.*

ELECTRO-MAGNETISM AND THE CONSTRUCTION OF DYNAMOS. Vol. I. By Prof. Dugald C. Jackson. 9s. 6d.

THE MANUFACTURE OF ELECTRIC LIGHT CARBONS—*see page 13.*

PRACTICAL DIRECTIONS FOR WINDING MAGNETS FOR DYNAMOS. By Carl Hering. 3s. 6d.

ELEMENTS OF CONSTRUCTION FOR ELECTRO-MAGNETS. By Count du Moncel. Translated by C. J. Wharton. 4s. 6d.

THE STEAM ENGINE INDICATOR, AND INDICATOR DIAGRAMS—*see page 13.*

ELEMENTARY PRINCIPLES OF ELECTRIC LIGHTING. By Alan A. Campbell Swinton. Enlarged and revised. Crown 8vo, cloth, 1s. 6d.

ELECTRIC LIGHTING AND POWER DISTRIBUTION : An Elementary Manual for Students. By W. Perren Maycock. In three parts. 2s. 6d. each.

DYNAMO CONSTRUCTION : A Practical Handbook for Engineer Constructors and Electricians in Charge. By John W. Urquhart. Second Edition. 7s. 6d.

A GUIDE TO ELECTRIC LIGHTING. By S. R. Bottone. Second Edition. 1s.

CONTINENTAL ELECTRIC LIGHT CENTRAL STATIONS. By Killingworth Hedges. 15s.

DYNAMO-ELECTRIC MACHINERY : A Text-Book for Students of Electro-Technology. By Silvanus P. Thompson, B.A., D.Sc., M.I.E.E., F.R.S. Just published. Fourth Edition, revised and enlarged. Cloth, 864 pages, 29 folding plates, 498 illustrations in text. 24s. post free.

THEORETICAL ELEMENTS OF ELECTRO-DYNAMIC MACHINERY. Vol. I. By A. Kennelly, F.R.A.S. 4s. 6d.

ELECTRIC LIGHTING FROM CENTRAL STATIONS. By Prof. George Forbes. 1s.

THE DYNAMO-TENDER'S HANDBOOK. By F. B. Badt. 4s. 6d.

PRINCIPLES OF DYNAMO-ELECTRIC MACHINES. By Carl Hering.
Practical Directions for Designing and Constructing Dynamos. With an Appendix containing several
Articles on Allied Subjects, and a Table of Equivalents of Units of Measurement. Cloth, 279 pages,
59 illustrations, 10s. 6d.

CENTRAL STATION MANAGEMENT AND FINANCE. By Horatio A.
Foster. 7s. 6d.

CENTRAL STATION BOOK-KEEPING. By Horatio A. Foster. 7s. 6d.

NOTES ON DESIGN OF SMALL DYNAMOS. By George Halliday. 2s. 6d.

ELECTRICITY IN OUR HOMES AND WORKSHOPS. By S. F. Walker.
Second Edition. Cloth, 5s.

ELECTRIC LIGHT INSTALLATIONS. By Sir David Salomons, Bart. In
Three Volumes. Vol. I., 5s. : The Management of Accumulators. Vol. II., 7s. 6d.: Apparatus, Engines,
Dynamos and Motors, Instruments, Governors, Switches and Switch Boards, Fuses, Cut-Outs, Connec-
tors and Minor Apparatus, Arc Lamps, Practical Applications. Vol. III., 5s. : Application.

HOUSE LIGHTING BY ELECTRICITY. By Angelo Fahie. Paper, 1s.; cloth, 2s.

DOMESTIC ELECTRIC LIGHTING. By E. C. de Segundo. 1s.

A HANDBOOK OF ELECTRICAL TESTING. By H. R. Kempe, M.I.E.E.
Fifth Edition, revised and enlarged. 18s.

ARTIFICIAL LIGHTING IN RELATION TO HEALTH. By R. E. B.
Crompton. 6d.

ELECTRIC LIGHTING SPECIFICATIONS, for the Use of Engineers and
Architects. By E. A. Merrill. 6s.

ELECTRICAL ENGINEERING: For Electric Light Artizans and Students.
By W. Slingo and A. Brooker. New and revised edition. Cloth, gilt, 12s.

PRECAUTIONS TO BE ADOPTED ON INTRODUCING THE ELECTRIC
LIGHT. By Killingworth Hedges. 2s. 6d.

ECONOMIC VALUE OF ELECTRIC LIGHT AND POWER. By A. R.
Foote. 4s.

ELECTRIC LIGHT: Its Production and Use. By John W. Urquhart. Fifth
Edition, carefully revised, with large additions. 7s. 6d.

RULES AND REGULATIONS RECOMMENDED FOR THE PREVENTION
OF FIRE RISKS FROM ELECTRIC LIGHTING, issued by the Society of Telegraph-Engineers. 8vo.
sewed, 6d.

THE PHŒNIX FIRE OFFICE RULES FOR ELECTRIC LIGHT INSTAL-
LATIONS AND ELECTRICAL POWER INSTALLATIONS. By Musgrave Heaphy, C.E. Eighteenth
Edition, 8vo, sewed, 6d.

ELECTRIC SHIP LIGHTING: For the use of Ship Owners and Builders,
Engineers, &c By John W. Urquhart. 7s. 6d.

ELECTRIC LIGHTING FOR MARINE ENGINEERS. By S. F. Walker. 5s.

ELECTRIC LIGHT FITTING: A Handbook for Electrical Engineers. By John
W. Urquhart. Fully Illustrated. 5s.

COMPREHENSIVE INTERNATIONAL WIRE TABLE. By W. S.
Boult. Full particulars of 469 Conductors (4 gauges), Single Wires and Cables, in English, American,
and Continental Units. Price 6s. 9d., post free.

COLLIERY LIGHTING BY ELECTRICITY. By S. F. Walker. 2s. 6d.

A PRACTICAL TREATISE ON THE INCANDESCENT LAMP. By J. H.
Randell. 2s. 6d.

ELECTRIC LAMPS AND ELECTRIC LIGHTING—*see page 5.*

PRACTICAL ELECTRIC LIGHT FITTING. By F. C. Allsop. 5s.

INCANDESCENT ELECTRIC LIGHTING. By L. H. Latimer. With additions
by C. J. Field and J. W. Howell. 2s.

TREATISE ON INDUSTRIAL PHOTOMETRY, with Special Application to
Electric Lighting. By Dr. A. Palaz. Translated by G. W. and M. R. Patterson. 12s. 6d.

WIRING SLIDE RULE (Trotter's Patent). By which can be found at once.—
1. Size of Cable; 2. Length of Cable; 3. Current Cable will carry; 4. Current Density; 5. Maximum
Current; 6. Resistance in Ohms; 7. Sectional Area in Square Inches. Full printed instructions are
supplied with each rule. For the pocket. Price 2s. 6d.; post free, 2s. 7d.

WOODHOUSE AND RAWSON WIRING TABLES. Price—Paper, 1s.,
post free ; mounted and glazed, 1s. 6d., post free ; in neat cloth case for pocket, 2s. 6d., post free ;
mounted and glazed and bound in cloth for pocket, 2s. 6d., post free. Printed directions how to use
the Tables are issued with each copy.

THE DISTRIBUTION OF ELECTRICITY. By Prof. George Forbes, M.A.,
F.R.S.E. 1s.

DEVELOPMENTS OF ELECTRICAL DISTRIBUTION. By Prof. George
Forbes, F.R.S. 1s.

MAY'S POPULAR INSTRUCTOR FOR THE MANAGEMENT OF
ELECTRIC LIGHTING PLANT. An indispensable Handbook for persons in charge of Electric Lighting
plants, more particularly those who have had little or no technical training. Pocket size, price 2s. 6d. ;
post free, 2s. 8d.

PRACTICAL ELECTRIC LIGHTING. By A. Bromley Holmes. Fourth
Edition. Cloth, 3s. 6d.

STANDARD TABLES FOR ELECTRIC WIREMEN. By Charles M. Davis. 5s.

MAY'S TABLE OF ELECTRIC CONDUCTORS. Showing the relations
between—(1) The sectional area, diameter of conductors, loss of potential, strength of current, and length
of conductors ; (2) The economies of incandescent lamps, their candle-power, potential, and strength of
current ; (3) The sectional area, diameter of conductors, and strength of current per square inch. For
office use, printed on cardboard, with metal edges and suspender, price 2s. ; post free, 2s. 2d. ; for the
pocket, mounted on linen, in strong case, 2s. 6d.; post free, 2s. 8d.

ELECTRIC LIGHT ARITHMETIC. By R. E. Day. 2s.

UNIVERSAL WIRING COMPUTER. By Carl Hering. 5s.

MAY'S BELTING TABLE. Showing the relations between—(1) The number
of revolutions and diameter of pulleys and velocity of belts ; (2) The horse-power, velocity, and square
section of belts ; (3) The thickness and width of belts ; (4) The square section of belts at different strains
per square inch. For office use, printed on cardboard, with metal edges and suspender, price 2s.; post
free, 2s. 2d. ; for the pocket, mounted on linen, in strong case, 2s. 6d.; post free, 2s. 8d.

THE GALVANOMETER AND ITS USES. A Manual for Electricians and
Students. By C. H. Haskins. Second edition. Illustrated. 18mo., 8s. 6d.

TRANSFORMERS : Their Theory, Construction, and Application Simplified.
By Caryl D. Haskyns. 4s. 6d.

INCANDESCENT WIRING HANDBOOK. With Tables. By F. B. Badt.
4s. 6d.

HISTORY OF THE TRANSFORMER. By J. Uppenborn. Translated from the
German. 3s.

ELECTRIC TRANSMISSION OF ENERGY, and its Transformation, Subdivision and Distribution. By Gisbert Kapp. Fourth Edition. 10s. 6d.

ELECTRIC LIGHT CABLES AND THE DISTRIBUTION OF ELEC-
TRICITY. By Stuart A. Russell, Assoc.M.Inst.C.E. 107 illustrations 7s. 6d.

ELECTRICITY AS A MOTIVE POWER. By Count du Moncel and Frank
Geraldy. Translated by C. J. Wharton. 7s. 6d.

ELECTRO-MOTORS : How Made and How Used. By S. R. Bottone. Second
Edition. 3s.

ELECTRIC MOTIVE POWER—*see page 14.*

THE ELECTRIC LIGHT POPULARLY EXPLAINED. By A. Bromley
Holmes. Sixth Edition, 1s.

HOW TO MAKE A DYNAMO. By Alfred Crofts. Fourth Edition. 2s.

MOTIVE POWER AND GEARING—*see page 14.*

THE DYNAMO: HOW MADE AND HOW USED. By S. R. Bottone. With
39 illustrations Eighth Edition. 2s. 6d.

HOW TO MANAGE A DYNAMO. By S. R. Bottone. 1s.

ELECTRIC TRANSMISSION HANDBOOK. By F. B. Badt. 4s. 6d.

AMERICAN ELECTRIC STREET RAILWAYS : Their Construction and
Equipment. By Killingworth Hedges. 12s. 6d.

ELECTRIC RAILWAYS, RECENT PROGRESS IN. By Carl Hering. 5s.

SECONDARY BATTERIES AND THE ELECTRICAL STORAGE OF ENERGY.
By Dr. Oliver Lodge. 1s.

ELECTRIC TRANSFORMATION OF POWER. By Philip Atkinson. 7s. 6d.

ELECTRO-MOTORS : The Means and Apparatus employed in the Transmission of Electrical Energy and its Conversion into Motive Power. By John W. Urquhart. 7s. 6d.

THE CHEMISTRY OF THE SECONDARY BATTERIES OF PLANTÉ AND FAURE. By J. H. Gladstone and A. Tribe. 2s. 6d.

THE STORAGE OF ELECTRICAL ENERGY, and Researches in the Effects created by Currents Combining Quantity with High Tension. By G. Planté. Translated from the French by Paul Bedford Elwell. With 89 illustrations, 8vo, cloth, 12s.

THE ELECTRIC MOTOR AND ITS APPLICATIONS. By T. C. Martin and J. Wetzler. Greatly enlarged edition. Quarto, 360 pages, 275 illustrations. 12s. 6d., post free 13s. 6d.

ELECTRIC BATTERIES, ELEMENTARY TREATISE ON. By Alfred Niaudet. Translated by L. M. Fishback. Sixth Edition, 12s. 6d.

DOMESTIC ELECTRICITY FOR AMATEURS. By E. Hospitalier. Translated by C. J. Wharton. 6s.

THE ELECTRIC RAILWAY IN THEORY AND PRACTICE. By Oscar T. Crosby and Louis Bell, Ph.D. Fully illustrated. 11s.

HOW TO WIRE BUILDINGS. By Augustus Noll. Second Edition. 6s.

PRACTICAL ELECTRICAL NOTES AND DEFINITIONS, for the use of Engineering Students and Practical Men. By W. Perren Maycock. Second Edition. 3s.

PRIMARY **BATTERIES.** By H. S. Carhart, A.M. 67 illustrations. 6s.

ELECTRICAL ENGINEERING LEAFLETS (HOUSTON AND KENNELLY'S). In three Grades—Elementary, Intermediate, and Advanced—of 35 Leaflets each. Price of single Leaflets, 6d., post free. Subscription price for any one grade, 12s. 6d. ; for any two Grades, 23s. ; for all three Grades. 33s. 6d. Further particulars on application.

LEÇONS SUR L'ELECTRICITÉ. By Eric Gerard. 19s.

L'ÉCLAIRAGE À PARIS. By H. Maréchal. 16s. 6d.

LES ACCUMULATEURS ÉLECTRIQUES. By A. Bandsept. 103 pages. 2s.

INDUCTEURS DYNAMO-ÉLECTRIQUES ET PYRO-ÉLECTRIQUES. By A. Bandsept. 6d.

TRAITÉ DES PILES ÉLECTRIQUES ; Piles hydro-électriques—Accumulateurs—Piles thermo-électriques et pyro-électriques. By Donato Tommasi, Docteur-ès-Sciences. 8vo., 670 pages, with 160 illustrations. 10s.

TRAITÉ ÉLÉMENTAIRE DE LA PILE ÉLECTRIQUE. By Alfred Niaudet. Third Edition. Revised by Hippolyte Fontaine, and followed by a Notice on Accumulators by E. Hospitalier. Illustrated, 8vo, 6s. 6d.

ÉCLAIRAGE À L'ÉLECTRICITÉ. By Hippolyte Fontaine. Third Edition, entirely re-written, with 326 illustrations, 8vo, 13s.

DIE DYNAMOELEKTRISCHE MASCHINE : Eine Physikalische Beschreibung für den Technischen Gebrauch. By Dr. O. Frölich. 8s.

UNTERSUCHUNGEN UEBER DIE AUSBREITUNG DER ELEKTRISCHEN KRAFT. By Prof. Dr. H. Hertz. Price 6s.

TELEGRAPHY AND TELEPHONY.

PRACTICAL NOTES FOR ELECTRICAL STUDENTS—*see page 13.*

HANDBOOK OF PRACTICAL TELEGRAPHY. By R. S. Culley, M.Inst.C.E. Eighth Edition. 17 plates and 135 woodcuts, 8vo, 16s.

ELECTRICITY AND THE ELECTRIC TELEGRAPH. With numerous illustrations. By George B. Prescott Eighth Edition, 2 vols., 8vo, cloth, £1. 16s.

ELECTRICAL ENGINEERING FORMULÆ—*see page 9.*

THE ELECTRIC TELEGRAPH : Its History and Progress. With Descriptions of some of the Apparatus. By R. Sabine, C.E., F.S.A. Limp cloth, 3s.

1, 2 and 3, Salisbury Court, Fleet Street, London, E.C.

SUBMARINE CABLE LAYING AND REPAIRING—*see page 14.*

RISE AND EXTENSION OF SUBMARINE TELEGRAPHY. By Willoughby
Smith. 21s.

COMMERCIAL AND RAILWAY TELEGRAPHY, MODERN SERVICE OF,
Designed for Students and Operators. Compiled and prepared by J. P. Abernethy. Seventh edition,
carefully revised, 427 pages, many illustrations, 8s. 6d.

MODERN PRACTICE OF THE ELECTRIC TELEGRAPH. By Frank L. Pope.
Fourteenth Edition, with numerous wood engravings, 8vo, cloth, 12s. 6d.

HANDBOOK OF THE ELECTRO-MAGNETIC TELEGRAPH. By A. E.
Loring. 2s.

"THE ELECTRICIAN" PRIMERS—*see page 8.*

TELEGRAPHIC CONNECTIONS : Embracing Recent Methods in Quadruplex
Telegraphy. By Charles Thom and W. H. Jones. 7s. 6d.

PRACTICAL GUIDE TO THE TESTING OF INSULATED WIRES AND
CABLES. By Herbert Laws Webb. 4s. 6d.

INSTRUCTIONS FOR TESTING TELEGRAPH LINES, and the Technical
Arrangements in Offices. By Louis Schwendler. 2 vols. Demy 8vo, 21s.

TELEGRAPHY. By W. H. Preece, C.B., F.R.S., M.I.C.E., and Sir J.
Sivewright, K.C.M.G., M.A. Ninth Edition, revised and enlarged. 6s.

TABLES TO FIND THE WORKING SPEED OF CABLES, comprising also
Data as to Diameter, Capacity, and Copper Resistance of all Cores. By A. Dearlove. 2s.

ON A SURF-BOUND COAST : Cable-laying in the African Tropics. By
Arthur P. Crouch, B.A. 7s. 6d.; New Edition, 5s.

TELEPHONES : THEIR CONSTRUCTION AND FITTING. By F. C. Allsop. 5s.

A HANDBOOK OF ELECTRICAL TESTING. By H. R. Kempe. Fifth
edition, revised and enlarged. 18s.

TELEPHONE LINES AND THEIR PROPERTIES. By Prof. W. J. Hopkins. 6s.

THE TELEPHONE HANDBOOK AND PRACTICAL GUIDE TO TELE-
PHONIC EXCHANGE. By Joseph Poole. 3s. 6d.

THE TELEPHONING OF GREAT CITIES, AND AN ELECTRICAL PARCEL
EXCHANGE SYSTEM. By A. R. Bennett. 1s.

MANUAL OF THE TELEPHONE. By W. H. Preece and A. J. Stubbs.
Over 500 pages and 334 illustrations. 15s.

THE TELEPHONE, THE MICROPHONE, AND THE PHONOGRAPH. By
Count du Moncel. Third Edition, 5s.

THE MAGNETO HAND TELEPHONE. By Norman Hughes. 3s. 6d.

PHILIPP REIS, INVENTOR OF THE TELEPHONE. A Biographical Sketch.
By Silvanus P. Thompson, B.A., D.Sc. With portrait and wood engravings. 8vo, cloth, 7s. 6d.

ELECTROMAGNETIC THEORY—*see page 4.*

ELECTRICAL ENGINEERING LEAFLETS (HOUSTON AND KENNELLY'S).
In three Grades—Elementary, Intermediate, and Advanced—of 35 Leaflets each. Price of single Leaflets,
6d., post free. Subscription price for any one Grade, 12s. 6d. ; for any two Grades, 23s. ; for all three
Grades, 33s. 6d. *Further particulars on application.*

TRAITÉ DE TÉLÉGRAPHIE ELECTRIQUE. By H. Thomas. With 702
illustrations. £1.

TRAITÉ DE TÉLÉGRAPHIE SOUS-MARINE. By E. Wünschendorff,
Ingénieur des Télégraphes, Directeur de Télégraphie Militaire. With 400 illustrations, 8vo, 33s.

LA TÉLÉPHONIE. By E. Piérard. 7s

ELECTRO-CHEMISTRY & ELECTRO-METALLURGY.

"THE ELECTRICIAN" PRIMERS—*see page 8.*

ELECTRO PLATING : A Practical Handbook on the Deposition of Copper,
Silver, Nickel, Gold, Aluminium, Brass, Platinum, &c., &c., By John W. Urquhart. Third Edition, 5s.

ELECTRO-METALLURGY : Practically treated. By Alexander Watt, F.R.S.A.
Ninth Edition. 12mo, cloth boards, 4s.

THE ART OF ELECTRO-METALLURGY, INCLUDING ALL KNOWN PRO-
CESSES OF ELECTRO-DEPOSITION. By G. Gore, LL.D., F.R.S. With 56 illustrations, fcp. 8vo, 6s.

THE ART OF ELECTROLYTIC SEPARATION OF METALS—*see page 11.*

A TREATISE ON ELECTRO-METALLURGY. By Walter G. M'Millan. 10s. 6d.

ELECTRO-DEPOSITION : A Practical Treatise on the Electrolysis of Gold, Silver, Copper, Nickel, and other Metals and Alloys. By Alexander Watt, F.R.S.S.A. With numerous illustrations. Third Edition. Crown 8vo, cloth, 9s.

ELECTRO-PLATER'S HANDBOOK. By G. E. Bonney. Second Edition. 3s.

ELECTRO-CHEMISTRY—*see page 11.*

ELECTRO-DEPOSITION OF METALS. By G. Langbein 25s.

THE USES OF ELECTROLYSIS IN SURGERY. By W. E. Steavenson, M.D., M.R.C.P. Cloth, 5s.

TRAITÉ THÉORIQUE ET PRATIQUE D'ÉLECTROCHIMIE. By Donato Tommasi, Docteur-ès-Sciences. Large 8vo, 1,200 pages. 30s.

ELECTRICAL ENGINEERING FORMULÆ—*see page 9.*

ELECTRICAL ENGINEERING LEAFLETS (HOUSTON AND KENNELLY'S). In three Grades—Elementary, Intermediate, and Advanced—of 35 Leaflets each. Price of single Leaflets, 6d., post free. Subscription price for any one Grade, 12s. 6d. ; for any two Grades, 23s. ; for all three Grades, 33s. 6d. *Further particulars on application.*

ELECTRICAL INSTRUMENTS, BELLS, &c.

"THE ELECTRICIAN" PRIMERS—*see page 8.*

ELECTRICAL INSTRUMENT MAKING FOR AMATEURS. A Practical Handbook. By S. R. Bottone. Fifth Edition. 3s.

INDUCTION COILS AND COIL MAKING. By F. C. Allsop. 3s. 6d.

INDUCTION COILS : A Practical Manual for Amateur Coil Makers. By G. E. Bonney. 3s.

ELECTRICAL ENGINEERING FORMULÆ—*see page 9.*

DYNAMOMETERS AND THE MEASUREMENT OF POWER. By Prof. J. J. Flather. 8s. 6d.

PRACTICAL ELECTRIC BELL FITTING. By F. C. Allsop. 3s. 6d.

THE BELL-HANGER'S HANDBOOK. By F. B. Badt. 4s. 6d.

ELECTRIC BELL CONSTRUCTION. By F. C. Allsop. 3s. 6d.

ELECTRIC BELLS, AND ALL ABOUT THEM. By S. R. Bottone. Fourth Edition. 3s.

ELECTRICAL APPARATUS FOR AMATEURS. By Various Authors. 1s.

REFERENCE BOOKS, &c.

"THE ELECTRICIAN" ELECTRICAL TRADES' DIRECTORY AND HANDBOOK—*see page 4.*

WILLING'S BRITISH AND IRISH PRESS GUIDE. A concise and comprehensive Index to the Press of the United Kingdom, with Lists of the principal Colonial and Foreign Journals. 1s.

HAZELL'S ANNUAL. A Cyclopædic Record of Men and Topics of the Day. 3s. 6d.

STATESMAN'S YEAR BOOK : A Statistical and Historical Annual of the States of the World. Revised after Official Returns. 10s. 6d.

WHITAKER'S ALMANACK. 2s. 6d.

SELL'S DIRECTORY OF REGISTERED TELEGRAPHIC ADDRESSES : From Official Lists supplied by the authority of the Postmaster General. Containing upwards of 40,000 firms. 21s.

KELLY'S POST OFFICE AND OTHER DIRECTORIES. (*Particulars and lowest prices on application.*)

ELECTRICAL ENGINEERING FORMULÆ—*see page 9.*

A POCKET-BOOK OF ELECTRICAL RULES AND TABLES, for the use
of Electricians and Engineers. By John Munro, C.E., and Andrew Jamieson, C.E., F.R.S.E. Tenth
Edition. Revised and enlarged. Pocket size, leather, 8s. 6d., post free.

THE ELECTRICAL ENGINEER'S POCKET-BOOK OF MODERN RULES,
FORMULÆ, TABLES, AND DATA. By H. R. Kempe, M.I.E.E. 250 pages, numerous illustrations.
Price 5s., post free.

USEFUL RULES AND TABLES FOR ENGINEERS AND OTHERS. By
Prof. Rankine. With Appendix by Andrew Jamieson, C.E. 10s. 6d.

A MANUAL OF RULES, TABLES AND DATA FOR MECHANICAL
ENGINEERS. By D K. Clark. Cloth, 16s.; half-bound, 20s.

MATHEMATICAL TABLES. By the Rev. Prof. Galbraith, M.A., and the
Rev. Prof. Haughton, D.C.L., F.R.S. 3s. 6d.

FACTORY ACCOUNTS: Their Principles and Practice. By E. Garcke and J.
M. Fells. Fifth Edition. 6s.

LAXTON'S PRICE BOOK. Published Annually. 4s.

LOCKWOOD'S BUILDERS' PRICE BOOK. Edited by Francis T. W. Miller. 4s.

ENGINEER'S YEAR BOOK of Formulæ, Rules, Tables, Data, and Memoranda.
By H. R. Kempe. 8s.

GRIFFIN'S ELECTRICAL ENGINEER'S PRICE-BOOK. Edited by H. J.
Dowsing. 8s. 6d.

**HOUSTON'S DICTIONARY OF ELECTRICAL WORDS, TERMS,
AND PHRASES.** New and greatly enlarged Edition. 560 pages, and nearly 600 illustrations. Price
21s., post free.

LAW RELATING TO ELECTRIC LIGHTING. By G. S. Bower and W. Webb.
Entirely re-written. 8vo, 12s. 6d.

DICTIONARY OF CHEMISTRY (Watts'). Revised and entirely rewritten by
M. M. Pattison Muir, M.A., F.R.S.E., and B. Forster Morley, M.A., D.Sc., assisted by eminent contributors. In 4 vols. Vols. I. and II., 42s. each; Vol. III., 50s; Vol. IV., 63s.

A DICTIONARY OF APPLIED CHEMISTRY. By T. E. Thorpe, assisted
by Eminent Contributors. Vols. I. and II., £2. 2s. each; Vol. III., £3. 3s.

THE GRAPHIC ATLAS AND GAZETTEER OF THE WORLD. 128 Map
Plates and Gazetteer of 280 pages. Crown 4to, cloth, 12s. 6d.; half-morocco, 15s.; full morocco 21s.

CHAMBERS'S CONCISE GAZETTEER OF THE WORLD. 768 pages. 6s.

DICTIONARY OF METRIC AND OTHER USEFUL MEASURES. By
Latimer Clark, F.R.S., &c. Crown 8vo, 6s.

BLACKIE'S MODERN CYCLOPÆDIA OF UNIVERSAL INFORMATION.
Edited by Charles Annandale, M.A., LL.D. Eight vols., cloth, 6s. each; half-morocco, 8s. 6d. each.

SPONS' DICTIONARY OF ENGINEERING. Three volumes. £5. 5s.

CASSELL'S STOREHOUSE OF GENERAL INFORMATION. In volumes. 5s.

CASSELL'S CONCISE CYCLOPÆDIA. 7s. 6d.

"TECHNICAL INDEX" OF THE "ENGINEERING MAGAZINE."—
(*Particulars on application.*)

PRACTICAL DICTIONARY OF MECHANICS. By Edward H. Knight.
Three volumes. With 15,000 illustrations. £3. 3s. Supplementary volume, £1. 1.

HAYDN'S DICTIONARY OF DATES. Twentieth edition; brought down to
the Autumn of 1892. Cloth, 18s.; half-calf, 24s.

DIRECTORY OF DIRECTORS: A list of Directors of Public Companies, with
the concerns with which they are connected. 12s. 6d.

STOCK EXCHANGE YEAR-BOOK: A Digest of Information relating to the
Origin, History, and Present Position of the Public Securities and Joint Stock Companies known to the
Markets of the United Kingdom. 18s.

UNIVERSAL ELECTRICAL DIRECTORY. 4s.

TABLES AND FORMULÆ FOR PLUMBERS, ARCHITECTS, SANITARY
ENGINEERS, &c. By J. Wright Clarke. With Electrical Tables and Memoranda, by W. Hibbert.
Second Edition. Waistcoat pocket size. 1s. 6d.

DIGEST OF THE LAW OF ELECTRIC LIGHTING, &c.—*see page 8.*

ELECTRICAL TABLES AND FORMULÆ FOR THE USE OF TELE-
GRAPH INSPECTORS AND OPERATORS. Compiled by Latimer Clark and Robert Sabine. With wood
engravings, crown 8vo, cloth, 12s. 6d.

TABLES AND FORMULÆ FOR ELECTRIC STREET RAILWAY EN-
GINEERS. Compiled by E. A. Merrill. 4s. 6d.

LIVES OF THE ELECTRICIANS. First Series: Profs. TYNDALL, WHEAT-
STONE AND MORSE. By William T. Jeans. Crown 8vo, cloth, 6s.

CASSELL'S NEW POPULAR EDUCATOR. Eight volumes. 5s. each.

CASSELL'S TECHNICAL EDUCATOR. Four volumes. 5s. each.

BARLOW'S TABLES OF SQUARES, CUBES, SQUARE ROOTS, CUBE
ROOTS. Reciprocals of all Integer Numbers up to 10,000. Post 8vo, cloth, 6s.

CHAMBERS' ENCYCLOPÆDIA. Ten volumes. Cloth, £5; half morocco
or half calf, £7. 10s.; and in better bindings up to £15.

A B C FIVE-FIGURE LOGARITHMS. By C. J. Woodward, B Sc., Cloth, 2s. 6d.

FOUR-FIGURE MATHEMATICAL TABLES. By J. T. Bottomley. 2s. 6d.

ILLUSTRATIONS OF THE C.G.S. SYSTEM OF UNITS. With Tables of
Physical Constants. By Prof. J. D. Everett. 5s.

LOGARITHMIC AND TRIGONOMETRICAL TABLES FOR APPROXI-
MATE CALCULATION, &c. By J. T. Bottomley, M.A. 1s.

MATHEMATICAL TABLES. By James Pryde, F.E.I.S. 4s. 6d.

SPONS' ENGINEERS' DIARY. Published annually. 3s. 6d.

SPONS' ENGINEERS' PRICE-BOOK. 7s. 6d.

ELECTRICAL TABLES AND MEMORANDA. By Silvanus P. Thompson,
D.Sc., and Eustace Thomas. Waistcoat-pocket size, 1s.; post free, 1s. 1d.

A POCKET-BOOK OF USEFUL FORMULÆ AND MEMORANDA FOR CIVIL
AND MECHANICAL ENGINEERS. By Sir G. L. Molesworth, K.C.I.E., and R. B. Molesworth, M.A.
Twenty-third Edition. 6s.

THE MECHANICAL ENGINEER'S POCKET-BOOK. By D. K. Clark. 9s.

TABLES, MEMORANDA, AND CALCULATED RESULTS, for Mechanics,
Engineers, Builders, Contractors, &c. By Francis Smith. Fifth Edition, revised and enlarged.
Waistcoat pocket size, 1s. 6d.

ENGINEER AND MACHINISTS' POCKET-BOOK. By Charles H. Haswell,
18s.

REID'S PATENT INDEXED READY RECKONER. With fore-edge Index.
Third Edition. 2s. 6d.

MISCELLANEOUS.

WIRE: ITS MANUFACTURE AND USES. By J. Bucknall Smith. 7s. 6d.

MODERN PHYSICS, CONCEPTS AND THEORIES OF. By J. B. Stallo.
Third Edition. 5s.

NATURAL PHILOSOPHY. By the Rev. Prof. Galbraith, M.A., and the Rev.
Prof. Haughton, D.C.L., F.R.S. 3s. 6d.

PROTECTION OF BUILDINGS FROM LIGHTNING. By Dr. Oliver J.
Lodge, F.R.S. 1s.

ELEMENTARY PHYSICS, EXAMPLES IN. With Examination Papers. By
W. Gallatly, M.A. 4s.

NATURAL PHILOSOPHY: An Elementary Treatise. By Prof. A. Privat
Deschanel. Translated and added to by Prof. J. D. Everett, D.C.L., F.R.S. Twelfth Edition. Cloth 18s.
Also in four parts.—I. MECHANICS, HYDROSTATICS, &c.; II. HEAT; III. ELECTRICITY AND MAGNETISM;
IV. SOUND AND LIGHT. 4s. 6d. each.

ELEMENTARY TEXT-BOOK OF PHYSICS. By Prof. Everett, D.C.L., F.R.S. Seventh Edition, 3s. 6d.

ELECTRICAL EXPERIMENTS. By G. E. Bonney. 2s. 6d.

THE PHONOGRAPH : and How to Construct It. By W. Gillett. 5s.

ELECTRICAL ENGINEERING AS A PROFESSION, AND HOW TO ENTER IT. By A. D. Southam. 3s. 6d.

AID BOOK TO ENGINEERING ENTERPRISE. By Ewing Matheson, M.I.C.E. New edition. 850 pages. £1. 1s.

ENGINEERING CONSTRUCTION IN IRON, STEEL, AND TIMBER. By W. H. Warren. 16s. net.

PRACTICAL PHYSICS. By R. T. Glazebrook and W. N. Shaw. 6s.

MANUAL OF PHYSICS. By William Peddie, F.R.S.E. 7s. 6d.

EXAMPLES IN PHYSICS. By D. E. Jones, B.Sc. Fcap. 8vo, 3s. 6d.

EXPERIMENTAL SCIENCE. By G. M. Hopkins. 680 engravings. Cloth, 15s.

LIGHTNING CONDUCTORS: THEIR HISTORY, NATURE, AND MODE OF APPLICATION. By Richard Anderson, F.C.S., F.G.S. Third Edition, 12s. 6d.

EXPERIMENTAL CHEMISTRY. By Stockhardt. Edited by C. W. Heaton, F.C.S. 5s.

AIDS TO CHEMISTRY. By C. E. Armand Semple, B.A., M.B. In four parts. Parts I. to III., 2s. and 2s. 6d. ; Part IV., 1s. and 1s. 6d.

A SHORT MANUAL OF ANALYTICAL CHEMISTRY FOR LABORATORY USE. By John Muter, Ph.D., M.A., F.C.S. 6s. 6d.

LIGHTNING CONDUCTORS AND LIGHTNING GUARDS. By Prof. Oliver J. Lodge, D.Sc., &c. With numerous illustrations. 15s.

PRACTICAL CHEMISTRY, INCLUDING ANALYSIS. By J. E. Bowman and Prof. C. L. Bloxam. Eighth Edition. With 90 engravings, fcap. 8vo, 5s. 6d.

CHEMICAL THEORY FOR BEGINNERS. By L. Dobbin, Ph.D., and J. Walker, Ph.D., D.Sc. 2s. 6d.

OUTLINES OF GENERAL CHEMISTRY. By Prof. W. Ostwald. Translated by Dr. J. Walker. 10s. net.

LESSONS IN ELEMENTARY CHEMISTRY. By Sir H. E. Roscoe, M.P., F.R.S. Sixth Edition, 4s 6d.

A TREATISE ON CHEMISTRY. By Sir H. E. Roscoe, M.P., F.R.S., and C. Schorlemmer. Vols. I. and II. : Inorganic Chemistry. Vol. I., The Non-Metallic Elements, 21s. Vol. II., Metals, in two parts, 18s. each. Vol. III. : Organic Chemistry. Parts 1, 2, 4, and 6, 21s. each ; Parts 3 and 5, 18s. each.

THE PRINCIPLES OF CHEMISTRY. By Prof. D. Mendeléeff. Translated by George Kamensky. Two Volumes. 36s.

PRACTICAL CHEMISTRY AND QUALITATIVE ANALYSIS. By Prof. Frank Clowes, D.Sc. 7s. 6d.

INORGANIC CHEMISTRY. By Prof. Edward Frankland, D.C.L., &c., and Prof. F. R. Japp, M.A., &c. 24s.

A SYSTEM OF INORGANIC CHEMISTRY. By Prof. William Ramsey, Ph.D., F.R.S. 15s.

ELEMENTARY SYSTEMATIC CHEMISTRY. By Prof. William Ramsey, Ph.D., F.R.S. 4s. 6d.

PRACTICAL CHEMISTRY, EXERCISES IN. By A. G. Vernon Harcourt, M.A., and H. G. Madan, M.A. Fourth Edition, 10s. 6d.

CHEMISTRY. By Edward Aveling, D.Sc. 6s.

METAL TURNING. By a Foreman Pattern-Maker. 4s.

ALUMINIUM : Its History, Occurrence, Properties, Metallurgy, and Applications, Including its Alloys. By J. W. Richards. Second Edition. 511 pages, £1. 1s.

PRACTICAL TREATISE ON THE STEAM-ENGINE. By Arthur Rigg. £1. 5s.

THERMODYNAMICS OF THE STEAM ENGINE AND OTHER HEAT ENGINES. By Prof. C. H. Peabody. 21s.

AN ELEMENTARY TREATISE ON STEAM. By Prof. John Perry. 4s. 6d.

STEAM ENGINES AND OTHER HEAT ENGINES. By Prof. J. A. Ewing, M.A., F.R.S. 15s.

STEAM ENGINE, THE. By George C. V. Holmes. 6s.

STEAM. By Prof. William Ripper. 2s. 6d.

STEAM AND THE STEAM-ENGINE. By Andrew Jamieson, M.Inst.C.E. Text-Book, 3s. 6d. ; Elementary Manual, 3s. 6d.

A HANDBOOK OF THE STEAM ENGINE. By John Bourne, C.E. 9s.

RECENT IMPROVEMENTS IN THE STEAM ENGINE. By John Bourne. 6s.

THE STEAM ENGINE: A Treatise on Steam-Engines and Boilers. By D. K. Clark. Two vols. 50s. net.

THE STEAM ENGINE, THE THEORY AND ACTION OF. By W. H. Northcott, C.E. 3s. 6d.

THE STEAM ENGINE. By the Rev. Prof. Galbraith, M.A., and the Rev. Prof. Haughton, D.C.L., F.R.S. 3s. 6d.

TEXT-BOOK ON THE STEAM ENGINE. By Prof. T. M. Goodeve, M.A. Eleventh Edition, 6s.

ON GAS ENGINES. By Prof. T. M. Goodeve, M.A. 2s. 6d.

HEAT ENGINES OTHER THAN STEAM. By H. Graham Harris. 1s. 6d.

THE STEAM ENGINE INDICATOR, AND INDICATOR DIAGRAMS— *see page 13.*

USES OF PETROLEUM IN PRIME MOVERS. By Prof. W. Robinson. 1s.

GAS ENGINES : their Theory and Management. By William Macgregor. 8s. 6d.

GAS ENGINE, THE. By Dugald Clerk. 7s. 6d.

THEORY OF THE GAS ENGINE. By Dugald Clerk. 2s.

GAS AND PETROLEUM ENGINES. By Prof. William Robinson. 14s.

A PRACTICAL TREATISE ON THE CONSTRUCTION OF HORIZONTAL AND VERTICAL WATERWHEELS. By William Cullen. Second Edition. 4s.

THE TURBINE MANUAL AND MILLWRIGHT'S HANDBOOK. By Charles Louis Hett. 2s.

HEAT AS A FORM OF ENERGY. By Prof. R. H. Thurston. 5s.

HEAT. By Prof. P. G. Tait. 6s.

CONVERSION OF HEAT INTO WORK. By Dr. William Anderson, F.R.S., Third Edition. 6s. Cheaper Edition. 2s. 6d.

THE THEORY OF HEAT. By J. Clerk Maxwell, M.A. Edited by Lord Rayleigh. 4s. 6d.

AN ELEMENTARY TREATISE ON HEAT. By William Garnett, M.A., D.C.L. Fifth Edition, 6s.

HEAT A MODE OF MOTION. By John Tyndall, D.C.L. 12s.

A COLLECTION OF EXAMPLES ON HEAT AND ELECTRICITY. By H. H. Turner. 2s. 6d.

PRINCIPLES OF MECHANICS. By Prof. T. M. Goodeve. 6s.

ELEMENTARY DYNAMICS. By E. J. Gross, M.A. 5s. 6d.

A TREATISE ON ELEMENTARY DYNAMICS. By William Garnett, M.A., D.C.L. Fifth Edition, 6s.

"THE ELECTRICIAN" PRIMERS—*see page 8.*

A TREATISE ON DYNAMICS. By W. H. Besant, D.Sc., F.R.S. Second Edition, 7s. 6d.

MODERN MECHANISM, exhibiting the latest Progress in Machines, Motors and the Transmission of Power. Edited by Park Benjamin, LL.B., Ph.D. 30s.

ELEMENTARY MECHANICS. By Oliver J. Lodge, D.Sc. Lond. Revised Edition, 208 pages. Cloth, 3s.

APPLICATIONS OF DYNAMICS TO PHYSICS AND CHEMISTRY. By Prof. J. J. Thomson. 7s. 6d.

TEXT-BOOK OF MECHANICAL ENGINEERING. By W. J. Lincham. 10s. 6d., net.

THE CONSERVATION OF ENERGY. By Balfour Stewart. Seventh Edition. 5s.

AN ELEMENTARY TEXT-BOOK OF APPLIED MECHANICS. By David Allan Low. 2s.

MECHANICS. By the Rev. Prof. Galbraith, M.A., and the Rev. Prof. Haughton, D.C.L., F.R.S. 3s. 6d.

PRACTICAL MECHANICS AND MACHINE DESIGN, NUMERICAL EXAMPLES IN. By R. G. Blaine. 2s. 6d.

THERMODYNAMICS, LESSONS ON. By R. E. Baynes, M.A. 7s. 6d.

MECHANICS. By Edward Aveling, D.Sc. 6s.

DYNAMICS OF A PARTICLE, TREATISE UPON. By Prof. P. G. Tait and W. J. Steele. 12s.

MOTIVE POWER AND GEARING—*see page 14.*

MODERN SHAFTING AND GEARING. By M. P. Bale. 2s. 6d.

THE TRANSMISSION OF ENERGY. By Prof. O. Reynolds, M.A., F.R.S. 1s.

HELICAL GEARS. By a Foreman Pattern-Maker. 7s. 6d.

APPLIED MECHANICS. By Prof. J. W. Cotterill. Third Edition. 18s.

LESSONS IN APPLIED MECHANICS. By Prof. J. H. Cotterill and J. H. Slade. 8s. 6d.

HYDRAULIC MOTORS: Turbines and Pressure Engines. By G. R. Bodmer, Assoc.M.Inst.C.E. With 179 illustrations. 14s.

THE MECHANICS OF MACHINERY. By Prof. A. B. W. Kennedy. 12s. 6d.

MACHINE DRAWING AND DESIGN. By Prof. William Ripper. In one Vol., 25s. ; or in five parts, 4s. 6d. each.

AN INTRODUCTION TO MACHINE DRAWING AND DESIGN. By David Allan Low (Whitworth Scholar). With 65 illustrations and diagrams, crown 8vo, 2s.

MANUAL OF MACHINE DRAWING AND DESIGN. By David Allan Low and Alfred Wm. Bevis. 7s. 6d.

MACHINE DRAWING. By G. Halliday. First Course. Part II. ELECTRICAL ENGINEERING, 2s. 6d. Second Course, 6s.

THE MECHANIC'S WORKSHOP HANDYBOOK: A Practical Manual on Mechanical Manipulation. By Paul N. Hasluck, A.I.M.E. Crown 8vo, cloth, 2s.

PRACTICAL MECHANICS. By Prof. Perry, M.E. 3s. 6d.

ELEMENTS OF MACHINE DESIGN. By Prof. W. C. Unwin. Part I., General Principles, 6s. Part II., Chiefly on Engine Details, 4s. 6d.

DRAWING FOR MACHINISTS AND ENGINEERS. (Cassell's Technical Manuals.) 4s. 6d.

ELECTRICITY : WHAT TO DO WITH IT, AND HOW TO DO IT— *In the Press.*

ENGINEER'S SKETCH BOOK of Mechanical Movements, Devices, Appliances. Contrivances, and Details employed in the Design and Construction of Machinery for every purpose. Nearly two thousand illustrations. 7s. 6d.

SCIENTIFIC AND TECHNICAL PAPERS of WERNER VON SIEMENS. Translated from the German by E. F. Bamber. Vol. I., Scientific Papers and Addresses. Vol. II., Applied Science. 14s. each.

PERSONAL RECOLLECTIONS OF WERNER VON SIEMENS. Translated by W. C. Coupland. 15s.

POPULAR LECTURES AND ADDRESSES ON VARIOUS SUBJECTS IN PHYSICAL SCIENCE. By Sir William Thomson (Lord Kelvin). In three vols. Vols. I. and III. ready, 7s. 6d. each.

JOURNAL OF THE INSTITUTION OF ELECTRICAL ENGINEERS (formerly the Society of Telegraph-Engineers and Electricians). Issued at irregular intervals.

THOMAS A. EDISON AND SAMUEL F. B. MORSE. By Dr. Denslow and J. Marsh Parker. 1s.

CHATS ON INVENTION. By John Martin. 1s.

FORTY YEARS AT THE POST OFFICE, 1850-1890. By F. E. Baines, C.B. Two Vols., 21s.

ROMANCE OF ELECTRICITY. By John Munro. Crown 8vo, 5s.

POPULAR LECTURES ON SCIENTIFIC SUBJECTS. By Prof. H. von Helmholtz. Two Volumes. 7s. 6d. each.

WATER OR HYDRAULIC MOTORS. By Philip R. Björling. 208 illustrations. 9s.

SCIENCE LECTURES AT SOUTH KENSINGTON. Containing Lectures by Celebrated Scientists. Two Volumes. 6s. each.

ELECTRICAL ENGINEERING FORMULÆ—see page 9.

PIONEERS OF SCIENCE. By Dr. O. J. Lodge. 7s. 6d.

PROF. CLERK MAXWELL, A LIFE OF. By Prof. L. Campbell, M.A., and W. Garnett, M.A. Second Edition. 7s. 6d.

GLEANINGS IN SCIENCE: A Series of Popular Lectures on Scientific Subjects. By Rev. G. Molloy. 7s. 6d.

ACHIEVEMENTS IN ENGINEERING DURING THE LAST HALF CENTURY. By Prof. Vernon Harcourt. 5s.

SPINNING TOPS. By Prof. J. Perry. 2s. 6d.

SOAP BUBBLES, and the Forces which Mould Them. By C. V. Boys. 2s. 6d.

FORTY LESSONS IN ENGINEERING WORKSHOP PRACTICE. (Polytechnic Series.) 1s. 6d.

SCIENCE FOR ALL. Edited by Dr. Robert Brown, M.A., F.L.S. Five volumes. 9s. each.

THEORY OF LIGHT. By T. Preston. 15s. net.

THEORY OF SOUND. By Lord Rayleigh. Vols. I. and II., 12s. 6d. each; Vol. III. in preparation.

THE MATERIALS OF CIVIL ENGINEERING. By Prof. R. H. Thurston. Part I. Non-Metallic Materials, 10s. 6d.; Part II. Iron and Steel, 18s.; Part III. Non-Ferrous Metals and Alloys (Second Edition), 12s. 6d.

THE SLIDE RULE: A Practical Manual. By C. N. Pickworth. 2s.

HISTOIRE D'UN INVENTEUR: Exposé des Decouvertes et des Travaux de M. Gustave Trouvé. By Georges Barral. With portrait of M. Trouvé and 260 illustrations.

MÉCANIQUE MOLÉCULAIRE. By A. Bandsept. 1s. 6d.

MANUEL D'ELECTROLOGIE MÉDICALE. By Gustave Trouvé. 7s.

PRODUCTION ET UTILISATION RATIONNELLES DE LA CHALEUR INTENSIVE. By A. Bandsept. 2s. 6d.